The Gregory Breit
Centennial Symposium

The Gregory Breit Centennial Symposium

Editors

Vernon W. Hughes
Francesco Iachello
Dimitri Kusnezov

Yale University, USA

World Scientific
Singapore • New Jersey • London • Hong Kong

Published by

World Scientific Publishing Co. Pte. Ltd.

5 Toh Tuck Link, Singapore 596224

USA office: 27 Warren Street, Suite 401-402, Hackensack, NJ 07601

UK office: 57 Shelton Street, Covent Garden, London WC2H 9HE

British Library Cataloguing-in-Publication Data
A catalogue record for this book is available from the British Library.

THE GREGORY BREIT CENTENNIAL SYMPOSIUM

ISBN-13 978-981-02-4553-5
ISBN-10 981-02-4553-X

Gregory Breit

July 14, 1899 – September 11, 1981

Born July 14, 1899
Nickolaev, Russia
A.B., Johns Hopkins University, 1918
A.M., Johns Hopkins University, 1920
Ph.D., Johns Hopkins University, 1921
National Research Council Fellow, University of Leyden, 1921-22
National Research Council Fellow, Harvard University, 1922-23
Assistant Professor of Physics, University of Minnesota, 1923-24
Mathematical Physicist, Carnegie Institute in Washington, D.C., 1924-29
Resident, Technische Hochschule, Zurich, Switzerland, 1928
Research Associate, Carnegie Institute in Washington, 1929-44
Professor of Physics, New York University, 1929-34
Member, Division of Physical Sciences, National Research Council, 1932-33, 1938-41
Professor of Physics, University of Wisconsin, 1934-47
Visiting Member Institute for Advanced Study, Princeton, 1935-36
Councillor, American Physical Society, 1935-38
Naval Ordinance Laboratory, Washington Navy Yard, October, 1940-January, 1941
Information Chief Coordinator, Fast Neutron Project, Metallurgical laboratory (Manhattan Project), University of Chicago, 1942
Member, Applied Physics Laboratory, Johns Hopkins University, 1942-43
Head Physicist, Ballistic Research Laboratory, Aberdeen Proving Grounds, Maryland, 1943-45
Professor of Physics, Yale University, 1947-68, (Donner Professorship, 1958-68)
Distinguished Professor of Physics, State University of New York at Buffalo, 1968-1977
Retired, Oregon, 1977-1981

Member, National Academy of Sciences, 1939
Fellow, American Academy of Arts and Sciences, 1951
Honorary Doctor of Science, University of Wisconsin, 1954
Benjamin Franklin Medal, 1964
National Medal of Science, 1968
Associate Editor, Physical Review, 1927-29, 1939-41, 1954-56, 1961-63
Associate Editor, Proceedings of the National Academy of Sciences, 1958-60
Associate Editor, Il Nuovo Cimento, 1964-1968

CONTENTS

The Legacy of Gregory Breit

Banquet Addresses

Frontiers of Nuclear Physics

PREFACE

On October 22-26, 1999, the Gregory Breit Memorial Symposium was held at Yale University in New Haven, Connecticut. This volume contains the Proceedings of that Symposium.

About 100 scientists attended the Symposium from the international community, Yale and surrounding universities, and the country as a whole. The Symposium commenced with welcoming addresses by Yale President Richard Levin and the Physics Department Chair Charles Baltay. The first day was devoted to the Legacy of Gregory Breit. Talks were presented by scientists from many countries on subjects introduced and developed by Gregory Breit in a variety of fields, atomic physics, nuclear physics and particle physics. The second day was devoted to the presentation of the Frontiers on Nuclear Physics at the dawn of the 21st Century.

We are grateful to the Physics Department of Yale University for providing the infrastructure and technical services in support of the Symposium and for a grant to support the outside speakers.

We owe a great deal of gratitude to the Conference Secretary, Diane Altschuler, and to the staff of the Physics Department, Laurelyn Celone, Marguerite Scalesse, for their help.

Vernon W. Hughes
Francesco Iachello
Dimitri Kusnezov

GREGORY BREIT CENTENNIAL SYMPOSIUM

Friday, October 29, 1999
Sloane Physics Laboratory, 217 Prospect Street, Room 57

The Legacy of Gregory Breit

8:45 a.m.	**Registration**
Session Chairman	**Vernon W. Hughes** Sterling Professor of Physics Emeritus, Yale University
9:30 a.m.	Welcome **Richard C. Levin** President, Yale University
	Introductory Remarks about Gregory Breit **Vernon W. Hughes**
	Quantum Electrodynamics and the Lamb Shift in Muonic Hydrogen **Toichiro Kinoshita** G. Smith Professor of Physics Emeritus, Cornell University
10:45 a.m.	Coffee
11:15 a.m.	Measurement of the Breit Interaction by Laser Spectroscopy **Massimo Inguscio** Professor of Physics, University of Florence, Italy Director, European Laboratory for Non-Linear Spectroscopy
12:00 noon	Lunch – Sage Hall, School of Forestry, 205 Prospect Street
Session Chairman	**Thomas W. Appelquist** Eugene Higgins Professor of Physics, Yale University

1:30 p.m.	Isotope Shifts **Ernst W. Otten** Professor of Physics, University of Mainz, Germany
	The Fermion - Fermion Interaction **Gerald E. Brown** Distinguished Professor of Physics, State University of New York, Stony Brook
	Nuclear Collisions: The Ultra Relativistic Frontier **Lawrence D. McLerran** Senior Scientist, Brookhaven National Laboratory
3:45 p.m.	Coffee
Session Chairman	**Samuel Devons** Professor of Physics Emeritus, Columbia University
4:15 p.m.	The Discovery of Supersymmetry in Nuclei **Francesco Iachello** J.W. Gibbs Professor of Physics and Chemistry, Yale University
	One Hundred Years of the Quantum:The Glory and the Shame **John A. Wheeler** Joseph Henry Professor of Physics Emeritus, Princeton University
6:30 p.m.	**Symposium Banquet** – The New Haven Lawn Club, 193 Whitney Avenue
	Martin J. Klein, Presiding Eugene Higgins Professor of Physics and the History of Science Emeritus, Yale University
	On Gregory Breit **McAllister Hull** Provost and Professor of Physics Emeritus, University of New Mexico

Remarks from a Family Member *
Ralph W. G. Wyckoff

*Due to a death in his family Ralph Wyckoff was unable to be present. His remarks were read by Martin Klein

Saturday, October 30, 1999
Luce Hall, 34 Hillhouse Avenue, Room 101

Frontiers of Nuclear Physics

Session Chairman	**Michael E. Zeller** Henry Ford II Professor of Physics, Yale University
9:00 a.m.	Color, Spin, and Flavor Forces in Quantum Chromodynamics **Robert L. Jaffe** Professor of Physics, Massachusetts Institute of Technology
9:45 a.m.	Coffee
10:15 a.m.	Nuclear Structure with Stable and Radioactive Beams **Richard F. Casten** Professor of Physics, Yale University Director, Wright Nuclear Structure Laboratory Equation of State of Nuclear Matter **C. Konrad Gelbke** University Distinguished Professor, Michigan State University Director, National Super-Conducting Cyclotron Laboratory
11:45 a.m.	Lunch- Kline Biology Tower, Dining Hall, 219 Prospect Street

Session Chairman	**Charles Baltay** Eugene Higgins Professor of Physics, Yale University Chairman, Physics Department
1:30 p.m.	Electric Dipole Moments as a Test of Time Reversal Symmetry **Norman F. Ramsey** Eugene Higgins Professor of Physics Emeritus, Harvard University Nuclear Astrophysics **Claus Rolfs** Professor of Physics, University of Bochum, Germany
3:00 p.m.	Coffee
3:30 p.m.	Physics with Relativistic Heavy Ions **John W. Harris** Professor of Physics, Yale University
4:15 p.m.	End of Symposium

INTRODUCTORY REMARKS ABOUT GREGORY BREIT

VERNON W. HUGHES
Physics Department, Sloane Physics Laboratory, Yale University, New Haven CT 06520-8120 USA

On behalf of the Yale Physics Department I am glad to welcome you and happy so many are able to attend the Gregory Breit Centennial Symposium. President Levin will be with us at about 10:30 am and will make his welcoming remarks then. He apologizes that he is unable to be here at the opening of our Symposium.

This symposium has been organized by a committee of the Yale Physics Department consisting of R. Adair, Y. Alhassid, T. Appelquist, C. Baltay, D.Allen Bromley, R. Casten, J. Harris, P. Hohenberg, V.W. Hughes (Co-chairman) and F. Iachello (Co-chairman).

The initial stimulus for this Symposium came as an apparently innocent question I was asked one evening about 8 months ago at Brookhaven National Laboratory. Andrzej Czarnecki, a young particle theorist at BNL, who has among other things calculated the higher order electroweak contribution to the muon g-2 value and also the positronium lifetime, had been reading papers by Breit on atomic hyperfine structure and quantum electrodynamics. Andrzej asked if I knew when Breit was born. I didn't happen to know, and he said 1899. This had an immediate implication, especially in this year of the American Physical Society Centennial celebration. We checked the 1899 date by the following day – there was some concern whether it might have been 1898. The National Academy of Sciences biographical memoir by Mac Hull settled the question. I am happy to say that Andrzej will be at our symposium. Although I hadn't known the year Gregory was born, I knew the day in the year-- namely July 14, French Bastille Day. Those of you who knew Gregory will understand that the symbolism of this day is not what one would easily associate with Gregory Breit.

Gregory Breit was on the faculty of the Yale Physics Department as the Donner Professor of Physics from 1947 to 1968, when he retired.

The first day of our Symposium, entitled The Legacy of Gregory Breit, emphasizes the recent status of topics to which Gregory Breit made major, seminal contributions. These include quantum electrodynamics and atomic physics and several topics in nuclear physics, notably the isotope shift, the nucleon-nucleon interaction and heavy ion collisions. The topic of supersymmetry in nuclei is distantly related to Breit's research on isospin symmetry of neutron and proton forces. A highlight of the first day is the talk by John Wheeler, who spent a year as a postdoctoral fellow with Breit, on "One Hundred Years of the Quantum: The Glory and the Shame." The second day of our symposium, entitled "Frontiers of Nuclear Physics," covers forefront and futuristic topics in nuclear physics.

Breit-Rabi Energy Level Diagram for Hydrogen Ground State

$$[\mathcal{H} = a\vec{I} \cdot \vec{J} + \mu_o g_J\, \vec{J} \cdot \vec{H} + \mu_o g_I \vec{I} \cdot \vec{H}]$$

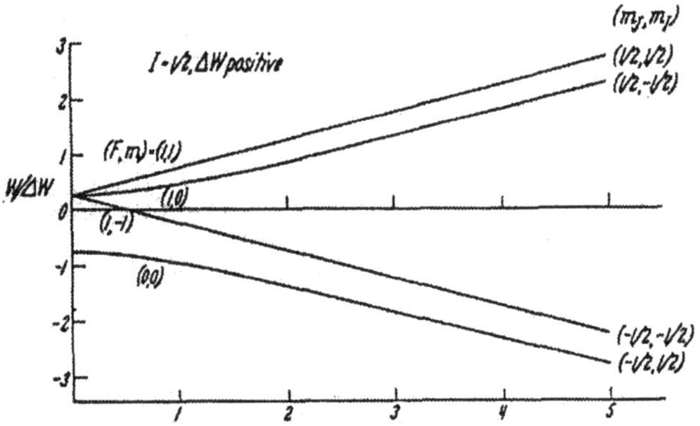

$$x = (g_J - g_I)\mu_o H / \Delta W$$

Transition Frequencies

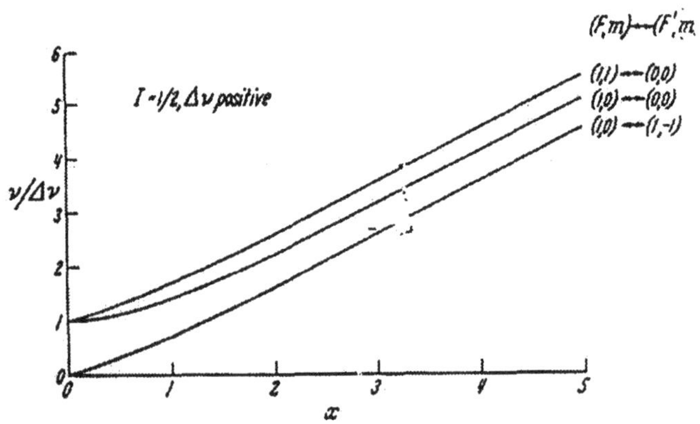

$$x = (g_J - g_I)\mu_o H / \Delta W$$

Figure 1. The Breit-Rabi energy level diagram and its consequences, taken from "Measurement of Nuclear Spin", G. Breit and I.I. Rabi, *Phys. Rev.* **38**, 2028 (1931).

There are several central and well-known topics in physics to which Breit's name is solidly attached. The important Breit-Rabi formula (Figure 1) for the energy levels of a hydrogen atom in its ground state as a function of magnetic field was developed at the time Breit was at New York University and Rabi was at Columbia. Rabi was building up his atomic and molecular beams laboratory and the interest was in nuclear spins and moments. The Breit-Rabi equation and its extensions to other atoms has been a cornerstone for precision atomic spectroscopy.

The electron-electron or Breit interaction (Figure 2) was developed by Breit based on the quantum electrodynamics field theory of the early 1930's. A principal application is to the fine structure of helium in its 2^3P_J states. Gerry Brown will discuss the Breit interaction from a theoretical point of view and Massimo Inguscio will describe the current status of measurement of *He* fine structure by laser spectroscopy.

Breit (Electron-Electron) Interaction

$$B = -\frac{e^2}{2r_{12}}\left[\alpha_1 \cdot \alpha_2 + \frac{(\alpha_1 \cdot r_{12})(\alpha_2 \cdot r_{12})}{r_{12}^2}\right]$$

For $He(1s\ 2p,\ ^3P_J)$

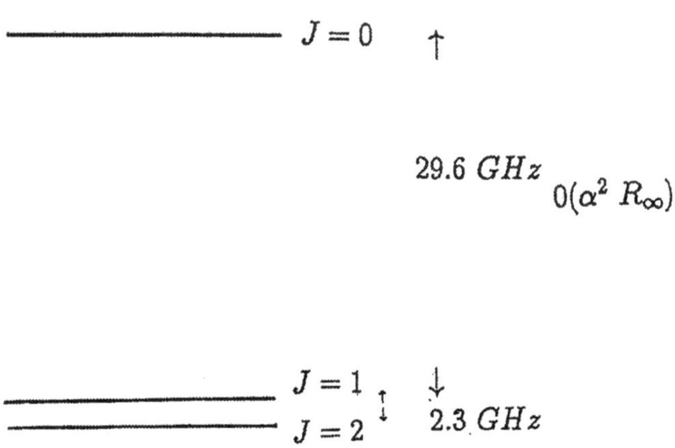

Figure 2. The Breit electron-electron interaction and its consequences in helium, taken from, "The Effect of Retardation on the Interaction of Two Electrons", G. Breit, *Phys. Rev.* **34**, 817 (1929), "Fine Structure of Helium as a Test of the Spin Interaction of Two Electrons", G. Breit, *Phys. Rev.* **36**, 383 (1930), "Dirac's Equation and the Spin-Spin Interaction of Two Electrons", G. Breit, *Phys. Rev.* **39**, 616 (1932).

Breit-Wigner Formula

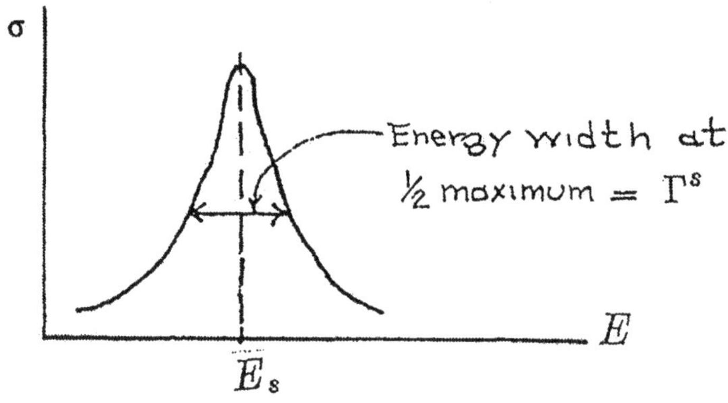

$$\sigma(\alpha,\beta) = (2l + 1)\,\pi\,\lambda\!\!\!^{-}{}_{\alpha}^{2}\,\frac{\Gamma_{\alpha}{}^{s}\,\Gamma_{\beta}{}^{s}}{(E - E_{s})^{2} + (\frac{1}{2}\Gamma^{s})^{2}}$$

Figure 3. The Breit-Wigner Formula, taken from "Capture of Slow Neutrons", G. Breit and E. Wigner, *Phys. Rev.* **49**, 519 (1936),

The Breit-Wigner resonance formula (Figure 3) was derived to understand the capture of slow neutrons by a heavy nucleus. An intermediate short-lived compound state is produced whose lifetime determines the width of the resonance in the capture cross section as a function of neutron energy. The Breit-Wigner formula has been a cornerstone of nuclear reaction theory.

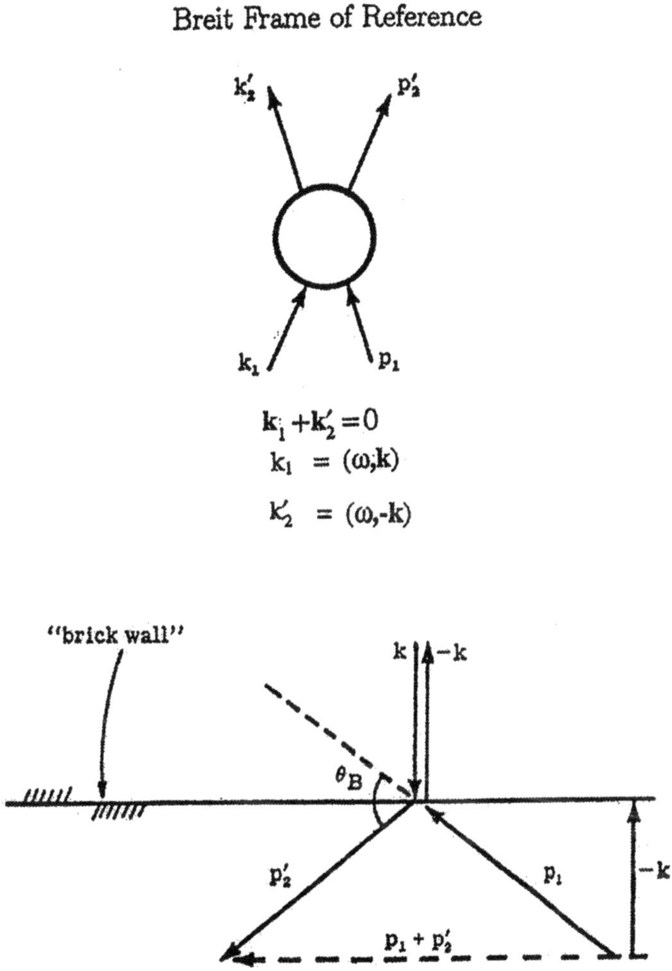

Figure 4. The Breit frame.

In addition to the laboratory and center of mass frames of reference, the Breit frame of reference (Figure 4) is often convenient to use in treating scattering problems. It was developed in Breit's studies of proton-proton scattering.

Breit made a major contribution to the discovery of the electron anomalous magnetic moment as indicated in Table 1. The measurement of the *hfs* interval in hydrogen provided the first discrepancy with a prediction of the quantum electrodynamics theory of the mid-1940s. Hence it was the first evidence of a radiative correction to QED, later understood within the renormalized theory of QED. Breit made the first suggestion that this discrepancy of $\Delta\nu$ (expt) with the Fermi theory might be due to an anomalous magnetic moment of the electron. He calculated this electron anomalous magnetic moment with the developing renormalized QED theory, but his calculation contained an error which Schwinger pointed out in his own correct calculation.

Some brief information about Breit's bibliography is given in Table 2. Breit's early training was in electrical engineering, as indicated in the first few references. His work with Tuve on propagation of radio waves was pioneering. Breit and Tuve also developed a high voltage electrostatic generator of 1 MeV for nuclear reaction studies. Early in his career Breit became interested in basic fundamental problems such as space quantization and the properties of a spinning electron.

We shall hear in this symposium of many other researches by Breit. They are included in his bibliography of some 330 papers, published over a period of 53 years. Breit wrote several major review articles, one on quantum theory of dispersion and another on the nucleon-nucleon interaction.

Despite all his accomplishments and active leadership in many fields of physics, Breit was basically a modest man. Professor Rabi told the following story. In 1939, the annual meeting of the National Academy of Sciences was held in April in Washington, DC and shortly thereafter the annual meeting of the American Physical Society took place in Washington at the Shoreham Hotel. At that National Academy meeting Breit was elected to be a member of NAS and was immediately told that he was elected. Then at the APS meeting the next day Rabi saw Breit sitting on a couch in the Shoreham Hotel, looking very sad and despondent. Rabi asked Breit, "Why do you look so sad? You have just received the honor of being elected to the Academy." Gregory replied: "But there are so many more worthy candidates." Rabi answered: "Don't worry, Gregory. You and I know it but nobody else does." The next year, 1940, Rabi was elected to the Academy.

Table 1.

Breit and the Electron Anomalous Magnetic Moment.

H hfs

H hfs measurement at Columbia in disagreement with theoretical value from Fermi formula by ~0.002.

$$\Delta\nu\,(\text{expt.}) = 1\,420.410(6)\ \text{MHz}$$
$$\Delta\nu\,(\text{theor.}) = 1\,416.97(54)\ \text{MHz}$$

"The hyperfine structure of atomic hydrogen and deuterium."

J.E. Nafe, E.E. Nelson, and I.I. Rabi, *Phys. Rev.* **71**, 914 (1947); J.E. Nafe and E.E. Nelson, *Phys. Rev.* **73**, 718 (1948).

H Lamb Shift

"Fine structure of the hydrogen atom by a microwave method."

W.E. Lamb, Jr. and R.C. Rutherford, *Phys. Rev.* **72**, 241 (1947).

Suggestion that electron has an anomalous magnetic moment

"Does the electron have an intrinsic magnetic moment?"'

G. Breit, *Phys. Rev.* **72**, 984 (1947).

Measurement of the electron anomalous magnetic moment

"The magnetic moment of the electron."

P. Kusch and H.M. Foley, *Phys. Rev.* **72**, 1256 (1947); *Phys. Rev.* **73**, 412 (1948).

Theory

"On quantum-electrodynamics and the magnetic moment of the electron."

J. Schwinger, *Phys. Rev.* **73**, 416L (1948).

Erratum: "Does the electron have an intrinsic magnetic moment?"

G. Breit, *Phys. Rev.* **73**, 1410 (1948).

"Some effects of the intrinsic magnetic moment of the electron."

G. Breit, *Phys. Rev.* **74**, 656 (1948).

"The effect of nuclear motion on the hyperfine structure of hydrogen."

G. Breit and R.E. Meyerott, *Phys. Rev* **72**, 1023 (1947).
G. Breit, G.E. Brown and G.B. Arfken, *Phys. Rev* **76**, 1299 (1949).

Table 2.

Early Publications

The calculation of detecting and amplifying properties of an electron tube from its static characteristics. *Phys. Rev.* **16**:387 (1920).

The propagation of fan shaped group of waves in a dispersing medium. *Phil. Mag.* **44**:1149 (1922).

The field radiated from two horizontal coils. *Bureau of Standards Scientific Paper* No. 431 (1922).

With P. Ehrenfest. A remarkable case of quantization. *Zeit. Physik* **9**:207 (1922), *Proc. Amsterdam Acad.* **25**:2 (1923).

The Heisenberg theory of the anomalous Zeeman effect. *Nature* **112:**396 (1923).

Note on the width of spectral lines due to collisions and quantum theory. *Proc. Nat. Acad. Sci.* **9**:244 (1923).

The electromagnetic mass and momentum of a spinning electron. *Proc. Nat. Acad. Sci.* **12**:451 (1926).

With M.A. Tuve. Radio evidence of the existence of the Kennelly-Heaviside layer. *J. Wash. Acad. Sci.* **16**:98 (1926).

With M.A. Tuve and O. Dahl. Effective heights of the Kennelly-Heaviside layer. *Proc. Inst. Radio Eng.* **16**:1236 (1928).

The magnetic moment of the electron. *Nature* **122**:649 (1928).

Quantum theory of dispersion. Parts I-V. *Rev. Mod. Phys.***4**:504 (1932).

Quantum theory of dispersion. Parts VI-VII. *Rev. Mod. Phys.* **5**:91 (1933).

* * * * * * * *

Some 330 papers

* * * * * * * *

On quantum theory, atomic physics, (hfs, fine structure), nuclear physics, (nuclear moments, isotope shift, nuclear forces, nuclear reactions, heavy ions), particle physics, (nucleon-nucleon interaction)

* * * * * * * *

Final Paper

With M. Tischler, S. Mukherje and G. Pappas. Magnetic-moment effects on tests of charge independence in nucleon-nucleon scattering. Proc. Nat. Acad. Sci. USA 70: 2178 (1973).

TOPICS FROM HIGH PRECISION TEST OF QUANTUM ELECTRODYNAMICS

TOICHIRO KINOSHITA
Newman Laboratory, Cornell University,
Ithaca, New York 14853, U. S. A.
E-mail: tk@hepth.cornell.edu

Professor Breit was one of the great pioneers who made important contributions to the development of quantum electrodynamics. At this Symposium celebrating his centennial I should like to talk on the current status of high precision test of quantum electrodynamics within the context of the Standard Model, or quantum mechanics in the extended sense.

1 Introduction

Let me begin by noting that the high precision test of QED is really a test of the Standard Model on systems in which the electromagnetic interaction plays the dominant role. Of course, non-QED effects cannot be ignored.

Before going into detail, it must be noted that the theory at present has several parameters such as mass m of the electron and elementary charge e (or the fine structure constant α) which cannot be fixed within QED. These parameters must be found elsewhere.

One problem is that no currently available value of α is accurate enough to test theory and the measurement of the electron magnetic moment anomaly a_e. From another perspective, however, this means that a_e is the best source of α available at present. Once α is determined from a_e, it can be used to examine how well α's based on other physics are measured. Comparison of α's measured by various methods is really a test of the internal consistency of quantum mechanics on which all these measurements are based.

In this talk I will discuss topics from two areas of QED in which Breit played an important role, namely, the lepton magnetic moment anomaly (Sec. 2) and the relativistic 2-body bound state (Sec. 3). A brief survey of high precision measurements of α available at present or in the near future is given in Sec. 4. The possible impact of high precision QED test in a broader context of general quantum physics is discussed in Sec. 5.

2 Lepton Magnetic Moment Anomaly

Within the context of current QED or the Standard Model the magnetic moment anomaly is the only calculable property of leptons in free space since other basic observables such as their masses and charges are not calculable and must be treated as external parameters. Thus I will discuss only the lepton magnetic moment anomaly. I skip the tau lepton magnetic moment anomaly which has not yet been precisely measured.

2.1 Electron

Breit was the first to suggest[1] that the anomalous results of hyperfine measurements[2,3] may be explained if the electron g value deviates slightly from 2, the Dirac value. This observation was soon confirmed by the experiment of Kusch and Foley.[4] Together with the discovery of the Lamb shift in the spectrum of hydrogen atom, this provided a timely stimulus to the renormalization theory of quantum electrodynamics (QED), which was just being developed. Schwinger demonstrated the power of QED by calculating the electron magnetic moment anomaly $a_e \equiv (g_e - 2)/2$ in the second-order covariant perturbation theory.[5]

This was the beginning of a long series of measurements and theoretical calculations in which the precision has been improved from 10^{-3} to 10^{-12} over 40 years. The latest results for the electron and positron anomalies obtained in Penning trap experiments are[6]

$$a_{e^-}(\text{exp}) = 1\ 159\ 652\ 188.4\ (4.3) \times 10^{-12},$$
$$a_{e^+}(\text{exp}) = 1\ 159\ 652\ 187.9\ (4.3) \times 10^{-12}, \qquad (1)$$

where e^- and e^+ refer to the electron and the positron, respectively. The numbers within parentheses stand for measurement uncertainties of $\pm\ 4.3 \times 10^{-12}$, which consists of several parts: the statistical error of 0.62×10^{-12}, the systematic error of 1.3×10^{-12} due to the uncertainty in a residual microwave power shift, and a large uncertainty of 4×10^{-12} assigned to a potential cavity-mode shift. This last error arises from a shift in the cyclotron frequency of the electron associated with image charges induced in the metallic Penning trap, an effect which depends on the cavity frequency modes and on the electron cyclotron frequency.[7]

Studies to improve the experimental precision for a_e focus on the understanding and control of this cavity influence on the cyclotron frequency. One approach is to reduce the Q of the cavity.[8] In another attempt Mittleman *et al.* produced and studied a many-electron (kiloelectron) cluster in the trap, which magnifies the shift of the cyclotron frequency.[9] Gabrielse *et al.* are studying the use of a cylindrical cavity where the cyclotron frequency shift can be calculated analytically and is hence under better control.[10] The eventual reduction of experimental uncertainty by about an order of magnitude is the goal of these experiments.

The current theoretical value of a_e can be written as[11,12]

$$a_e(\text{theory}) = 0.5\left(\frac{\alpha}{\pi}\right) - 0.328\,478\,965\,579\,\ldots\left(\frac{\alpha}{\pi}\right)^2 + 1.181\,241\,456\ldots\left(\frac{\alpha}{\pi}\right)^3 - 1.509\,8\,(384)\left(\frac{\alpha}{\pi}\right)^4 + 4.382\,(19)\times 10^{-12}, \tag{2}$$

The analytic values of the coefficients of the α and α^2 terms have been known for a long time. The analytic value of the α^3 term has been obtained only recently.[13] It is in excellent agreement with the latest numerical result[14]

$$1.181\,259\,(40), \tag{3}$$

which was obtained shortly before the analytic result became available.

The α^4 term requires the evaluation of 891 four-loop Feynman diagrams. This problem is so huge that analytic evaluation is prohibitively difficult even with the help of the fastest computers. Crude numerical evaluation of these integrals began around 1981.[15] It is only in the last few years that the calculation of this term began to move from a "qualitative" to a "quantitative" stage thanks to the development of massively-parallel computers. The last term of (2) consists of terms caused by vacuum-polarization loops due to muon and tau and by hadronic and weak interactions.[11,16,17,18,19]

To compare the theory of a_e with experiment, it is necessary to know the value of α. Currently the best non-QED measurement of α is that based on the quantum Hall effect: [20]

$$\alpha^{-1}(\text{q.Hall}) = 137.036\,003\,7\,(33) \qquad [2.4\times 10^{-8}], \tag{4}$$

where the number within the brackets represents the fractional precision. Substituting this value in (2), one finds

$$a_e(\text{q.Hall}) = 1\,159\,652\,153.4\,(1.2)\,(28.0)\times 10^{-12}, \tag{5}$$

where the numbers enclosed in parentheses are the uncertainty in the numerical integration result and that of α from (4), respectively.

Note that the intrinsic theoretical uncertainty in Eq. (5) is already quite small. Thus the overall uncertainty is dominated by the uncertainty of α given in (4). In other words, the insufficient precision of this α prevents us from testing the validity of QED to the extent allowed by the theory and measurement of a_e. This means, in turn, that comparison of theory and measurement of a_e will give a more precise value of α. The value of α determined from the average of a_{e^-} and a_{e^+} and the theory is

$$\begin{aligned}\alpha^{-1}(a_e) &= 137.035\ 999\ 58\ (14)\ (50)\\ &= 137.035\ 999\ 58\ (52) \qquad [3.8\times 10^{-9}],\end{aligned} \tag{6}$$

where the uncertainties on the first line are from the α^4 term and the measurement uncertainty of a_e given in (1), respectively.

2.2 Muon

The measurement of the muon anomalous magnetic moment $a_\mu = \frac{1}{2}(g_\mu - 2)$ at Brookhaven National Laboratory is making a rapid and steady progress. It is expected to achieve a precision of about 40×10^{-11}, more than 20 times better than that of the best previous result[21]

$$a_\mu(\text{exp}) = 1\ 165\ 923\ (8.5) \times 10^{-9} \qquad [7 \times 10^{-6}]. \tag{7}$$

Preliminary results obtained at the Brookhaven National Laboratory in the 1997 and 1998 runs are[22]

$$a_\mu^+(\text{exp}) = 11\ 659\ 251\ (150) \times 10^{-10} \qquad [13 \times 10^{-6}] \tag{8}$$

and[23]

$$a_\mu^+(\text{exp}) = 11\ 659\ 191\ (59) \times 10^{-10} \qquad [5 \times 10^{-6}], \tag{9}$$

respectively.

The primary purpose of the new muon $g-2$ experiment is to verify the presence of the electroweak effect of the Standard Model, which has been evaluated to two loop order. The Standard Model prediction of a_μ consists of five parts:

(i) Pure QED contribution

$$a_\mu(\text{QED}) = 116\ 584\ 705.6\ (1.7) \times 10^{-11}, \tag{10}$$

(ii) hadronic vacuum polarization contribution. This is obtained mainly from the measured hadronic production cross section in e^+e^- collisions. We quote here only the latest value that includes additional information obtained from the analysis of hadronic tau decay data: [17]

$$a_\mu(\text{had.1}) = 6\ 924\ (62) \times 10^{-11}, \tag{11}$$

(iii) higher order hadronic vacuum polarization effect: [18,24]

$$a_\mu(\text{had.2}) = -101\ (6) \times 10^{-11}, \tag{12}$$

(iv) hadronic light-by-light scattering contribution: [19]

$$a_\mu(\text{had.ll}) = -79.2\ (15.4) \times 10^{-11}, \tag{13}$$

(v) electroweak contribution of up to two loop orders: [25]

$$a_\mu(\text{weak}) = 153\ (3) \times 10^{-11}. \tag{14}$$

The sum of these contributions gives the prediction of the Standard Model:

$$a_\mu(\text{th}) = 116\ 591\ 602\ (65) \times 10^{-11}. \tag{15}$$

This is in good agreement with the measurements (7), (8), and (9). The uncertainty in (15) comes mainly from the hadronic vacuum-polarization contribution (11). It must be improved by at least a factor of two before we can extract useful physical information from the new high precision measurement of a_μ. Fortunately, this contribution can be calculated from the measured value of R ($= \sigma^{total}(e^+e^- \to \text{hadrons})/\sigma^{total}(e^+e^- \to \mu^+\mu^-)$ in e^+e^- collisions. Future measurements of R at VEPP-2M, VEPP-4M, DAΦNE and BEPS as well as analysis of the hadronic tau decay data [26,27,28], will reduce the uncertainty of this contribution to a satisfactory level.

The contribution of the hadronic light-by-light scattering effect (13) is smaller but is potentially a source of serious problem because it is difficult to express in terms of experimentally accessible observables; it must be evaluated by theoretical consideration. It has been estimated by two groups, within the framework of chiral perturbation theory and the $1/N_c$ expansion.[29,30] Recently the theory dependence of these calculations has been reduced [19] by improving a part of the calculation incorporating the measurements of the $P\gamma\gamma^*$ form factors [31], where P stands for π^0, η, and η'. The value quoted in (13) is the result of this work. Evaluation of these effects in lattice QCD will be particularly timely and interesting.

Although the QED contribution a_μ(QED) given in (10) has a relatively small uncertainty, it is being improved further by new numerical work which will reduce the uncertainty by an order of magnitude.

Because of the unusually high sensitivity of a precise experimental value of a_μ to possible physics beyond the Standard Model, theoretical predictions of the contributions to a_μ of these theories are also of great interest. In general any new particles or interactions which couple to the muon or to the photon contribute to a_μ, whose value then provides a sum rule for physics.[32] In comparison with experimental data from the higher energy colliders (LEP II, Tevatron, LHC), an a_μ value with a precision of 0.35 ppm, as projected for the current BNL experiment, provides a comparable or greater sensitivity to a composite structure of the muon or W boson and also to the new particles in supersymmetric theories (SUSY). For the muon a composite mass scale $\Lambda = 4$ TeV and for the W boson an anomalous magnetic moment $\kappa = 0.04$ would be detectable. Of course, any observation of physics beyond the Standard Model from a_μ would be indirect and would not by itself determine the process involved.

Of particular interest is possible SUSY contributions which arise from smuon-neutralino and sneutrino-chargino loops.[33] They can be significant if the supersymmetric particles are not too massive and if $\tan\beta \equiv v_2/v_1$ is large. In the large $\tan\beta$ limit the one-loop SUSY effect gives

$$a_\mu^{\rm SUSY} \simeq \frac{\alpha}{8\pi \sin^2\theta_W} \frac{m_\mu^2}{\tilde{m}^2} \tan\beta \qquad (16)$$

where $\tilde{m}$ represents a typical SUSY loop mass. If $\tan\beta \simeq 40$, a sensitivity of 1 ppm in the a_μ measurement probes $\tilde{m}$ at the 750 GeV level, which may be competitive with direct high energy collider searches. Other theories discussed thus far include muon internal structure[32], extra space-time dimensions[34] and possible violations of CPT and/or Lorentz invariance.[35]

3 Relativistic two-body bound states

It was Breit[36] who obtained the first relativistic correction to bound systems beyond that of the Dirac equation in the static Coulomb potential. Since Bethe and Salpeter gave a relativistic formulation of two-body systems within the framework of QED it has been customary to treat such a system starting from the Bethe-Salpeter equation. Various radiative, binding, and recoil correction terms have been evaluated in this scheme. In recent years, however, a new approach, called NRQED[37], has been developed, which is more readily applicable to bound state problems and enables us to calculate higher order radiative and binding corrections in a systematic manner as an expansion in powers of α and of the velocity of bound particles. (Note that NRQED is not a nonrelativistic approximation to QED. It is a systematic expansion of QED in the electron velocity and is fully equivalent to QED on re-summation.)

Obviously, it is impossible to cover all topics of relativistic two-body bound systems in this short talk. Instead I will select just a few topics from those of my current interest.

3.1 Muonium hyperfine structure

The hyperfine splitting of the muonium ground state can be calculated very precisely in QED or NRQED, being relatively free from the effect of hadronic and weak interactions. Together with the recent precise measurement: [38]

$$\Delta\nu(\text{exp}) \;=\; 4\,463\,302\,765\,(53)\ \text{Hz} \qquad [12\times 10^{-9}], \tag{17}$$

this calculation provides the most stringent test of bound-state QED.

The bulk of the hyperfine splitting is given by the Fermi formula

$$\begin{aligned} E_F &= \frac{16}{3}\alpha^2 c R_\infty \frac{m}{M}\left[1+\frac{m}{M}\right]^{-3} \\ &= 4\,453\,839\,075\,(511)\ \text{Hz}, \end{aligned} \tag{18}$$

where R_∞ is the Rydberg constant for infinite nuclear mass, and m and M are the electron and muon masses, respectively. Here the values for α, R_∞ are from Ref. [39], and that for M/m is from Ref. [38]:

$$\begin{aligned} \alpha^{-1} &= 137.035\,999\,76\,(50) && [3.8\times 10^{-9}], \\ R_\infty &= 10\,973\,\ 731.568\,549\,(83)\ \text{m}^{-1} && [7.6\times 10^{-12}], \\ \frac{M}{m} &= 206.768\,277\,(24) && [117\times 10^{-9}]. \end{aligned} \tag{19}$$

Many theoretical corrections to E_F have been calculated over the last 50 years, mostly starting from the Bethe-Salpeter equation, although some recent evaluations of higher-order terms have been carried out by the NRQED method. Taking advantage of the expansion in the small parameter m/M, they may be classified into non-recoil and recoil terms, both including radiative and binding effects. In addition there are contributions of hadronic vacuum-polarization and weak interaction effects. Thus one may write

$$\Delta\nu(\text{theory}) = \Delta\nu(\text{non-recoil}) + \Delta\nu(\text{recoil}) + \Delta\nu(\text{hadron}) + \Delta\nu(\text{weak}). \tag{20}$$

I would like to sketch the calculation of higher order terms of the muonium hyperfine structure. A purely radiative non-recoil term of order $\alpha(Z\alpha)$ and an approximate value of the $\alpha(Z\alpha)^2$ term (where Ze is the nuclear charge) have been known for some time.[40] Of course $Z = 1$ for the muon, but it is kept here in order to identify the origin of the correction terms. Note that each radiative photon on the electron line contributes a factor α, each on the muon line a factor $Z^2\alpha$, and each jumping from electron and muon a factor $Z\alpha$. These Z's are straightforward. The analysis of the bound state problem becomes complicated because the factor $Z\alpha$ (as well as $\ln(Z\alpha)$) also arises from the effect of binding on the velocity distribution of atomic electrons. Recently, radiative corrections of orders $\alpha(Z\alpha)^2$ and $\alpha^2(Z\alpha)$ have been evaluated with high precision.[41,42,43,44,45] Including these results and the Fermi term E_F itself, the non-recoil term of the muonium hyperfine structure can be written as

$$\begin{aligned}\Delta\nu(\text{non-recoil}) &= E_F(1+a_\mu)\left(1+b+a_e+\alpha(Z\alpha)\left(\ln 2-\frac{5}{2}\right)\right.\\ &- \frac{8\alpha(Z\alpha)^2}{3\pi}\ln(Z\alpha)\left[\ln(Z\alpha)-\ln 4+\frac{281}{480}\right]\\ &+ \left.16.9042(11)\frac{\alpha(Z\alpha)^2}{\pi}+0.7717(4)\frac{\alpha^2(Z\alpha)}{\pi}\right)\\ &= 4\ 464\ 098\ 346\ (511)\ \text{Hz}. \qquad (21)\end{aligned}$$

Here a_e and a_μ are the anomalous magnetic moments of the electron and muon, respectively, and b is the Breit term

$$b = \frac{1}{\beta(2\beta-1)} - 1, \quad \beta = (1-(Z\alpha)^2)^{1/2}. \qquad (22)$$

The factor $(1+a_\mu)$ in (21) is in accord with our definition (18) of E_F. The coefficient 16.9042 of the $\alpha(Z\alpha)^2$ term is different from 15.39 given in an earlier literature [40]. This is because the latter contains, besides the $\alpha(Z\alpha)^2$ term, terms of order $\alpha(Z\alpha)^3$ and higher. More recent evaluation of this term without expansion in $Z\alpha$ is given in Ref. [46]. For a detailed discussion see Ref. [43].

In order to match the precision of the new measurement [38], however, it is necessary to improve the theoretical prediction of terms of order α^4. This is because, when enhanced by the factor $\ln^2(m/\gamma)$ or $\ln(m/\gamma)$ where $m/\gamma = 1/Z\alpha$, corrections of order $\alpha(Z\alpha)^3$ and $\alpha^2(Z\alpha)^2$ contribute to the muonium hyperfine structure as much as the $\alpha^2(Z\alpha)$ term. (There is no $\ln(Z\alpha)$ enhancement for the α^4, $\alpha^3(Z\alpha)$ and $(Z\alpha)^4$ terms.)

The leading log contribution of the $\alpha(Z\alpha)^3$ term is known exactly.[41,44,47,48] The non-log term has also been estimated.[46] Altogether we have

$$\begin{aligned}\Delta\nu(\alpha(Z\alpha)^3) &= \alpha(Z\alpha)^3 E_F \ln\left(\frac{m}{\gamma}\right)\left[\left(5\ln 2-\frac{191}{16}\right)+\frac{13}{24}\right]\\ &- (12\pm 2)\alpha(Z\alpha)^3 E_F/\pi\\ &= -542\ (8)\ \text{Hz}, \qquad (23)\end{aligned}$$

where the first term inside the brackets is the contribution from the radiative photon and the second is due to the vacuum polarization. The uncertainty is an estimate based on the non-log part of the $\alpha(Z\alpha)^3$ term given in Ref. [46]. We are in the process of improving this estimate.[49]

For the $\alpha^2(Z\alpha)^2$ term the leading log part plus an estimate of non-log part gives [50,51]

$$\begin{aligned}
\Delta\nu(\alpha^2(Z\alpha)^2) &= \frac{\alpha^2(Z\alpha)^2}{\pi^2}E_F \ln^2\left(\frac{m}{\gamma}\right)\left(-\frac{4}{3}\right)\left(1-\frac{4\pi m}{\alpha M}\right) \\
&+ \frac{\alpha^2(Z\alpha)^2}{\pi^2}E_F \ln\left(\frac{m}{\gamma}\right)\left[-2\left(-\frac{4358}{1296}-\frac{10}{27}\pi^2+\frac{3}{2}\pi^2\ln 2\right.\right. \\
&\left. - \frac{9}{4}\zeta(3)\right) - \frac{3}{2}\left(\frac{197}{144}+\left(\frac{1}{2}-3\ln 2\right)\zeta(2)+\frac{3}{4}\zeta(3)\right) \\
&\left. + \left(1-\frac{4\pi m}{\alpha M}\right)\frac{8}{3}\left(-\ln 2+\frac{3}{4}\right)+\frac{1}{2}\left(-\frac{11}{9}-1+\frac{8}{15}\right)\right] \\
&+ \frac{\alpha^2(Z\alpha)^2}{\pi^2}E_F\left(1-\frac{4\pi m}{\alpha M}\right)(10\pm 2.5) \\
&= 193\ (24)\ \text{Hz}, \qquad (24)
\end{aligned}$$

where the factor m/M originates from the reduced mass effect. An error in an earlier literature[41] due to an incorrect treatment of the low energy contribution is rectified in (24).

The pure recoil and $\mathcal{O}(\alpha)$ radiative recoil corrections, except for the last term of order $(m/M)^2$ obtained recently[52], have been known for a while[40,53,54]

$$\begin{aligned}
\Delta\nu(\text{recoil}) &= E_F\left(-\frac{3Z\alpha}{\pi}\frac{mM}{M^2-m^2}\ln\frac{M}{m}\right. \\
&\left. + \frac{\gamma^2}{mM}\left[2\ln\frac{m_r}{2\gamma}-6\ln 2+\frac{65}{18}\right]\right) \\
&+ E_F\frac{\alpha(Z\alpha)}{\pi^2}\frac{m}{M}\left(-2\ln^2\frac{M}{m}+\frac{13}{12}\ln\frac{M}{m}+6\zeta(3)+\zeta(2)\right. \\
&- \frac{71}{72}+3\pi^2\ln 2+Z^2\left[\frac{9}{2}\zeta(3)+\frac{39}{8}-3\pi^2\ln 2\right] \\
&\left. + \frac{\alpha}{\pi}\left[-\frac{4}{3}\ln^3\frac{M}{m}+\frac{4}{3}\ln^2\frac{M}{m}+\mathcal{O}\left(\ln\frac{M}{m}\right)\right]\right) \\
&+ E_F\alpha(Z\alpha)\left(\frac{m}{M}\right)^2\left[\left(-6\ln 2-\frac{3}{4}\right)-\frac{17}{12}Z^2\right] \\
&= -795\ 227\ (2)\ \text{Hz}, \qquad (25)
\end{aligned}$$

where $\gamma \equiv Z\alpha m_r$, $m_r = mM/(m+M)$.

Further refinement of $\Delta\nu(\text{recoil})$ comes from evaluation of $(Z\alpha)^3(m/M)$ terms. One of the leading log terms of this contribution arises from the recoil hfs potential $V_{\rm hfs}$ (the first line of (25)) and the 1-photon exchange part of the Coulomb Green function (G_{1p}): [51]

$$\begin{aligned}\Delta\nu^{(1)}((Z\alpha)^3(m/M)) &= 2 < V_{\rm hfs}G_{1p}V_K > +2 < V_{\rm hfs}G_{1p}V_D > \\ &\simeq -\frac{3(Z\alpha)^3}{\pi}\frac{m}{M}E_F\ln\left(\frac{M}{m}\right)\ln\left(\frac{m}{\gamma}\right) \\ &= -210\ (43)\ \mathrm{Hz}, \qquad (26)\end{aligned}$$

where V_K and V_D are the k^4 kinetic energy and the Darwin term, respectively. The uncertainty is estimated assuming that it is a factor $\ln(Z\alpha)$ smaller than the leading term.

Another log term arises from the Salpeter recoil correction to the Lamb shift.[55] This contribution involves both relativistic and nonrelativistic regions. Correcting an error in an earlier literature [41] we find [44,50,51]

$$\begin{aligned}\Delta\nu^{(2)}((Z\alpha)^3(m/M)) &= (Z\alpha)^3\pi E_F\frac{m}{M}\ln^2\left(\frac{m}{\gamma}\right)\left[-\frac{2}{3}\right] \\ &+\frac{(Z\alpha)^3}{\pi}E_F\frac{m}{M}\ln\left(\frac{m}{\gamma}\right)\left[-2C_S+\frac{32}{3}\left(-\ln 2+\frac{3}{4}\right)\right] \\ &+\frac{(Z\alpha)^3}{\pi}E_F\frac{m}{M}(40\pm 22) \\ &= -194\ (59)\ \mathrm{Hz}, \qquad (27)\end{aligned}$$

where one factor of $Z\alpha$ is the radiative correction due to exchange of a transverse photon between the electron and the muon and C_S is given by

$$\begin{aligned}C_S = &-\frac{1}{9}+\frac{7}{3}(2\ln 2+3)-\frac{2}{1-(m/M)^2} \\ &\times\left(\ln(1+m/M)-(m/M)^2\ln(1+M/m)\right). \qquad (28)\end{aligned}$$

The hadronic vacuum polarization contributes [56]

$$\Delta\nu(\mathrm{hadron}) = \frac{\alpha(Z\alpha)}{\pi^2}\frac{mM}{m_\pi^2}(3.5988\pm 0.1045)E_F = 240\ (7)\ \mathrm{Hz}, \qquad (29)$$

where m_π is the charged pion mass.

The weak-interaction contribution due to the Z^0 exchange is [57,58,59]

$$\Delta\nu(\mathrm{weak}) = -G_F\frac{3\sqrt{2}mM}{8\alpha\pi}E_F \simeq -65\ \mathrm{Hz}, \qquad (30)$$

where G_F is the Fermi weak coupling constant.

Collecting all these contributions we obtain the best current estimate

$$\Delta\nu(\text{theory}) = 4\ 463\ 302\ 542\ (510)\ (33)\ \text{Hz}\ , \tag{31}$$

where the first and second errors come from those of M/m and α in (19).

The uncertainties in (23), (24), (26), and (27) are 8, 24, 43, and 59 Hz, respectively. For further improvement it is necessary to evaluate the non-logarithmic terms of the $\mathcal{O}(\alpha^4)$ and $\mathcal{O}(\alpha^3(m/M))$ contributions. At present, work is in progress to evaluate the complete $\alpha(Z\alpha)^3$ term.[49,51] The theoretical uncertainty from the not-yet calculated terms is probably less than 100 Hz.

3.2 Ortho-Positronium

Current interest in this system is inspired by the long-standing discrepancy between the Michigan measurements of the life-time of ortho-positronium [60,61]

$$\Delta\Gamma = 7.0514\ (14)\ \mu s^{-1}, \tag{32}$$

$$\Delta\Gamma = 7.0482\ (10)\ \mu s^{-1}, \tag{33}$$

and the QED calculation to order α. The current theoretical status may be expressed as [62]

$$\begin{aligned}\Delta\Gamma^{th} &= \Gamma_0 \left(1 - 10.286\ 606\ (10)\ \frac{\alpha}{\pi} + \frac{1}{3}\alpha^2 \ln\alpha + B\left(\frac{\alpha}{\pi}\right)^2 \right.\\ &\left. - \frac{3}{2}\frac{\alpha^2}{\pi}\ln^2\alpha + \cdots\right),\\ &\simeq (7.0382\ +\ 3.9\times 10^{-5}\ B)\ \mu s^{-1},\end{aligned} \tag{34}$$

where B stands for the non-logarithmic two-loop effect which is not yet fully known. To explain the discrepancy between (33) and (34) one would need $B \simeq 250$, which is nearly an order of magnitude larger than is expected from QED. However, recent Tokyo measurement [63]

$$\Delta\Gamma = 7.0398\ (25)\ (15)\ \mu s^{-1}, \tag{35}$$

where the first error is statistical and the second systematic, requires a much smaller B ($\simeq 40$), in better agreement with theoretical expectation.

A recent Michigan paper [64] shows that thermalization of positronium in gases takes longer than previously assumed. A new experiment is underway at Tokyo to improve on their earlier result.[65]

Inspired by the Michigan measurements[61] numerous theoretical works have been carried out to see whether such a large discrepancy can be understood as a large $\mathcal{O}(\alpha^2)$ binding effect. [62,66,67,68,69,70,71,72,73] None have produced large enough term, however. One early calculation [68] gave some hope for sizable (though not sufficiently large) contribution to B due to a one-photon virtual annihilation term:

$$\Delta B = 9.0074\ (9). \tag{36}$$

Recently, however, it was found that this term receives a substantial correction from soft virtual photons of momentum of order $m\alpha$.[73] With this correction (36) becomes

$$\Delta B = -0.8622\ (9). \tag{37}$$

This seems to be a warning signal that it may not be possible to generate a large correction term of order α^2 from the bound state effect after all. For a recent survey of the general positronium problem see Ref. [74].

3.3 Hydrogen atom

The theory of the hydrogen atom is of course very similar to that of muonium. The only difference, although crucial, is that the effect of hadronic interaction must be taken into account.

The hydrogen atom is currently a subject of very active experimental study. Developments in laser spectroscopy have enabled the exploitation of Doppler-free two-photon transitions between various states. Some of these works led to extremely precise values of the Rydberg constant: [75,76]

$$R_\infty = 10\ 973\ \ 731.568\ 639\ (91)\ \text{m}^{-1}\ [8.3 \times 10^{-12}], \tag{38}$$

$$R_\infty = 10\ 973\ \ 731.568\ 516\ (84)\ \text{m}^{-1}\ [7.7 \times 10^{-12}]. \tag{39}$$

The value of R_∞ given in (19) is derived mainly from these measurements.

Now that these measurements are so precise one can determine the proton charge radius from the measurement of the hydrogen Lamb shift. Assuming the validity of the QED bound state calculations, these measurements lead to the proton charge radius [77]

$$r_p \equiv < r^2 >^{1/2} = 0.883\ (14)\ \text{fm}. \tag{40}$$

This is somewhat larger than the value deduced from elastic electron-proton scattering experiment at Mainz [78]

$$r_p \equiv < r^2 >^{1/2} = 0.862\ (11)\ \text{fm}, \tag{41}$$

which was obtained by analyzing the scattering data in the first Born approximation. Recently it has been shown [79] that the small discrepancy between (40) and (41) may be accounted for by Coulomb and recoil corrections. (An earlier Stanford experiment[80] gave a smaller value (0.805 (11) fm). But this may not be taken seriously since it is based on an (incorrect) $Q^2 \to 0$ extrapolation of the dipole form factor fitted at higher Q^2.)

For a precision test of QED, better measurements of the elastic electron-proton scattering cross section are highly desirable. On the theoretical side, r_p is one of the fundamental quantities that can in principle be calculated from QCD. Unfortunately, lattice QCD calculations have thus far given only crude values for r_p.[81,82]

3.4 Muonic Hydrogen

This atom is very similar to the ordinary hydrogen atom. However, it has a new feature that the vacuum-polarization effect is enhanced by a huge factor $(m_\mu/m_e)^2$ and hence is much larger than the pure radiative corrections. The hadronic vacuum polarization is similarly scaled up.

The interest in muonic hydrogen arises from the fact that, if the theory is made sufficiently accurate, it not only provides another high precision test of QED but will lead to a very precise *experimental* determination of the proton charge radius r_p. This is exactly what one would need to reduce the theoretical uncertainty in ordinary hydrogen atom discussed in Sec. 3.3, making it much more useful for precision testing of QED.

An experiment is being prepared at PSI to measure the Lamb shift of the muonic hydrogen to a precision of 10 ppm.[83]

On the theoretical side, the largest theoretical uncertainty was the α^3 vacuum-polarization contribution. Recently we have evaluated this term by at least two independent methods, using the exact α^3 vacuum-polarization term and its Padé approximants.[84,85] Our result is

$$\begin{aligned}\Delta E^{(6)} &= 0.120\,045\,(12)\; m_r(Z\alpha)^2\left(\frac{\alpha}{\pi}\right)^3 \\ &= 0.007\,608\,(1)\ \text{meV}. \end{aligned} \tag{42}$$

The total Lamb shift $\mathcal{L}$ is

$$\begin{aligned}\mathcal{L} &= \Delta E(\text{lower order}) + \Delta E^{(6)} + \Delta E(\text{had.vp}) \\ &= 206.049 - 5.1975\, r_p^2 \\ &+ 0.007\,608\,(1) \\ &+ 0.011\,3\,(3) \\ &= 206.068 - 5.1975\, r_p^2\ (\text{in meV}),\end{aligned} \tag{43}$$

where ΔE(lower order) and ΔE(had.vp) are from Refs. [86,87].

The QED part of $\mathcal{L}$ has sufficient precision now. If $\mathcal{L}$ is measured to 10 ppm or ± 0.002 meV as planned, r_p^2 can be determined 10 times more accurately than it can at present. This will lead to a substantial improvement of the QED test for the ordinary hydrogen atom. It may also encourage a more precise lattice QCD calculation of r_p^2.

4 Various determinations of α

Currently the best measurements of α, with a relative standard uncertainty of less than 1×10^{-7}, are those based on the quantum Hall effect [20], the ac Josephson effect [39], the de Broglie wave length of a neutron beam [88], the muonium hyperfine structure[38], an absolute optical frequency measurement of the Cesium D_1 line [89], and the electron magnetic moment anomaly (6):

$$\begin{aligned}
\alpha^{-1}(\text{q.Hall}) &= 137.036\,003\,7(33) \qquad [2.4 \times 10^{-8}],\\
\alpha^{-1}(\text{acJ}) &= 137.035\,988\,0(51) \qquad [3.7 \times 10^{-8}],\\
\alpha^{-1}(\text{h/m}_\text{n}) &= 137.036\,011\,9\ (51) \qquad [3.7 \times 10^{-8}],\\
\alpha^{-1}(\mu\text{hfs}) &= 137.035\,993\,2\ (83) \qquad [6.0 \times 10^{-8}],\\
\alpha^{-1}(\text{optical}) &= 137.035\,992\,4\ (41) \qquad [3.0 \times 10^{-8}],\\
\alpha^{-1}(\text{a}_\text{e}) &= 137.035\,999\,58\ (52) \quad [3.8 \times 10^{-9}].
\end{aligned} \tag{44}$$

Precision determinations of silicon lattice spacing in the measurements [90,91] lead to somewhat more accurate values of $\alpha^{-1}(\text{h/m}_\text{n})$ than that of Ref. [88]. A preliminary result from measurements based on the atom beam interferometry of the Cesium atom has also been reported [92]

$$\alpha^{-1}(\text{C}_\text{s}) = 137.036\,002\,8\ (77) \qquad [5.6 \times 10^{-8}]. \tag{45}$$

The fine structure of the He atom will be another good source of high precision α when the current theoretical work is completed.[93] Note that measurements of the He atom fine structure are already much more precise than theory [94] and further improvements are in progress.[95]

Continuing theoretical work on a_e will soon reduce the intrinsic uncertainty of theory further by a factor of 3 to 4. If the experimental precision is improved by an order of magnitude [9,10], the precision in $\alpha(a_\text{e})$ will exceed 1 part in 10^9. In the near future determinations of α based on atom beam interferometry [96,97] and single electron tunneling [98] may also reach a precision of 10^{-9}, competing with the improved $\alpha(a_e)$.

How far can we go in testing quantum mechanics beyond the precision of 10^{-9} ? I do not know. I only point out that the theory of a_e will hit a very serious brick wall at the level of

$$\left(\frac{\alpha}{\pi}\right)^5 \simeq 0.68 \times 10^{-13}$$
$$\simeq 0.58 \times 10^{-10} a_e. \quad (46)$$

Even if the measurement of a_e achieves a precision of 10^{-10}, an improvement of $\alpha(a_e)$ much beyond $\sim 10^{-10}$ will not be possible unless the α^5 term is evaluated or estimated reliably. A very formidable task indeed !

5 Concluding remarks

I have already discussed in Sec. 2.2 the significance of the high precision measurement of a_μ as a test of the Standard Model and beyond.

Thus let me focus here on a_e as a tool to test the validity of quantum mechanics. The *intrinsic* uncertainty in the theoretical value of a_e listed in (6) or (44) is already quite small.

It is fortunate that many independent ways have become available for measuring α with high precision. This offers an opportunity to examine the theoretical bases of *all* these measurements on an equal footing. The precision of these measurements requires that the underlying theories must be valid to the same extent. Such theories must be based on quantum mechanics, or, more precisely, its extention including the relativistic effects, radiative corrections, and renormalization with respect to electroweak and strong interactions.

At present such a theoretical basis is fully established only for $\alpha(a_e)$ and for α determined by the muonium hyperfine structure and some atomic measurements. Although the principle of neutron de Broglie wavelength measurement looks very simple, it requires determination of the *free* neutron mass from nuclear physics, which can be fully justified only within the context of a renormalizable quantum field theory. The α determined in condensed matter physics has another unresolved problem. It is argued that, although the theories of the ac Josephson effect and the quantum Hall effect start from the condensed matter physics Hamiltonian with its usual simplifying approximations, their predictions may in fact be valid to the same or higher degree than that of $\alpha(a_e)$ because they are derived from the gauge invariance and single-valuedness of the wave function and are not dependent on specific approximation procedures practiced in condensed matter physics. It is important to note, however, that this assertion has not yet been proven rigorously: In particular, the theory of condensed matter physics in the present form is not renormalizable. The NRQED method of Caswell and Lepage [37] may provide a rigorous starting point for establishing a sounder basis for condensed matter physics.

Currently, the Standard Model is the simplest theory to represent extended quantum mechanics, and within its context all measurements of α which can be reduced to those of the charge form factor or the magnetic form factor at zero momentum transfer must give the same answer.

An expectation that the α's obtained from the charge form factor may be affected by short range interactions by $\sim (\alpha/\pi)^2(m_e/m_\rho)^2 \simeq 2.4 \times 10^{-12}$, where m_ρ is the ρ meson mass, is not realized. This effect cannot be detected since it is absorbed by charge renormalization, which applies universally to all measurements of charge form factor at threshold. The magnetic form factor, on the other hand, will be affected by the known short range forces by $\sim 1.7 \times 10^{-12}$, which contributes about 1.5 ppb to $\alpha(a_e)$. But this effect is already taken into account in defining $\alpha(a_e)$. Thus $\alpha(a_e)$ determined from the magnetic form factor must have the same value as α's derived from the charge form factor. This equality is not affected by short distance effects. This remark applies as well to α derived from the muonium hyperfine structure.

Effects beyond the Standard Model on $\alpha(a_e)$ can also be estimated using the measured a_μ insofar as the new interaction satisfies μ-e universality. Relative to the weak interaction, this effect will scale as $(m_W/m_X)^2$, where m_X is the mass scale of the new interaction. Such an effect will be too small to be significant at the present level of precision of $\alpha(a_e)$. Another useful constraint on a new interaction may come from a new measurement of the muon electric dipole moment. [99]

How well does the constancy of α stand up against the evidence ? The data (44) cast some doubt on this point, although it is not yet too serious. Further improvement of precision in the α value, both experimental and theoretical, will be needed to obtain a definitive answer to this question. If the discrepancy persists after the uncertainties are reduced, however, we may have to reexamine the theoretical basis of various determinations of α very closely.

In conclusion, comparison of α's provides a concrete means to quantify the extent of applicability of quantum mechanics in the extended sense. If quantum mechanics fails this test at some point, it would have a profound impact on the future of physics.

Acknowledgments

This work is supported in part by the U. S. National Science Foundation.

References

1. G. Breit, *Phys. Rev.* **72**, 984L (1947); **73**, 1410 (1948); **74**, 656 (1948).
2. J. E. Nafe, E. B. Nelson, and I. I. Rabi, *Phys. Rev.* **71**, 914 (1947).

3. J. E. Nagle, R. S. Julian, and J. R. Zacharias, *Phys. Rev.* **72**, 971 (1947).
4. P. Kusch and H. M. Foley, *Phys. Rev.* **72**, 1256 (1947).
5. J. Schwinger, *Phys. Rev.* **73**, 416L (1948).
6. R. S. Van Dyck, Jr., P. B. Schwinberg, and H. G. Dehmelt, *Phys. Rev. Lett.* **59**, 26 (1987).
7. L. S. Brown *et al.*, *Phys. Rev.* A **32**, 3204 (1985); *Phys. Rev. Lett.* **55**, 44 (1985).
8. R. S. Van Dyck, Jr., P. B. Schwinberg, and H. G. Dehmelt, in *The Electron*, ed. by D. Hestenes and A. Weingartshofer (Deventer: Kluwer, 1991), pp. 239 - 293.
9. R. Mittleman *et al.*, *Phys. Rev. Lett.* **75**, 2839 (1995).
10. G. Gabrielse and J. Tan, *Advances in Atomic, Molecular and Optical Physics*, Supplement **2** (1994), p. 267.
11. See references in the review article by T. Kinoshita, *Rep. Prog. Phys.* **59**, 1459 (1996).
12. V. W. Hughes and T. Kinoshita, *Rev. Mod. Phys.* **71**, S133 (1999).
13. S. Laporta and E. Remiddi, *Phys. Lett.* B **379**, 283 (1996).
14. T. Kinoshita, *Phys. Rev. Lett.* **75**, 4728 (1995).
15. For the literature prior to 1990, see T. Kinoshita, *Theory of the anomalous magnetic moment of the electron – Numerical approach,* in *Quantum Electrodynamics*, ed. by T. Kinoshita (World Scientific, Singapore, 1990), pp. 218 - 321.
16. A. Czarnecki and M. Skrzypek, *Phys. Lett.* B **449**, 354 (1999).
17. M. Davier and A. Höcker, *Phys. Lett.* B **435**, 427 (1998).
18. B. Krause, *Phys. Lett.* B **390**, 392 (1997).
19. M. Hayakawa and T. Kinoshita, *Phys. Rev.* D **57**, 465 (1998).
20. A. Jeffery *et al.*, *IEEE Trans. Instrum. Meas.* **46**, 264 (1997).
21. F. J. M. Farley and E. Picasso, *The Muon g - 2 Experiments*, in *Quantum Electrodynamics*, edited by T. Kinoshita (World Scientific, Singapore, 1990), pp. 479 - 559.
22. R. M. Carey *et al.*, *Phys. Rev. Lett.* **82**, 1632 (1999).
23. B. L. Roberts, hep-ex/0002005.
24. T. Kinoshita, B. Nižić, and Y. Okamoto, *Phys. Rev.* D **31**, 2108 (1985).
25. A. Czarnecki, B. Krause, and W. J. Marciano, *Phys. Rev. Lett.* **76**, 3267 (1996), *Phys. Rev.* D **52**, 2619 (1995); S. Peris, M. Perrottet, and E. de Rafael, *Phys. Lett.* B **355**, 523 (1995); T. V. Kukhto, E. A. Kuraev, A. Schiller, and Z. K. Silagadze, *Nucl. Phys.* B **371**, 567 (1992); G. Degrassi and G. F. Giudice, *Phys. Rev.* D **58**, 053007 (1998).
26. Z.-G. Zhao, *Proc. of the XIX International Symposium on Lepton and*

Photon Interactions at High Energies, Stanford, August 1999, ed. M. Peskin.
27. S. Bertolucci, *Proc. of the XIX International Symposium on Lepton and Photon Interactions at High Energies*, Stanford, August 1999, ed. M. Peskin.
28. J. Z. Bai *et al.*, *Phys. Rev. Lett.* **84**, 594 (2000).
29. M. Hayakawa, T. Kinoshita and A. I. Sanda, *Phys. Rev. Lett.* **75**, 790 (1995); *Phys. Rev.* D **54**, 3137 (1996).
30. J. Bijnens, E. Pallante and J. Prades, *Phys. Rev. Lett.* **75**, 1447 (1995); *Phys. Rev. Lett.* **75**, 3781 (1995); *Nucl. Phys.* B **474**, 379 (1996).
31. J. Gronberg *et al.*, *Phys. Rev.* D **57**, 33 (1998).
32. T. Kinoshita and W. J. Marciano, *Theory of the Muon Anomalous Magnetic Moment*, in *Quantum Electrodynamics*, ed. by T. Kinoshita (World Scientific, Singapore, 1990), pp. 419 - 478.
33. A. Czarnecki and W. J. Marciano, *Nucl. Phys. B* (Proc. Suppl.) **76**, 245 (1999).
34. M. L. Graesser, hep-ph/9902310 (1999); G. Shiu, R. Shrock, and S.-H. Henry Tye, *Phys. Lett.* B **458**, 274 (1999).
35. R. Jackiw and V. A. Kostelecký, *Phys. Rev. Lett.* **82**, 3572 (1999); R. Bluhm, V. A. Kostelecký, and C. D. Lane, *Phys. Rev. Lett.* **84**, 1098 (2000).
36. G. Breit, *Phys. Rev.* **34**, 553 (1929).
37. W. E. Caswell and G. P. Lepage, *Phys. Lett.* **167B**, 437 (1986).
38. W. Liu *et al.*, *Phys. Rev. Lett.* **82**, 711 (1999).
39. P. J. Mohr and B. N. Taylor, to be published in Rev. Mod. Phys.
40. J. R. Sapirstein and D. R. Yennie, in *Quantum Electrodynamics*, edited by T. Kinoshita (World Scientific, Singapore,1990), pp. 560 - 672.
41. T. Kinoshita and M. Nio, *Phys. Rev. Lett.* **72**, 3803 (1994).
42. M. I. Eides and V. A. Shelyuto, *Phys. Rev.* A **52**, 954 (1995).
43. T. Kinoshita and M. Nio, *Phys. Rev.* D **53**, 4909 (1996).
44. M. Nio and T. Kinoshita, *Phys. Rev.* D **55**, 7267 (1997).
45. K. Pachucki, *Phys. Rev.* A **54**, 1994 (1996).
46. S. A. Blundell, K. T. Cheng, and J. Sapirstein, *Phys. Rev. Lett.* **78**, 4914 (1997).
47. G. P. Lepage, private communication.
48. S. G. Karshenboim, *Z. Phys.* D **36**, 11 (1996).
49. M. Nio and T. Kinoshita, to be published.
50. S. G. Karshenboim, *Zh. Eksp. Teor. Fiz.* **103**, 1105 (1993), [*JETP* **76**, 541 (1993)].
51. T. Kinoshita and M. Nio, in "*Frontier Tests of QED and Physics of the*

Vacuum" eds. E. Zavattini, D. Bakalov, and C. Rizzo (Heron Press, Sofia 1998), pp. 151 - 167.
52. M. I. Eides, H. Grotch, and V. A. Shelyuto, *Phys. Rev.* D **58**, 013008 (1998).
53. M. I. Eides, S. G. Karshenboim, and V. A. Shelyuto, *Phys. Lett.* **177**, 425 (1988); **202B**, 572 (1988); M. I. Eides, S. G. Karshenboim, and V. A. Shelyuto, *Ann. Phys.* **205**, 231 (1991); **205**, 291 (1991).
54. M. I. Eides, S. G. Karshenboim, and V. A. Shelyuto, Phys. Lett. **B216**, 405 (1989); M. I. Eides and V. A. Shelyuto, *Phys. Lett.* B **146**, 241 (1984).
55. E. E. Salpeter, *Phys. Rev.* **87**, 328 (1952).
56. A. Karimkhodzhaev and R. N. Faustov, *Yad. Fiz.* **53**, 1012 (1991) [Eng. transl.: Sov. J. Nucl. Phys. **53**, 626 (1991)]; R. N. Faustov, A. Karimkhodzhaev, and A. P. Martynenko, *Phys. Rev.* A **59**, 2498 (1999).
57. M. A. B. Bég and G. Feinberg, *Phys. Rev. Lett.* **33**, 606 (1974); G. T. Bodwin and D. R. Yennie, *Phys. Rep.* **43C**, 267 (1978).
58. M. A. B. Bég and G. Feinberg, *Phys. Rev. Lett.* **35**, 130 (1974).
59. Unpublished note by M. Nio, 1995, private communications from J. Sapirstein, 1995, and a paper by M. I. Eides, *Phys. Rev.* A **53**, 2953 (1996).
60. C. I. Westbrook *et al.*, *Phys. Rev. Lett.* **58**, 1322 (1987); *Phys. Rev.* A **40**, 5489 (1989).
61. J. S. Nico *et al.*, *Phys. Rev. Lett.* **65**, 1344 (1990).
62. G. S. Adkins, *Phys. Rev. Lett.* **76**, 4903 (1996).
63. S. Asai, S. Orito, and N. Shinohara, *Phys. Lett.* B **357**, 475 (1995).
64. M. Skalsey *et al.*, *Phys. Rev. Lett.* **80**, 3727 (1998).
65. S. Asai, private communication.
66. A. P. Burichenko, *Yad. Fiz.* **56**, 123 (1993) [*Phys. At. Nucl.* **56**, 640 (1993)].
67. G. S. Adkins and Y. Shiferaw, *Phys. Rev.* A **52**, 2442 (1995); A. P. Burichenko and D. Yu. Ivanov, *Yad. Fiz.* **58**, 898 (1995) [*Phys. At. Nucl.* **58**, 832 (1995)].
68. G. S. Adkins and M. Lymberopoulos, *Phys. Rev.* A **51**, 2908 (1995).
69. I. B. Khriplovich and A. I. Milstein, *J. Exp. Theor. Phys.* **106**, 689 (1994) [*Sov. Phys. JETP* **79**, 379 (1994)].
70. P. Labelle, G. P. Lepage, and U. Magnea, *Phys. Rev. Lett.* **72**, 2006 (1994).
71. R. N. Faustov, A. P. Martynenko, and V. A. Saleev, *Phys. Rev.* A **51**, 4520 (1995).
72. V. Antonelli *et al.*, *Eur. Phys. J.* C5, 535 (1998).

73. G. S. Adkins and K. Melnikov, and A. Yelkhovsky, *Phys. Rev.* A **60**, 3306 (1999).
74. A. Czarnecki, hep-ph/9911455.
75. Th. Udem *et al.*, *Phys. Rev. Lett.* **79**, 2646 (1997).
76. C. Schwob et al., *Phys. Rev. Lett.* **82**, 4960 (1999).
77. K. Melnikov and T. van Ritbergen, hep-ph/9911277.
78. G. G. Simon *et al.*, *Nucl. Phys.* A **333**, 381 (1980).
79. R. Rosenfelder, nucl-th/9912031. See also S. G. Karshenboim, *Can. J. Phys.* **77**, 241 (1999).
80. D. J. Drickey and L. N. Hand, *Phys. Rev. Lett.* **9**, 521 (1962); L. N. Hand, D. J. Miller, and R. Wilson, *Rev. Mod. Phys.* **35**, 335 (1963).
81. T. Drapers, R. M. Woloshin, and K.-F. Liu, *Phys. Lett.* B **234**, 121 (1990).
82. D. B. Leinweber and T. D. Cohen, *Phys. Rev.* D **47**, 2147 (1993).
83. D. Taqqu et al., Hyperfine Interact. **101/102**, 599 (1996).
84. T. Kinoshita and M. Nio, *Phys. Rev. Lett.* **82**, 3240 (1999).
85. T. Kinoshita and M. Nio, *Phys. Rev.* D **60**, 053008 (1999).
86. K. Pachucki, *Phys. Rev.* A **53**, 2092 (1996).
87. J. L Friar, J. Martorell, and D. W. L. Sprung, nucl-th/9812053.
88. E. Krüger, W. Nistler, and W. Weirauch, *Metrologia* **36**, 147 (1999).
89. Th. Udem et al., *Phys. Rev. Lett.* **82**, 3568 (1999).
90. G. Basile et al., *Phys. Rev. Lett.* **72**, 3133 (1994); *IEEE Trans. Instrum. Meas.* **44**, 538 (1995).
91. K. Nakayama and H. Fujimoto, *IEEE Trans. Instrum. Meas.* **46**, 580 (1997).
92. B. C. Young, Stanford University Ph. D. thesis, 1997.
93. T. Zhang, Z.-C. Yan, and G. W. F. Drake, *Phys. Rev. Lett.* **77**, 1715 (1996); T. Zhang and G. W. F. Drake, *Phys. Rev.* A **54**, 4882 (1996).
94. D. L Shiner, R. Dixson and P. Zhao, *Phys. Rev. Lett.* **72**, 1802 (1994); D. L Shiner and R. Dixson, *IEEE Trans. Instrum. Meas.* **44**, 518 (1995).
95. F. Minardi *et al.*, *Phys. Rev. Lett.* **82**, 1112 (1999).
96. D. S. Weiss, B. C. Young, and S. Chu, *Phys. Rev. Lett.* **70**, 2706 (1993).
97. S. Chu, private communication.
98. M. W. Keller *et al.*, *SCIENCE* **285**, 1706 (1999).
99. Y. K. Semertzidis, in "*Frontier Tests of QED and Physics of the Vacuum*" eds. E. Zavattini, D. Bakalov, and C. Rizzo (Heron Press, Sofia 1998), pp. 369 - 376.

HELIUM FINE STRUCTURE, FROM BREIT INTERACTION TO PRECISE LASER SPECTROSCOPY

F. MINARDI AND M. INGUSCIO

European Laboratory for Non-linear Spectroscopy (LENS) and Istituto Nazionale di Fisica della Materia (INFM), Largo E. Fermi 2, I-50125 Firenze, Italy

E-mail: minardi@lens.unifi.it

1 Introduction: Helium atom and the fine structure constant α

Gregory Breit pioneered the issue of two interacting electrons within the frame of relativistic electrodynamics and, although his interest in this topic has been mainly confined to the early years of his long and various scientific career, his work has represented the starting point for the whole theory of two-electrons atoms. He attacked the problem of the correct form for the retarded interaction between two electrons as early as in 1929, only one year later that Dirac wrote down his famous equation[1].

The step from hydrogen to helium, from one to two electrons atoms, is of course a major one. Already in classical mechanics, the problem of three bodies interacting with a Coulomb potential doesn't allow any exact solution. Quantum mechanics has dealt with this problem since its very beginnings and after some 70 years Helium can now be considered in some sense a "simple" system. Simple systems are what we look for in order to measure the fundamental physical constants of nature. The fine structure constant α deserves a special attention since what is perhaps our most precise theory, quantum electrodynamics QED, is generally build in a perturbative fashion over this small adimensional number, α.

The Helium atom has been progressively become connected to α as its energy levels have been calculated to increasing degrees of precision, so that it seems now within reach the goal of obtaining a value for the fine structure constant α from the spectroscopic results.

2 Breit's two electron theory ...

Since his first paper in 1929[2], Breit aimed to a formulation of a two-particle equation that was correct to within terms $(v/c)^2$ for the energy. He was deliberately attempting to establish an equation for the 16-component Dirac wavefunction, to serve as the starting point for a reduction to "large" components, following the example set by Darwin with Dirac's equation. He was not the first to take over this challenging task: such an equation had already been published by Gaunt and Eddington[3], but G. Breit claimed that it was incorrect. A classical Hamiltonian taking into account retardation effects was derived by Darwin. Guided by the "correspondence" principle, G. Breit transposed this Hamiltonian into the Dirac theory in such a way as to recover the expected equations of motion for the position and momentum operators. He proposed as a wave equation:

$$\Big(i\hbar\frac{\partial}{\partial t} + e(A_0^I + A_0^{II}) - \frac{e^2}{r} + c(\boldsymbol{\alpha}^I \boldsymbol{p}^I + \boldsymbol{\alpha}^{II}\boldsymbol{p}^{II}) + (\alpha_4^I + \alpha_4^{II})mc^2 \qquad (1)$$
$$+\tfrac{e^2}{2}(\boldsymbol{\alpha}^I\boldsymbol{\alpha}^{II} r^{-1} + (\boldsymbol{\alpha}^I \boldsymbol{r})(\boldsymbol{\alpha}^{II}\boldsymbol{r})r^{-3})\Big)\,\psi = 0$$

where we denote with α^I, α^{II} the Dirac matrices of the two electrons, with $\boldsymbol{r}$ their distance and with A_0 the electrostatic potential.

That cast the milestone in the theory of two electrons systems, introducing the term that nowadays is called after him, the "Breit term":

$$B = \frac{e^2}{2r}(\boldsymbol{\alpha}^I\boldsymbol{\alpha}^{II} + (\boldsymbol{\alpha}^I\boldsymbol{r})(\boldsymbol{\alpha}^{II}\boldsymbol{r})r^{-2}) \qquad (2)$$

At that stage G. Breit still considered eq.(1) an approximated yet correct wave equation. He then started the reduction procedure and worked out the trivial zero-order Schrödinger Hamiltonian for the two electrons, including their Coulomb interaction, plus all the operators of order $(v/c)^2$, or equivalently $mc^2\alpha^4$:

$$WU(\boldsymbol{r}_1, \boldsymbol{r}_2) = (H_0 + H_1 + H_2 + H_3 + H_4 + H_5)U(\boldsymbol{r}_1, \boldsymbol{r}_2) \qquad (3)$$
$$W = E - 2mc^2$$

where $U(\boldsymbol{r}_1, \boldsymbol{r}_2)$ represents the large component wavefunction and:

$$H_0 = \frac{1}{2m}((\boldsymbol{p}^I)^2 + (\boldsymbol{p}^{II})^2) - e(A_0^I + A_0^{II}) + \frac{e^2}{r} \qquad (4)$$

$$H_1 = -\frac{1}{8m^3c^2}((\boldsymbol{p}^I)^4 + (\boldsymbol{p}^{II})^4) \qquad (5)$$

$$H_2 = -\frac{e^2}{2r(mc)^2}[\boldsymbol{p}^I\boldsymbol{p}^{II} + \frac{\boldsymbol{r}\,(\boldsymbol{r}\,\boldsymbol{p}^I)\boldsymbol{p}^{II}}{r^2}] \qquad (6)$$

$$H_3 = \frac{e\hbar}{2mc}[\boldsymbol{H}^I\boldsymbol{s}^I + \boldsymbol{H}^{II}\boldsymbol{s}^{II}] \qquad (7)$$

$$H_4 = -\frac{ie\hbar}{(2mc)^2}[\boldsymbol{p}^{\ I}\boldsymbol{\mathcal{E}}^{\ I} + \boldsymbol{p}^{\ II}\boldsymbol{\mathcal{E}}^{\ II}] \tag{8}$$

$$H_5 = \frac{e\hbar}{2(mc)^2}[(\boldsymbol{\mathcal{E}}^{\ I} \times \boldsymbol{p}^{\ I} + \frac{2e}{r^3}\boldsymbol{r} \ \times \boldsymbol{p}^{\ II})\boldsymbol{s}^{\ I} + \\ +(\boldsymbol{\mathcal{E}}^{\ II} \times \boldsymbol{p}^{\ II} + \frac{2e}{r^3}\boldsymbol{r} \ \times \boldsymbol{p}^{\ I})\boldsymbol{s}^{\ II}] \tag{9}$$

$$H_6 = 4(\frac{e\hbar}{2mc})^2\langle(\boldsymbol{\nabla}^{\ II}\boldsymbol{s}^{\ II})(\boldsymbol{\nabla}^{\ I}\boldsymbol{s}^{\ I})r^{-1}\rangle \tag{10}$$

$$H_7 = \frac{e^4}{8mc^2}[\frac{3 - 8(\boldsymbol{s}^{\ I}\boldsymbol{s}^{\ II})}{r^2} + \frac{(\boldsymbol{s}^{\ I}\boldsymbol{r}\ )(\boldsymbol{s}^{\ II}\boldsymbol{r}\ )}{r^4}] \tag{11}$$

(we used the same notation of G. Breit's original papers, except for the spin operators therein denoted with $-\boldsymbol{\sigma}\ /2$; the notation $\langle\rangle$ indicates that differential operators apply to what is inside $\langle\rangle$).

Quite soon, G. Breit doubted of the last term proprtional to e^4 and not vanishing in the limit $\hbar \to 0$. Actually, in the following years he made an effort to find the correct formulation of his theory and to check it against experimental data. As a result, he acknowledged that the term (11) is wrong and must be removed.

Already at that time, Helium levels were regarded as the obvious check for any theory. To obtain quantitative predictions to compare with experimental data, G. Breit established a set of approximate atomic wavefunctions. Extending the work by Hylleras, who considered only S states, G. Breit showed that the six-variables Schrödinger equation for the two interacting electrons within a central Coulomb potential can be reduced to a three-variables equation for any state of assigned orbital angular momentum[4]. He introduced three Euler's angles and wrote the two-electrons wavefunction as a linear combination of products of *angle functions*, depending only on the Euler's angles, that are fully determined by the value of the angular momentum, and *distance functions*, that in turn depend only on r_1, r_2 and r. Though the reduced equation in (r_1, r_2, r) doesn't allow for an exact solution, yet it is amenable of a more practical treatment by means of variational techniques. In his paper, G. Breit worked out explicitely the case of $L = 1$, the P terms.

In a subsequent paper[5], significantly entitled "The fine structure of He as a test of the spin interactions of two electrons" G. Breit employed the Ritz variational method to find the approximate "distance functions" for the $2P$ state and derived the fine structure separations with an accuracy that he estimated to be about 10%. At that time spectroscopic values accurate to $10^{-3}R_\infty\alpha^2$ were available. The comparison with these data showed an acceptable agreement, provided that the doubtful term was neglected. Anyway, he concluded that the precision of theoretical results and experimental data was still too low to discriminate between competitive theories, such as his own and the one by Gaunt.

It is also of interest that G. Breit recognized that the Breit's term was useful only to calculate the energy corrections but the eigenfunctions should be obtained from the Schrödinger wave equation.

3 ... and Developments

In the late 50's Araki[6] and Sucher[7] independently calculated the next terms of the series, they worked out all the terms $O(m_ec^2\alpha^5, m_ec^2\alpha^5 \log\alpha)$, which are due mainly to the electron's anomalous magnetic moment. More than a decade later, in 1972, another big step forward was done by Douglas and Kroll[8], who completed the derivation of all contributions $O(m_ec^2\alpha^6)$.

Parallel to the derivation of the operators, the calculus of the wavefunctions was pursued by variational methods. As we already mentioned, it was started in 1928 by Hylleras, who first proposed for the S states trial wavefunctions in the form:

$$\psi(r_1, r_2) = \sum_{i,j,k} a_{ijk} r_1^i r_2^j r^k e^{-\alpha r_1 - \beta r_2} + b_{ijk} r_1^i r_2^j r^k e^{-\alpha r_2 - \beta r_1}. \quad (12)$$

However, it was the advent of digital computation that allowed a significant advancement, culminating, in the late '50s, with the important work of Pekeris and coworkers, who obtained the non-relativistic energies of the ground and low-lying excited states within 10^{-4} cm^{-1}.

In the last two decades, relevant improvements have been achieved in particular by the group of G. W. F. Drake that, doubling the Hylleras basis set, employed trial wavefunctions of the type[9]:

$$\psi(r_1, r_2) = \sum_{i,j,k} a_{ijk}\chi_{ijk}(\alpha_A, \beta_A) + b_{ijk}\chi_{ijk}(\alpha_B, \beta_B) + (\text{exchange } r_1 \leftrightarrow r_2),$$

$$\chi_{ijk}(\alpha, \beta) = r_1^i r_2^j r^k e^{-\alpha r_1 - \beta r_2}. \quad (13)$$

Due to faster convergence of the series (13), Drake and coworkes obtain the non-relativistic energies within 10^{-15} for the lowest levels. Obviously, the calculation of He levels must be pushed beyond non-relativistic values. The relavistic and QED corrections are taken into account in a perturbative fashion as a power series of α: currently all the operators up to $mc^2\alpha^x, \alpha^x \log\alpha$ have been derived and the ultimate precision in the energies is limited to about 10^{-9} by the higher order, uncalculated terms. However, very accurate variational wavefunctions are not an academic exercise as they are required to evaluate matrix elements of the higher order operators.

For energy differences, such as fine structure separations, the theory takes advantage of several cancellation and allows for predictions with a smaller absolute uncertainty. As an example, the Helium $2\,^3P$ state is splitted into a 3-component fine structure multiplet (see Fig. 1) and the calculation of its separations have been recently extended to include all the $mc^2(\alpha^7, \alpha^7 \log \alpha)$ operators [10,11], with an extimated precision of the order of 10 kHz, i.e. 0.3 ppm.

4 Experiments

4.1 The work of W. Lamb

In 1929, when G. Breit's first paper on this subject was published, the $2\,^3P_1$ and $2\,^3P_2$ levels had just been experimentally resolved for the first time, and fine structure intervals were known to within 1 %. Later on, from '30s to '50s, several groups repeated these measurements with a comparable accuracy and all the results were found in agreement within 1 %, i.e. 0.01 cm^{-1} for the $2\,^3P_0 - 2\,^3P_1$ splitting. Also the separations of the $3\,^3P$ had been determined with a similar accuracy.

But it was only in the late '50s that sub-MHz spectroscopy of Helium fine structure was started by W. E. Lamb Jr., who proposed and implemented a new method based on direct excitation of the $J = 2$ to $J = 1$ and $J = 1$ to $J = 0$ transitions in the microwave domain[12]. Helium atoms were subject to bombardment of electrons with energy just above the $n\,^3P$ excitation threshold. The light emitted by radiative decay was monitored as the sample was illuminated by polarized microwave radiation. The resonance was detected as a change in the radiation pattern of the emitted fluorescence due to the change of atomic orientation induced by the microwave field. A moderate magnetic field of 30 to 60 mT was applied for convenience and it was varied to tune the atomic transitions in resonance with the fixed-frequency microwave field. The calibration of the field intensity against the coils current was done via NMR before and after the measures to the desired level of accuracy but the field inhomogeneity across the interaction volume was identified as the major source of uncertainty. The experiment was first performed on the $3\,^3P$ level because the fluorescence at 388.9 nm is more efficiently detected by photomultipliers than the 1083 nm light emitted in the decay of the $2\,^3P$ level.

A throughout reading of W. Lamb's papers is certainly instructive, as all the aspects of the experiment are discussed: the general principles, the underlying physics involved, the technical details of the experimental realization, the data acquisition and reduction, the analysis of the error sources. The Lamb experiences led to an improvement of a factor, respectively, 7, 2700 and 60 in the accuracy of the $3\,^3P_2 - 3\,^3P_1$, $3\,^3P_1 - 3\,^3P_0$ and $2\,^3P_1 - 2\,^3P_2$ values. As he remarked,

> The results are much more precise than existing data and should be of use in theoretical work on helium fine structure.

It may be worth noticing that the reported $3\,^3P$ intervals equal, within the uncertainty, the most recent values[13], while the $2\,^3P$ interval is only 1.5 standard deviation larger[a].

4.2 The Yale microwaves experiments

A second major advancement in Helium fine structure spectroscopy was realized by V. Hughes and collaborators with a series of experiments that started in the late '60s and continued for more than a decade[14,15,16,17]. In the first issue of the "Comments on Atomic and Molecular Physics", V. Hughes foresaw the relevance of Helium fine structurefor the determination of the fine structure constant α[18]:

> One of the most promising possibilities for determining α with high precision (perhaps 1 ppm or better) is provided by the fine structureintervals in the $2\,^3P_J$ state of helium. Both the intervals $J = 0$ to $J = 1$ and $J = 1$ to $J = 2$ have been measured with a precision of about 2 ppm. Although the precise calculation of the fine structure intervals is a difficult two-electrons problem, it seems to be a tractable problem with modern digital computing machines.

The technique adopted resembles that introduced by W. Lamb: a magnetic-resonance atomic beam technique. To circumvent the low quantum efficiency of fluorescence detection, the magnetization variations induced by the microwaves were directly observed with a double Stern-Gerlach selector followed by an atom detector. Indeed, since the $2\,^3S_1$ metastable state carries an internal energy of about 20 eV, metastable helium atoms can be detected with a quantum efficiency close to unity. The preparation in the $2\,^3S_1$ level with a well-defined magnetization was obtained by a trajectory selection operated after the passage through a magnetic field gradient. Then, atoms were excited to the $2\,^3P$ level by a discharge lamp and were simultaneously illuminated by the microwave field causing a magnetization variation. A second

---

[a] A detailed recognition of the $2\,^3P$ fine structure measurements is given below.

magnetic selector allowed only the desired atoms to impinge on the atomic detector. Like in Lamb's, in these experiments the main issues were related to the presence of the bias magnetic fields of 10 to 200 mT: a remarkable work has been done both to assure the necessary stability and homogeneity and to have an accurate calibration of the intensity in the interaction region.

The lineshape distortion, consequent to the way the resonances were sweeped (i.e. by fixing the microwave frequency and varying the magnetic field), and the second order Zeeman shift required a careful analysis, so that also the $2\,^3P\,g_J$ was measured.

V. Hughes and collaborators measured all the three intervals separately, since the mixing of fine structure levels induced by the magnetic fields allowed even $\mathcal{J}=2$ to $\mathcal{J}=0$ transitions (we note as $\mathcal{J}$ a level that connect to the J manifold at zero magnetic field). Their results[b]

$$
\begin{aligned}
2\,^3P_2 - 2\,^3P_1 &= (2291.196 \pm 0.005)\ \text{MHz}^{14,15} \\
2\,^3P_1 - 2\,^3P_0 &= (29616.864 \pm 0.036)\ \text{MHz}^{16} \\
2\,^3P_2 - 2\,^3P_0 &= (31908.040 \pm 0.020)\ \text{MHz}^{17}
\end{aligned}
$$

improved by almost two order of magnitudes the experimental accuracy.

4.3 Recent developments

A completely different approach was taken in a recent experiment, completed in 1995 at Yale University[19]. The development of laser spectroscopy made advantageous to go back to fine structure measurements in the optical domain, i.e. as frequency differences of optical transitions. Shiner and coworkers measured the wavelength of the different fine structure components of the 1083 nm $2\,^3S_1 - 2\,^3P$ transitions. The detection scheme was again based on a double Stern-Gerlach apparatus, but the demagnetization was caused by the laser beams. The major source of uncertainty was attributed to the interferometric technique employed to measure the laser wavelength. The results

$$
\begin{aligned}
2\,^3P_2 - 2\,^3P_1 &= (2291.173 \pm 0.003)\ \text{MHz} \\
2\,^3P_1 - 2\,^3P_0 &= (29616.962 \pm 0.003)\ \text{MHz} \\
2\,^3P_2 - 2\,^3P_0 &= (31908.135 \pm 0.003)\ \text{MHz},
\end{aligned}
$$

showed quite a disagreement with those of V. Hughes and coworkers, their differences being 3 to 4 times larger than the combined standard deviation.

---

[b] According to a private communication of V. Hughes, we deliberately disregard the sentence of Ref. 17, in which the authors suggest to use the difference of the $2\,^3P_2 - 2\,^3P_0$ minus $2\,^3P_2 - 2\,^3P_1$ as a more precise value for the $2\,^3P_1 - 2\,^3P_0$ interval.

At York University, Toronto, microwave measurements have been revived by Storry and Hessels [20], that could benefit of lasers to excite the $2\,^3P$ level, whereas Hughes and coworkers had to use lamps. The 13 kHz experimental uncertainty of a first measurement was a factor 4 larger with respect to Shiner *et al.* Very recently, this group has been published a new microwave measurement of the $2\,^3P_1 - 2\,^3P_2$ interval[21]. With uncertainty of only 1.4 kHz, this represents a valuable test for the QED calculations. Unfortunately, the relative uncertainty is not as good as the previous best measurements since the investigated interval is one order of magnitude smaller then the other two (see Fig. 1).

Finally, we wish to mention that, at Harvard University, another laser spectroscopy experiment was carried out in a discharge cell: the uncertainty of this unpublished measures is 8 kHz[22].

5 The Florence experiments

In Florence we have chosen an approach that combines laser spectroscopy with the direct frequency measures of the microwave experiments[23]. We take advantage of the obvious consideration that to obtain the fine structure separations there's no need to precisely know the optical transitions frequencies but just their differences. Thus, if we dispose of two laser frequencies whose difference can be accurately controlled, we may use one as a fixed reference and tune the second across the atomic resonances, as illustrated by Fig. 1.

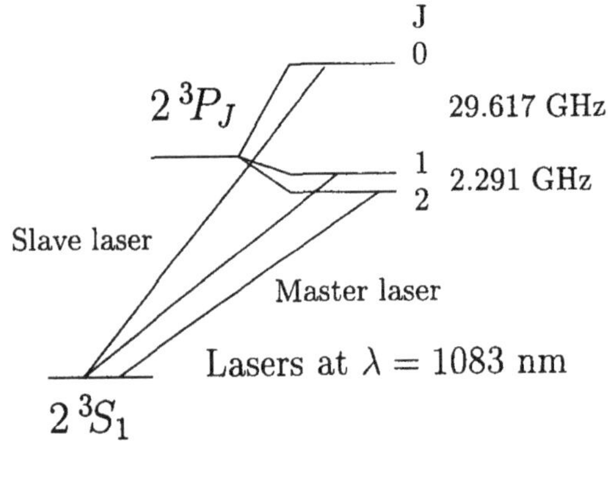

Figure 1. Scheme of helium energy levels relevant for our experiment.

Infact, our approach reverts to an etherodyne technique, where all the transitions are measured with respect to the same the reference frequency, that can take any arbitrary but stable value. In the experimental realization we obtain the two frequencies by phase-locking two diode lasers (master and slave), i.e. phase-locking their beat note to a microwave oscillator [24].

Diode lasers are very compact and convenient sources, although their emission is generally broader in frequency than the one of solid state lasers, it can be narrowed by reflecting part of the emitted light back into the laser chip. This widely adopted technique, named "optical feedback", is implemented by mounting in front of the diode laser a partially reflecting element, such as a beam splitter or a diffraction grating (extended-cavity configuration). Figure 2 shows quite clearly the effect of the optical feedback.

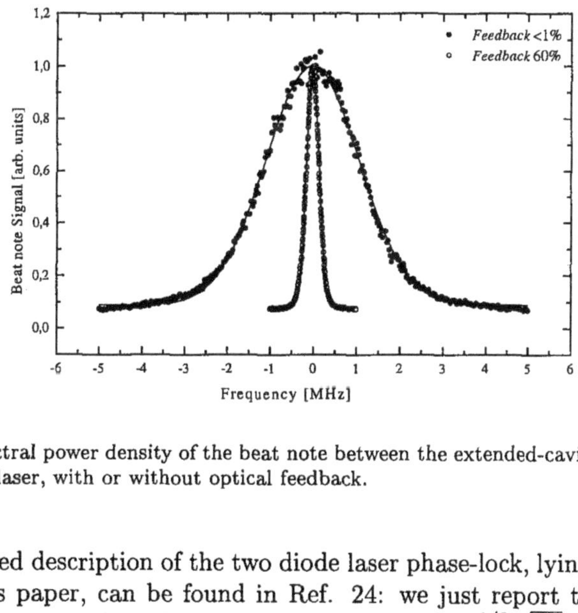

Figure 2. Spectral power density of the beat note between the extended-cavity master laser and the slave laser, with or without optical feedback.

A detailed description of the two diode laser phase-lock, lying outside the scope of this paper, can be found in Ref. 24: we just report the measured root Allan variance of the frequency difference, $0.1 \cdot \tau^{-1/2}\sqrt{\text{Hz}}$.

We detected the fluorescence emitted by a beam of $2\,^3S_1$ metastable atoms as they cross at right angle the laser light. We increase the signal-to-noise ratio by means of standard frequency modulation: a third harmonic demodulated lineshape is shown in Fig. 3.

Perhaps, the most significant difference with respect to previous experiments is that no bias magnetic field is present: a cylinder of high magnetic permettivity metal (mu-metal) embeds the interaction region to shield it against

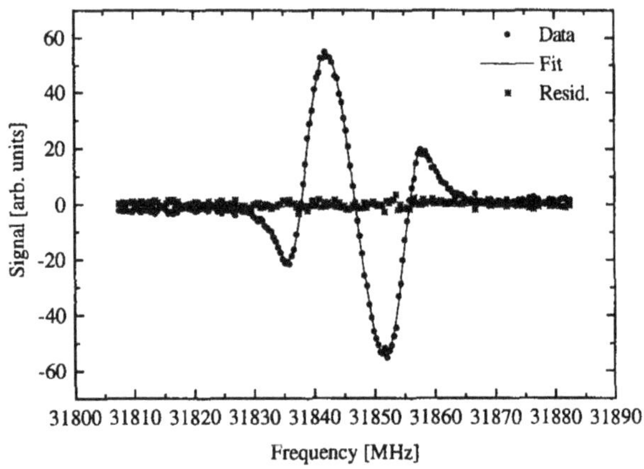

Figure 3. Third harmonic demodulated signal of the $2\,^3S_1 \rightarrow 2\,^3P_2$ transition, fitted with the expected function.

dc and ac magnetic fields while a couple of Helmholtz coils inserted therein allow for the fine cancellation. In general, by selecting the transitions connecting $m = 0$ states, the first order Zeeman effect is (in principle) eliminated but at the current level of accuracy even the second order effect is relevant: e.g. the separation between $2\,^3P_0\; m = 0$ and $2\,^3P_1\; m = 0$ decreases by 0.18 MHz in a field of 3 mT. Moreover, in our experience a magnetic field transverse to the laser polarization of 16 μT gave rise to a 0.05 MHz shift of the measured ν_{01} separation. This shift is too large to be explained by simply admitting a linear Zeeman shift combined with the residual ellipticity of the laser polarization. As a consequence, we choose to work with zero magneti field. To minimize it *in situ*, we made use of the non-linear Hanle effect: in presence of optical pumping toward non-coupled sublevels of the ground state, the fluorescence signal drops, unless a magnetic field recycles the sublevels populations. Thus, we tuned the Helmholtz coils current in search for the minimum of signal. We then measured the magnetic field to be below 0.1 μT in a cube of 1 cm^3 around the interaction region. The related systematic error, with the conservative assumption that it scales linearly with the residual field, is less than 0.3 kHz.

Surprisingly, however, we found that the major limitation to our experimental accuracy was due to the recoil imparted by photons to atoms. The radiation forces are of course well-known, they are responsible for the laser cooling, nevertheless their effect in precision spectroscopy is often disregarded as the interaction time between light and particles is short. However, we found that these forces cause a lineshape asymmetry and a consequent shift of the linecenter[25]. We can understand at least the origin of this shift with simple qualitative arguments, while for a precise characterization a numerical analysis is required.

Basically, the shift occur becuase the velocity and position distribution of atoms is modified by the interaction with the laser light. In our experimental configuration, the laser was arranged in a standing wave with two linearly-polarized, counterpropagating beams, shined orthogonally to the atomic beam. Let's consider what happens when the laser is detuned below resonance (red detuning). The velocity-dependent component of the force is damping, therefore the transverse velocity distribution of the atoms is cooled. Since the saturated fluorescence we detect is proportional to the number of atoms which have nearly zero transverse velocity, cooling yields an increase of the signal. Moreover, in the standing wave the position-dependent component of the force pushes the atoms toward the high intensity region, i.e. the antinodes, leading to a further increase of the signal. The opposite occurs for blue-detuning. For cycling transitions, the number of photons scattered by any atom is limited only by its transit time across the laser beam: in our experiment, the shift of the cycling $2\,^3S_1 - 2\,^3P_2$ line was of the order of 0.2 MHz, about one tenth of the natural width. On the other hand, the $2\,^3S_1$ – $2\,^3P_J$ $(J = 0, 1)$ are "open" transitions, i.e. transitions for which the selection rules make at least one of the Zeeman $2\,^3S_1$ sublevels uncoupled to the laser light ("dark" levels). For this two transitions optical pumping toward dark levels reduces the number of scattered photons and, as a consequence, the lineshift turns out to be almost one order of magnitude smaller. Furthermore, we take advantage of the partial cancellation in the frequency differences.

A detailed theoretical analysis of the shift has been given in Ref.26, where the authors use a simplified two level model and compare its results with the experimental data for the "closed" transition. For the fine structure experiment, we implemented a numerical solution of the optical Bloch equations and, after we verified that the predicted shifts were in agreement with the observed values for different interaction times[25], we corrected our $2\,^3P_0 - 2\,^3P_1$ measured value by 11.7(9) kHz.

Our final result shows clearly how the obtained accuracy depends on whether or not the closed transition $2\,^3S_1$ – $2\,^3P_2$ enters the measured interval[23,27]:

$$2\,^3P_2 - 2\,^3P_1 = (2291.169 \pm 0.015)\ \text{MHz}$$
$$2\,^3P_1 - 2\,^3P_0 = (29616.9497 \pm 0.0020)\ \text{MHz}$$
$$2\,^3P_2 - 2\,^3P_0 = (31908.124 \pm 0.015)\ \text{MHz}.$$

In Fig. 4 we compare our best result with those previously mentioned.

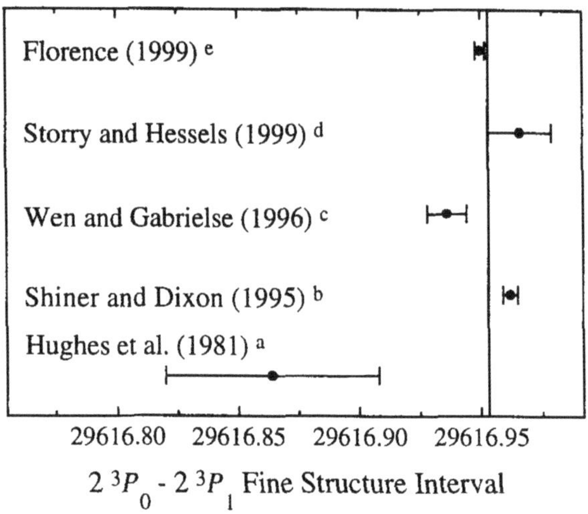

Figure 4. Measured values of the $2\,^3P_0 - 2\,^3P_1$ interval and their weighted average; a: Ref. 16, b: Ref. 19, c: Ref. 22, d: Ref. 20, e: Ref. 23

5.1 A new reference

To further improve our final accuracy we have built a brand new stabilization system for the master laser. So far, we have been locking the master laser to the saturated absorption signal of a Helium cell discharge. Fig. 5 shows the short term frequency stability of the reference laser as obtained from the error signal of the stabilization loop. In addition, we estimated a drift of the order of 20 kHz/hour, that is not readable from the error signal, probably due to the large rf fileds of the discharge and to the slow pressure variations inside the cell.

Although we have adopted a suitable acquisition sequence to cancel out the effect of this drift on the measured fine structure splittings, a new, more

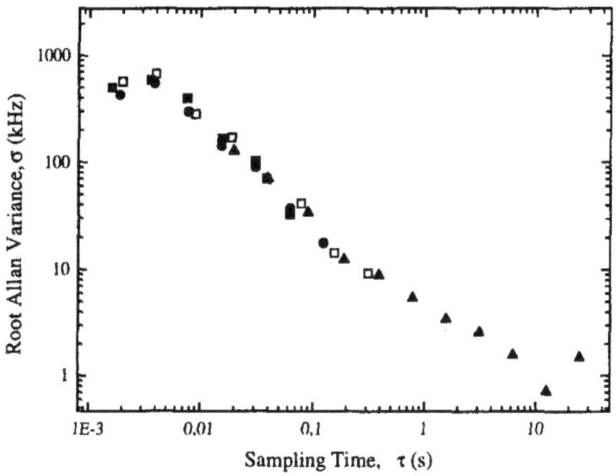

Figure 5. Frequency stability of our reference laser.

stable reference is necessary to reduce the statistical and hence systematic uncertainty. In the search for it, we have been inspired by the work by J. Hall and coworkers[28] that stabilize a Nd:YAG 1064 nm laser on the rich spectrum of the iodine molecule I_2 at half its wavelength, 532 nm. The idea is to frequency double our 1083 nm laser and lock it to the saturated lines of I_2 (see Fig. 6).

The first step is to frequency double our 1083 nm laser. Unfortunately, its output power is about 10 mW, a factor 10 to 100 less than that of a Nd:YAG laser. To increase the second harmonic generation (SHG) efficiency, we employ an Yb^+-doped fibre-amplifier[29], that allows to gain 30 dB in the 1083 nm power, yielding up to 1 W. In addition, after a first experiment with a lithium niobate crystal embedded by an enhancement cavity[30], we use a periodically-poled KTP ($KTiOPO_4$) crystal, which has the advantages of a larger SHG efficiency and an easier phase-matching. Iodine lines have been detected with a signal-to-noise of the order of 10^3 with an integration time of 1 s, and as a preliminary result, the frequency stability of the locked diode laser is already improved by almost one order of magnitude with respect to the He cell discharge[31].

This scheme is promising, as it can be obviously extended to other frequency difference measurements, such as ^{3}He hyperfine splittings and ^{3}He - ^{4}He isotope shifts. But perhaps more important is the perspectve of a link to the I_2 lines at 532 nm, that now represents a secondary frequency standard.

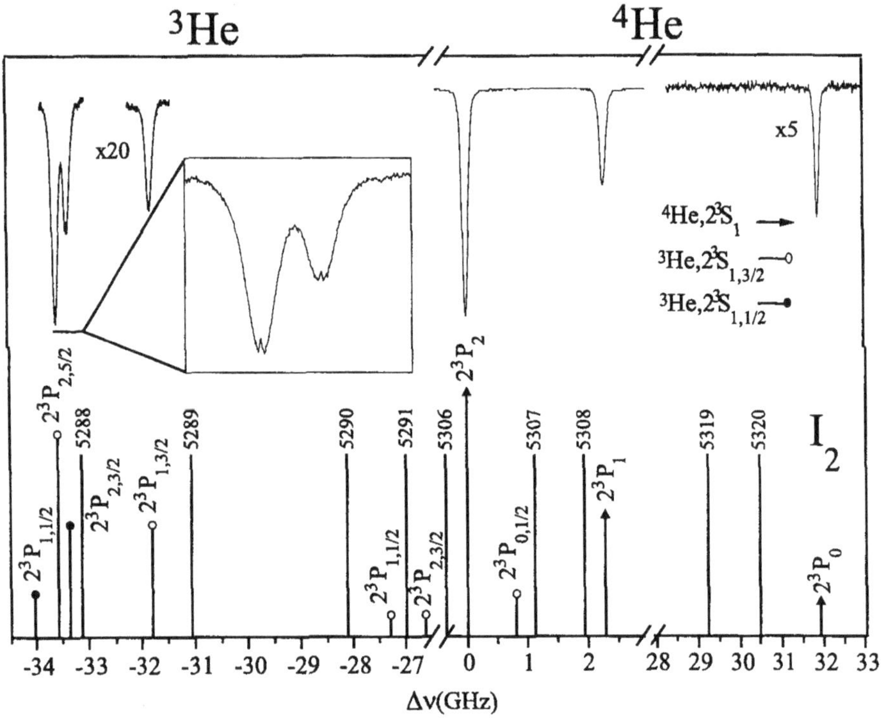

Figure 6. Coincidence between twice the frequency of ^{3}He and ^{4}He lines and the I_2 spectrum (courtesy of P. Cancio Pastor)

6 Fine structure constant determinations and perspectives.

Among the we several determinations of α yielded by different physical systems currently available, the only one that goes beyond 5 ppb of accuracy comes from comparing the measured[32] and the calculated[33] values of the electron magnetic anomaly. A new determination promising to improve this level of accuracy is expected from the relation:

$$\alpha^2 = \frac{2R_\infty}{c}\frac{h}{M_{\mathrm{Cs}}}\frac{M_{\mathrm{Cs}}}{m_e}$$

At Stanford, atomic interferometry is used to measure the ratio of Planck's constant to Cesium mass[34] h/M_{Cs} via the Cesium D_1 line frequency[35], while the Cesium-to-electron mass ratio has already been measured with ppb accuracy[36]. The Rydberg constant is known better than 10^{-2} ppb[?,37].

The Helium $2\,^3P$ fine structure can compete with these determinations only if the experimental uncertainty will decrease by an order of magnitude and the theory will come to a similar accuracy. As for the electron magnetic anomaly, it should be pointed out that the theory involves an heavy and complex work, on both analytic and numerical side. Up to now, this formidable task has been undertaken mainly by one single group, namely that of G. Drake and coworkers. It seems therefore desirable to have an indepedent check of the calculations from a different group[38].

On the experimental side, reducing the uncertainty by another order of magnitude is certainly a challenge. But a few group are already at work.

Acknowledgments

In the years, many people have contributed to the work above reported and we are indebted with all of them: F. Marin, F. S. Pavone, P. De Natale, M. Prevedelli, G. Bianchini, P. Cancio Pastor and G. Giusfredi.

References

1. P. A. M. Dirac, *Proc. Roy. Soc. (London)* **A117**, 610 (1928).
2. G. Breit, *Phys. Rev.* **34**, 553 (1929).
3. Gaunt, *Proc. Roy. Soc.* **A122**, 153 (1929); Eddington, *Proc. Roy. Soc.* **A122**, 358 (1929).
4. G. Breit, *Phys. Rev.* **35**, 569 (1930).
5. G. Breit, *Phys. Rev.* **36**, 383 (1930).
6. H. Araki, *Progr. Theoret. Phys. Japan* **17**, 619 (1957).
7. J. Sucher, *Phys. Rev.* **109**, 1010 (1958).
8. M. Douglas and N. M. Kroll, *Ann. Phys.* **82**, 89 (1974).
9. G. W. F. Drake, in *Long-Range Casimir Forces: Theory and Recent Experiments on Atomic Systems*, eds. F. S. Levin and D. A. Micha, (Plenum, New York, 1989).
10. Tao Zhang, *Phys. Rev. A* **53**, 3896 (1996);
11. T. Zhang, Z.-C. Yan, and G. W. F. Drake, *Phys. Rev. Lett.* **77**, 1715 (1996).
12. W. E. Lamb Jr., *Phys. Rev.* **105**, 559 (1957); I. Wieder and W. E. Lamb Jr., *Phys. Rev.* **107**, 125 (1957).
13. D. J. Yang, P. McNicholl and H. Metcalf, *Phys. Rev. A* **33**, 1725 (1986).

14. F.M.J. Pichanick, R.D. Swift, C.E. Johnson, and V.W. Hughes, *Phys. Rev.* **169**, 55 (1968).
15. S.A. Lewis, F.M.J. Pichanick, and V.W. Hughes, *Phys. Rev. A* **2**, 86 (1970).
16. A. Kponou, V.W. Hughes, C.E. Johnson, S.A. Lewis, and F.M.J. Pichanick, *Phys. Rev. A* **24**, 264 (1981).
17. W. Frieze, E.A. Hinds, V.W. Hughes, and F.M.J. Pichanick, *Phys. Rev. A* **24**, 279 (1981).
18. V. W. Hughes, *Comm. At. Mol. Phys.* **1**, 5 (1969).
19. D. L. Shiner, R. Dixson, and P. Zhao, *Phys. Rev. Lett.* **72**, 1802 (1994); D. L. Shiner, R. Dixson, *IEEE Trans. Instrum. Meas.* **44**, 518 (1995).
20. C. H. Storry and E. A. Hessels, *Phys. Rev. A* **58**, R8 (1998).
21. C. H. Storry, M. C. George, and E. A. Hessels, *Phys. Rev. Lett.* **84**, 3274 (2000).
22. J. Wen, *Ph. D. Thesis*, Harvard University, 1996 (advisor: G. Gabrielse).
23. F. Minardi, G. Bianchini, P. Cancio Pastor, G. Giusfredi, F. S. Pavone and M. Inguscio, *Phys. Rev. Lett.* **82**, 1112 (1999).
24. M. Prevedelli, P. Cancio, G. Giusfredi, F. S. Pavone, M. Inguscio, *Opt. Comm.* **125**, 231 (1996).
25. F. Minardi, M. Artoni, P. Cancio, M. Inguscio, G. Giusfredi, and I. Carusotto, *Phys. Rev. A* **60**, 4164 (1999).
26. M. Artoni, I. Carusotto and F. Minardi, *Phys. Rev. A* in press.
27. P. Cancio Pastor, M. Artoni, G. Giusfredi, F. Minardi, F. S. Pavone, M. Inguscio, in *Atomic Physics 16*, eds. W. E. Baylis and G. W. F. Drake, (American Institute of Physics, Woodbury NY 1999). Note that, in this reference, we have miscalculated the correction due to the microwave oscillator calibration for the $2\,^3P_2 - 2\,^3P_1, 2\,^3P_2 - 2\,^3P_0$ values. The present value correct this mistake, the difference being within the overall uncertainty.
28. M. L. Eickhoff and J. L. Hall, *IEEE Trans. Instrum. Meas.* **44**, 155 (1995).
29. P. De Natale, G. Giusfredi, P. Cancio, D. Mazzotti and M. Inguscio, in *Laser Spectroscopy XIV*, eds. R. Blatt, J. Eschner, D. Leibfried and F. Schmidt-Kaler, (World Scientific, Singapore, 1999).
30. D.J.E. Knight, F. Minardi, P. De Natale, and P. Laporta, *Eur. Phys. J. D* **3**, 211 (1998).
31. P. Cancio Pastor, P. Zeppini, A. Arie, P. De Natale, G. Giusfredi, G. Rosenman and M. Inguscio, *Opt. Comm.* **176**, 453 (2000).

32. R. S. Van Dyck, Jr. P. B. Schwinberg, and H. G. Dehmelt, *Phys. Rev. Lett.* **59**, 26 (1987).
33. T. Kinoshita, *IEEE Trans. Instrum. Meas.* **46**, 108 (1997).
34. B. C. Young, Ph.D thesis, Stanford 1997 (unpublished).
35. Th. Udem, J. Reichert, J. Holzwarth, and T. W. Hänsch, *Phys. Rev. Lett.* **82**, 3568 (1999).
36. M. J. Bradley, J. V. Porto, S. Rainville, J. K. Thompson, and D. E. Pritchard *Phys. Rev. Lett.* **83**, 4510 (1999) and references therein; a preliminary α value with a 30 ppb uncertainty is reported in this paper.
37. C. Schwob *et al.*, *Phys. Rev. Lett.* **82**, 4960 (1999).
38. Recently, K. Pachucki has begun to work on Helium: K. Pachucki, *J. Phys. B: At. Mol. Opt. Phys.* **31** (1998), 2489.

THE ISOTOPE SHIFT

E. W. OTTEN
Institut für Physik, Universität Mainz,
D-55099 Mainz
E-mail: Ernst.Otten@uni-mainz.de

The first half of the paper deals with the discovery of the nuclear volume effect in the isotopic shift of spectral lines and the development of its interpretation over the years emphasizing Breit's leading role. With the advent of tunable lasers and mass separators on-line with accelerators experiments could be extended to isotopes far off stability yielding a wealth of systematic data which had large impact on basical problems of nuclear structure and nuclear theory. This is discussed in the second half of the paper.

1 Introduction

That spectral lines should exhibit an isotopic shift (IS), this was evident from the first days of Bohr's model of the atom and Soddy's discovery of isotopes, both dating back to 1912. The difference in nuclear mass $M^A - M^{A'}$ between 2 isotopes A, A' causes an increase of the reduced mass of the electron with mass number. In hydrogen-like spectra this leads to a general blue shift of any spectral line of frequency ν by

$$\delta\nu_m^{AA'} = \nu^A - \nu^{A'} = \frac{M^A - M^{A'}}{M^A \cdot M^{A'}} m_e \nu^A \approx N \frac{A - A'}{A \cdot A'}, \tag{1}$$

where the right side is an approximation in terms of the mass numbers instead of the masses themselves. $N = \nu^A m_e/(a.u.)$ is the so-called normal mass shift constant, (a.u.) the atomic mass unit. The mass shift was easily discovered in the isotopes of light elements like hydrogen, lithium and neon, where it is largest due to the denominator in (1). In complex spectra, particularly in those of transitional elements, the motion of a pretty large number of valence electrons may be correlated, hence pulling coherently on the nucleus and causing a mass shift up to an order of magnitude larger than the normal one. A satisfactory calculation of this so-called specific mass shift is still a challenge for the theory of atomic spectra!

Though still burning, the problem of specific mass shift is not in the center of the current interest in isotope shift, but regarded as a nasty pertubation to another isotopic effect, namely, the so-called nuclear volume shift or field shift. It accounts for the change of the Coulomb interaction with an extended nuclear charge as compared to that with a point-like charge. This volume shift yielded already in the thirties, before the age of accelerators, a first insight into nuclear structure. Still in our days, the nuclear volume shift is the only source which gives us direct information on the nuclear mean squared charge radii $\langle r^2 \rangle_{ch}$ of unstable nuclei.

Gregory Breit was among the first theoreticians who searched for a quantitative explanation of the effect. He tackled the atomic problem of the electron nucleus interaction and then tried to understand the observed nuclear volume shifts in terms of nuclear theory. He published his first paper on IS in 1932 [1] and the last one in 1958 [2]. Despite of his widespread interests he was obvioulsy quite faithful to the problem.

In the following I will recall the history of the discovery of the nuclear volume shift and then discuss the first harvest of results which reflects the status as of the late fifties, when nuclear physics had matured already to a grown up science. Due to the advance of tunable lasers, the last third of the centruy saw a rapid development of new powerful spectroscopic methods with fabulous increase of precision and sensitivity. This opened the way to systematic maesurements of nuclear radii and moments far off stability. New phenomena showed up in the data, and old problems in the systematics of nuclear volume shift found satisfactory explanation in modern theory. A survey of this second generation of research in IS as well as of present trends.

2 The discovery of the nuclear volume effect in isotpe shift

Almost simultaneously Hermann Schüler [3] and Hans Kopfermann [4] discovered an IS in heavy elements which turned out to be much larger then could be expected from the mass shift (1). In addition, it showed the wrong sign, that means a red shift for the heavier isotope. Schüler found it in thallium isotopes; the decisive sentence in his paper reads: *"Eine Erweiterung des Strukturschemas ist nun dadurch gegeben, daß die Niveaus für die Isotopen 203 und 205, beide mit $i = 1/2$ nicht mehr zusammenfallen, sondern gegeneinander verschoben sind!"*. [A generalization of the scheme is now given by the fact that the levels of isotopes 203 and 205 both having $i = 1/2$ do not coincide anymore, but are shifted relative to each other.] Kopfermann investigated lead isotopes, once in their usual isotopic mixture and once so-called "uranium lead" which is found in ores of uranium as decay product; it is but pure ^{206}Pb. Kopfermann's experiment was also the first spectroscopic experiment with separated isotopes. This made the effect appear very clearly (see Fig. 1). Kopfermann's comment in this very short note reads: *"Auffallend ist (analog wie beim Lithium) die starke Isotopiverschiebung zwischen den Linien des ^{206}Pb und ^{208}Pb."* [A striking fact (analogue to Li) is the strong isotope shift between the lines of lead 206, and 208.] Furthermore, the analysis shows the center of gravity of the hfs components of 207 to lie closer to 206 than to 208 - the first manifestation of the effect of the odd-even staggering of the IS.

Figure 1: First IS measurement with separated isotopes [7]. Uranblei is pure ^{206}Pb. On the left natural Pb. Shown are photographs of subsequent Perot Fabry orders.

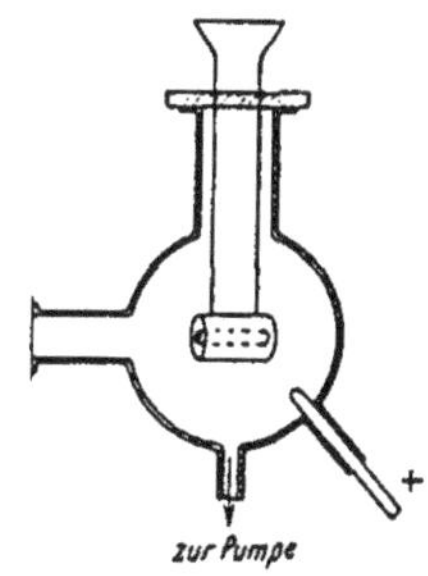

Figure 2: A hallow cathode, the spectroscopic tool of the thirties [7].

The spectrosopic method chosen by both authors is shown in Fig. 2. It was standardized already by that time and remained principally unchanged for a generation: A plasma burns into a hollow cathode (Schüler'sche Hohlkathode) and atomizes and excites the probe therein. The cathode is liquid air-cooled in order to reduce the Doppler width. The spectral light then passes a highly resolving Perot-Fabry-etalon and before entering a monochromator with a photo plate in its focal plane.

In their short notes [3,4], neither Schüler nor Kopfermann gave an explanation of the effect. But in the full papers published thereafter, first suggestions are given in the right direction. Schüler and Keystone [5] write on the thallium results: *"Die bei einigen Linien beobachteten Isotopenverschiebungen sind vermutlich auf irgendwelche Verschiedenheiten der elektrischen Atomkernfelder der Tl-Isotopen zurückzuführen".* [The isotope shifts observed on some lines can presumably be traced down to differences of the electric nuclear fields on the Tl isotopes.] In 1932 D.A. Jackson - an important independent scholar of his time who was faithful to this field for his lifetime - has confirmed already Schüler's measurements and writes on the thallium isotope shift [6]: *"Sie muß vielmehr auf irgendeinem Unterschied der Kernstruktur beruhen, der eine geringe Verschiedenheit des Kraftgesetzes in der Nachbarschaft des Kernes bedingt."* [It appears to be connected to some difference in nuclear structure, causing in turn a slight change of the power law in the region of the nucleus.] In the same issue Kopfermann comments his findings in lead isotopes as follows[7]: *"Man ist beim augenblicklichen Stand der Forschung mehr geneigt, den Isotopieverschiebungseffekt als einen typischen Kerneffekt zu deuten (siehe z.B. W. Pauli, Diskussionsbemerkungen der Züricher Vorträge 1.c) und zwar in dem Sinne, daß man als Ursache der Termverschiebung eine Verschiedenheit der elektrischen Felder der einzelnen Isotopenkerne annimmt".* [At the present state of research one tends to interpret the isotope shift effect as a typical nuclear effect - in a sense that one supposes the term shift to be caused by a difference of electric fields of the respective isotopic nuclei (see e.g. W. Pauli, discussion remarks of the Zurich talks 1.c).] The consensus between the experimentalists towards the right explanation is obvious.

3 Early theory of isotope shift

But what did Pauli whom Kopfermann quotes really say in Zurich in May 1931, when he responded to the just discovered isotope shift? Kopfermann attended this meeting. But in another secondary source a summary of this session by Bretscher and Guth[8] we read: *"Nach Pauli und Peierls lassen sie sich nicht deuten durch die Annahme nur wenig verschiedener Kernvolumina für die beiden (Tl-)isotope".* [According to Pauli and Peierls, "they cannot be interpreted by the assumption of slightly differing nuclear volumes for both Tl isotopes".] How to interpret this negative statement? Maybe that it reflects Pauli's notorically critical mind in the simple sense that neither he nor Peierls managed to reproduce the effect qunatitatively by theoretical calculations. If this interpretation is correct, then they were at least in good company. Because shortly after, Breit and Rosenthal as well as Racah entered the scene and published independently from each other papers on the solution of the Dirac equation for an electron moving in the field of homogeneously or surface charged spherical nuclei of radius R_0 and $R_0 + \delta r$, respectively, where δr is the isotopic increase[1,9]. It turned out that at high Z the relativistic mass gain of the electron increases the probability to be within the nucleus by a pretty large factor, as compared to the non-relativistic Schrödinger case. Both papers finally calculate a nuclear volume shift an order of magnitude larger than the observed one. Since the relativistic correction was calculated quite correctly, however, the problem was not lying with the shape of the electronic wave function near the origin but with its size, that means its normalization factor between its unscreened inner and its screened outer parts. The calculations of the normalization factors were performed at that time by help of the Thomas Fermi model. Obviously they lacked the precision of today's numerical procedures for such problems.

In Breit and Rosenthal's paper we find an interesting note added in proof which puts the nuclear physics of that time in the right perspective: *"Since the paper was written we learned of a new nuclear theory of Heisenberg according to which the nucleus is supposed to consist of protons and neutrons."* And further: *Our thanks are due to Prof. Werner Heisenberg for informing us of the new developments before publication and for a helpful discussion on the interaction of electrons with the nucleus."*

A few months later Breit gets everything settled in another paper on IS[10]:
i) The main point is that he switches from aborigine to semi-empirical calculation of the normalization factor, i.e. the nonrelativistic density of a valence electron at the origin, given by

$$| \psi_{(0)} |^2_{nr} = \frac{Z_i Z_0}{\pi (a_0 n^*)^3} \tag{2}$$

with Z_i = unscreened inner charge number, Z_0 = charge number of the ionic core, a_0 = Bohr radius and n^* = effective quantum number as determined from Rydberg-Ritz series. This and similar formulas had served before already for calculating fine- and hyperfine splittings and carry the names of their famous inventors: Goudsmit, Fermi, Segré, Landé. Breit writes: *"With Goudsmit's extension of Landé's formula for $\langle r^{-3} \rangle$ it is possible to explain the order of magnitude of the isotope displacement in Hg, Tl, Pb arc and spark spectra on the hypothesis of small changes in nuclear radii. The nuclear radius is supposed to be proportional to the 1/3 power of the atomic weight. The effective nuclear charge is supposed to be distributed with a roughly uniform density through the interior of the nucleus."*
ii) Breit stresses the dominance of s-states in field shift: *"The larger displacements are due to the addition or removal of a 6s or 7s electron ..."*
iii) Breit clears up the electronic configurations in Tl-spectra and thereby the confusing interpretations of their IS.
iv) Breit introduces the concept of screening of s-shells by low lying states with $\ell > 0$: *"These changes are presumably caused by differences in screening of the two 6s-electrons from the nucleus as the valence electron is excited from the 6p-state to the ionization limit."*

On that basis isotopic shifts were analyzed in the decades to follow.

Todays status of theory: Leaving for a moment the historical line, I would like to present in a few lines the nuclear volume effect in its present form: The Coulomb energy between a nuclear charge distribution $\rho_N(\vec{r})$ and an electronic one $\rho_e(\vec{r}')$ may be written in the alternative forms

$$E_c = \int \frac{\rho_N(\vec{r})\rho_e(\vec{r}')}{|\vec{r}-\vec{r}'|} d\tau d\tau' = \begin{cases} \int V_N(\vec{r}')\rho_e(\vec{r}')d\tau' \\ \int V_e(\vec{r})\rho_N(\vec{r})d\tau \end{cases} , \tag{3}$$

where the first integration leads either to the potential of the nuclear charge distribution acting on the electronic one or vice versa. For tutorial reasons we prefer the latter one, assuming that the electron density is, to first order, constant over the nuclear volume. This leads to a parabolic electron potential $V_e(\vec{r})$ inside the nucleus. The resulting Coulomb energy then consists of two terms

$$E_c = -Ze^2\langle r_e^{-1}\rangle + \frac{2\pi}{3}Ze^2 \mid \psi_e(0) \mid^2 \langle r_N^2\rangle_{ch} . \tag{4}$$

The first would apply for a point-like nucleus and the second gives a correction proportional to the mean squared charge radius of the nucleus times the electronic density at the origin. The virtue of this presentation is its independence of a particular nuclear model. In the IS of a spectral line the first term in (4) cancels in the difference of Coulomb energy leading to a nuclear volume shift

$$\delta\nu_{vol}^{AA'} = -\frac{2\pi}{3}Ze^2\Delta \mid \psi(0) \mid^2 \delta\langle r_N^2\rangle_{ch}^{AA'} = F\delta\langle r_N^2\rangle_{ch} . \tag{5}$$

$\Delta \mid \psi(0) \mid^2$ is the difference of electronic charge density at the origin between the lower and the upper electronic state. For heavy elements a correction factor has to be applied which takes into account relativity as well as the next order of pertubation, namely, the change of the electronic wave function inside the nucleus by the nuclear charge distribution itself. Bodmer [11] has generalized Racah's [9] and Breit's [1] calculations and has shown that for any nuclear model the IS depends only on radial moments of the nuclear charge distribution

$$\delta\nu_{vol}^{AA'} \propto \langle (r_N^A)^{2\rho}\rangle_{ch}\delta\langle r_N^2\rangle_{ch}^{AA} \tag{6}$$

with $\rho = (1-\alpha^2 Z^2)^{1/2}$. According to Seltzer [12] still higher radial moments contribute with a few percent of the effect in the heaviest elements. Hence $\delta\langle r^2\rangle$ should be replaced by a power series

$$\delta\langle r_N^2\rangle_{ch}^{AA'} \longrightarrow \lambda^{AA'} = \delta\langle r_N^2\rangle_{ch}^{AA'} + C_4\delta\langle r_N^4\rangle_{ch}^{AA'} + \dots \quad . \tag{7}$$

How do we treat nowadays the normalization factor of the electronic wave function at the origin? Still by Goudsmit's semi-empirical formula (2)? The answer is yes, by parts. On the other hand, we have now for quite a number of cases reliable relativistic ab origine calculations at hand. As an example,

Fig. 3 shows the case of the magnetic hyperfine coupling constant A(7s) and the corresponding electronic F-factor of the field shift (5) for the ground state and the resonance line, respectively, of francium, calculated by Ann-Marie Mårtensson-Pendrill[13]. The graph shows successive refinements of the theory, starting from the Dirac-Fock result which is normalized to 1. The calculation of the A factor can be checked by experiments since the magnetic moment has been measured independently. The agreement is within the experimental error bars of 2 %.

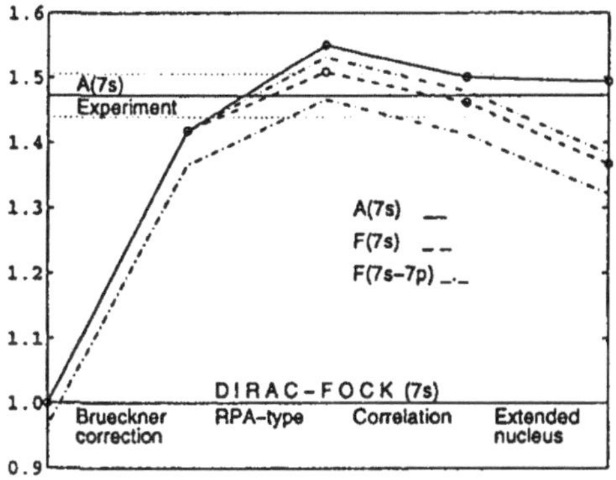

Figure 3: Recent calculations of hfs and IS constants in Fr [14].

Other sources on nuclear radii: At that point I may also mention which other sources yield today information on the nuclear charge distribution. These are in first line:

- Spectra of myonic atoms from which one determines a total, particular, state-dependent radial moment of the nucleus, the so-called "Barret moment" which can be related to $\langle r^2 \rangle$ [14].

- Form factor measurements in elastic electron scattering which can resolve also details of the radial charge distribution.

Both methods have also been applied for IS measurements for a limited number of cases, restricted of course to stable nuclei. These cases are very valuable for checking or calibrating the optical IS.

4 Survey of first generation results

Looking back again to the early thirties, it is striking to find all the big names to be engaged in IS and hyperfine structure of atomic spectra. Why is that so?

The answer is quite simple: At that time the atomic spectrum is one of the very few, narrow windows to the still totally dark world of nuclear structure.

After basic understanding of the nuclear volume effect had been achieved in these early years, a long period of systematic data collection of volume shifts as well as on spins and nuclear moments via the hyperfine structure started. In 1958, Reviews of Modern Physics published an issue with a collection of articles on the status of nuclear structure. In this issue Breit [15] summarizes the theoretical and Brix and Kopfermann [16] the experimental status of IS. In the latter article we find a synopsis of known volume shifts presented in the so-called Brix-Kopfermann-plot (see Fig. 4). As a function of neutron number it shows the volume shift between neighbouring even even isotopes normalized to the so-called standard shift. The latter is the volume shift expected for a homogenously charged spherical nucleus with a radius R expanding as

$$R_0 = r_0 A^{1/3} \tag{8}$$

with $r_0 = 1.2$ fm. The plot indicates already most of the salient features which govern IS still today where we dispose of a much richer and better data set. I will discuss them one by one:

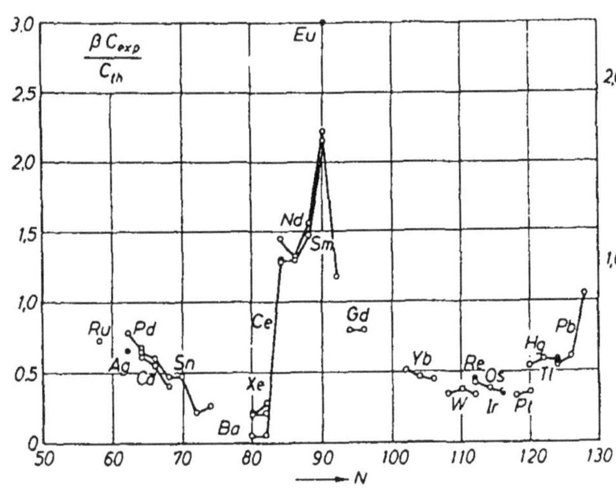

Figure 4: Brix-Kopfermann-plot of differential IS for $\Delta N = 2$.

General IS discrepancy: The majority of data, in particular those obtained in regions of relatively stable nuclear structure and shape, exhibit an IS only one half of what is expected for an imcompressible liquid drop expanding like (8). The explanation of this so-called IS discrepancy gave theoreticians a lot of headache for many decades. Thirty years ago, when Greogry Breit's 70. birthday was celebrated, Gerald Brown put these headaches in the following words[17]: *"When the neutron is in the center of the nucleus it tends to push protons out because of the near imcompressibility of nuclear matter. When on the surface of the nucleus, it tends to pull protons out, because of the attractive neutron proton force. So far so good! But the quantitative description of these effects, beginning from first or second principles, is not easy and that is a problem I wish to deal with today."* His calculations then lead him to the following conclusions: *"Calculations using density independent forces give quite wrong results, even with the wrong sign, for the isotope shift, chiefly because the attractive contribution from the interior of the nucleus wins out and the added neutron then tends to pull the protons in."* In fact, the first breakthroughs came when density dependent forces were introduced into the Hartree-Fock codes. For instance, the calculations by Angeli et al.[18] have shown that the addition of neutrons in an isotopic chain leads to a build-up of a neutron skin rather than diluting the nuclear charge distribution homogeneously according (8).

In this context many people have raised the question whether the IS is an effect of nuclear core polarization by the outer valence neutrons. Brown comes to the conclusion that a low lying giant monopole resonance would lead to such a polarization, but then rejects the idea with the statement: *"No breathing mode excitations of nuclei have been found"*. Nevertheless, his conclusion points to the old problem of nuclear compressibility: *"The isotope shift gives practically the only information on the compressibility of finite nuclei"*.

A parameterized gross nuclear model like Meyer's and Swiadetzky's droplet model[19,20], on the other hand, can incorparate the IS discrepancy by allowing for different proton- and neutron radii. It has therefore become a standard in phenomenological IS analysis (see e.g. [21]). A deviation from the droplet prediction is then interpreted as being caused by a change in nuclear shape rather than volume.

Shape effect of IS: The last remark in the foregoing paragraph leads to another important feature in Fig. 4, namely, the anomalously large IS observed in rare earth elements. It has a sharp peak for the pair N = 88, 90. Schüler and Schmidt[22], were the first to observe this large effect in the pair ^{150}Sm, ^{152}Sm. This happened a year before the same authors discovered the nuclear quadrupole splitting in atomic hyperfine structure and the large quadrupole moments of europium isotopes which also increase by a factor of more than 2 between N = 88 and 90[23] (compare Fig. 12). For a long time this anomalous peak in the IS remained unexplained, until Brix and Kopfermann[24,25] resumed the topic after the war and dared the idea that also spin 0 nuclei like the even

Sm isotopes, mentioned above, might have a quadrupole deformation. It is called an intrinsic deformation because it cannot lead to a quadrupole splitting due to I = 0, but still can contribute to the IS since a deformed nucleus has a larger second radial moment $\langle r^2 \rangle$ than a spherical one of equal volume. In an intrinsic coordinate system of a deformed, but axially symmetric nuclear drop one may define its surface by a radius vector which is (in modern terms [20]) an expansion in spherical harmonics $Y_{i0}(\theta)$

$$R_d = R_0 n[1 + \sum_i \beta_i Y_{i0}(\theta)] \tag{9}$$

with deformation parameters β_i for each multipole. If R_0 stands for a spherical nucleus of same volume, then n is a normalization factor close to 1 which slightly depends on the β_i [20]. We may then split up $\delta\langle r^2 \rangle$ in good approximation into two components

$$\delta\langle r^2 \rangle = \delta\langle r^2 \rangle_s + \langle r^2 \rangle_s \frac{5}{4\pi} \sum_i \delta\langle \beta_i^2 \rangle\,. \tag{10}$$

The first one results from a change of nuclear volume, calculated as if the nucleus were spherical. The second one is due to a change of deformation parameters β_i. The forefactor $\langle r^2 \rangle_s$ is the mean squared radius of a spherical nucleus of same volume.

The echo of theoreticians to Brix and Kopfermann's idea was split between scepticism and stimluation. Among the stimulated ones we count Bohr and Mottelson who founded their collective nuclear model on the Brix-Kopfermann-effect as one of its cornerstones. In the paper from 1950 on "Spectroscopic Isotope Shift and Nuclear Polarization" [26], Breit and collaborators comment the new situation as follows: *"Kopfermann's suggestion that these anomalies are caused by the deviation of the nuclear charge distribution from spherical symmetry is not literally tenable in the form stated by him, because nuclei of spin 0 must have spherical symmetry. The essential idea is applicable, however..."* and later *"Another interpretation of Kopfermann's viewpoint brought out in a discussion by Dr. A. Bohr is that the nucleus has an intrinsic angular momentum and quadrupole moment with respect to an axis definable by internal coordinates"*.

The discussion on intrinsically deformed nuclei went on for years but we see Breit on the right side from the beginning. In modern terms we would characterize this effect as an example of spontaneous symmetry breaking. But in these days, such concepts were not so common among physicists, except for the analogue cases of molecules where such symmetry breakings were known quite well as Jahn-Teller-effect [27]. So we may call it a Nuclear Jahn-Teller-effect [28].

IS at shell closures: The fifties were the golden age not only of the collective nuclear model but also of the shell model. It was a burning question in these years, therefore, whether one could see a shell closure effect also in the IS. The synopsis in Fig. 4 shows two positive examples, indeed, a big jump in cerium when crossing the shell closure at N = 82 and a much smaller, but sill significant one in lead isotopes when crossing N = 126. One is inclined to ascribe the former one to an onset of deformation, already, whereas the interpretation of the latter remains open. We will come back to these questions when we see the much richer data of today.

Odd-even staggering and nuclear polarization: Another phenomenon of IS, well documented already in the early days, was the odd-even staggering. It is not shown in Fig. 4 because that is restricted to even isotopes. Fig. 5 is giving a striking example of this effect in mercury and thallium isotopes. It is taken from Tomlinson and Stroke's systematic measurements [29] on long-lived radioactive isotopes. These experiments form the on-set of the second generation of IS measurements which is devoted to long isotopic chains extending far off stability. We see that the staggering is repeated also in isotones, and, moreover, that the i = 13/2 isomers in light Hg-isotopes are unstaggered, apparently.

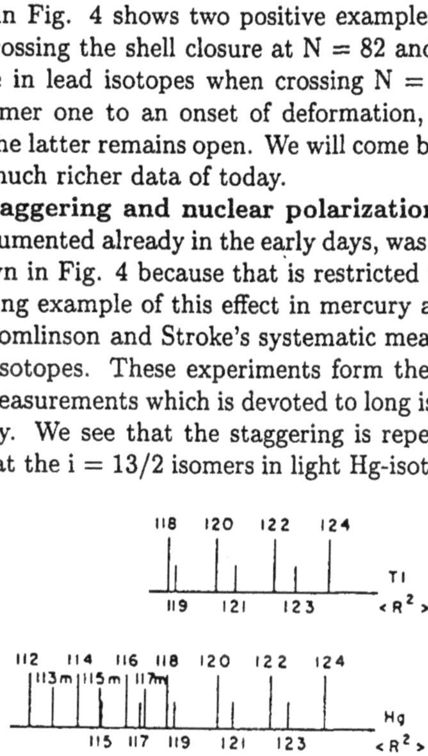

Figure 5: IS in Hg and Tl also showing examples of odd-even-staggering.

Several more or less plausible explanations of the odd-even staggering were tried already in the early days. But none of them found a solid theoretical approval. In their already mentioned, great paper on isotope shift from 1950 [26] Breit et al. also analyze odd-even-staggering. In particular, they raise the question whether a nuclear polarization by the valence electron which would strengthen the binding might play a role: *"It appears natural, to investigate the order of magnitude of polarization effects of a nucleus by atomic electrons. Because such effects should lower the energy and because the observed density of nuclear levels for odd isotopes is usually larger than for even ones".* They have developed a pertubation theory for this effect which formed the basis of all later papers on the subject. But the question on the character and size of the relevant nuclear matrix elements remained open at that time. Still Breit spots the right solution for polarization effects at the horizon, namely, giant resonances, which were proposed just at that time: *"Returning to the general subject of orders of magnitude of polarization needed for the explanation of the isotope effect, it is of interest that Goldhaber and Teller consider it likely that the (γn) reactions indicate the possibility of exciting nuclear "dipole vibrations" of an exceptionally strong type".*

We see that Breit has raised in this seminal paper all the big questions on IS right. But experimental input on nuclear structure and excitation data was still too poor to answer them. Hoffmann et al.[30] have finalized later the calculations on the effect of nuclear polarization in IS. They show that the giant resonances contribute significantly to the total binding of a valence electron, but that their isotopic change is too small and smooth to be discernable in the IS, except when an abrupt change of nuclear structure occurs like in the aforementioned shape transition at N = 90.

Coming back to the odd-even-staggering Reehal and Sorensen[31] have given the present standard explanation in terms of a deformation effect. They point out that the odd particle blocks pairing vibrations and thus leads to a decrease of the mean squared deformation $\langle \beta_2^2 \rangle$ (compare (10)). Their success in matching the experiments in quantitative terms remained rather limited on the other hand. More recently, Zawischa et al.[32,33] have extended the model to include neutron proton cluster pairing and receive better agreement, in some cases they have investigated. The problem is certainly not exhausted from the side of theory.

Reversed odd-even-staggering and octupoles: Far off stability in the region of neutron-rich radium[34], francium[35], and radon[36] isotopes one has identified in the eighties a region of rather pronounced octupole deformation. Interesting enough, the odd-even-staggering in this region is found to be reversed, i.e., the odd N isotope lies closer to its heavier neighbour than to its lighter. This unexpected observation corraborates calculations by Leander and Sheline[37] who suggest that octupole deformation should be more pronounced in odd than in even nuclei. In this context it is interesting to note that Talmi applied a core polarization model by a pure j^n neutron configuration[38]. His formulas imply a regular odd-even-staggering for the case of an even multipole and an inverted one for odd multipoles (compare also[21]).

5 Second generation of IS physics

Around 1960 the harvest of isotope shifts and nuclear moments in the stable valley was about finished. Moreover, modern experimental techniques like nuclear reactions, nuclear spectroscopy, electron scattering, spectroscopy of myonic atoms, etc., superseded the traditional atomic spectroscopy, giving more detailed information on nuclear structure. Therefore, atomic spectroscopists, still faithful to the field, had to climb out of the stable valley, up the steep slopes to the land of exotic nuclei which was out of reach for most of the competitors. The route of access was prepared in these years by isotope separators on-line (ISOL) with isotope production facilities like cyclotrons, reactors or heavy ion accelerators. Yield curves span over thirty isotopes and have maxima excessing 10^{10} ions/s[39].

On-line laser spectroscopy: Tunable lasers lead to a fabulous methodological creativity in optical spectroscopy during the last three decades. With its enormous flexibility, sensitivity and precision laser spectroscopy found soon a rich field of application in on-line separated exotic isotopes. Nowadays online hfs and IS measurements by laser spectroscopy cover almostthe entire range of an ISOL yield curve. Many different methods have been invented and successfully applied in this particular field. Quite often they are hybridized with other techniques like atomic and ionic beams, optical pumping and rf-spectroscopy, all kinds of particle detection etc. (for review see ref.[21]). They are quite subtle and complex but very effective. I will confine myself here to a short description of three of these methods which formed the mainstreams in the field over the years.

Laser induced optical pumping (LIOP): Alkalis are favourite elements for on-line mass separation as well as for almost any kind of atomic spectroscopy. The LIOP method is addressed to unstable isotopes of these elements and has yielded hfs and IS data for almost hundred of them from sodium to francium. Fig. 6 shows the scheme. The mass-separated ion beam is stopped in an oven and converted into an atomic beam which passes a six pole magnet where states with positive electronic spin projection are focussed onto the entrance of another mass spectrometer. This serves as a background-free particle detector. A tunable laser crosses the atomic beam in front of the six pole at right angles in order to suppress the Doppler width. When it hits a hyperfine component of the resonance line, optical pumping sets in by which the resonant hyperfine component is depopulated in favour of the other. The optical pumping effect is analyzed by the state selective six pole magnet. The experiments were performed by an Orsay-ISOLDE-Collaboration[35].

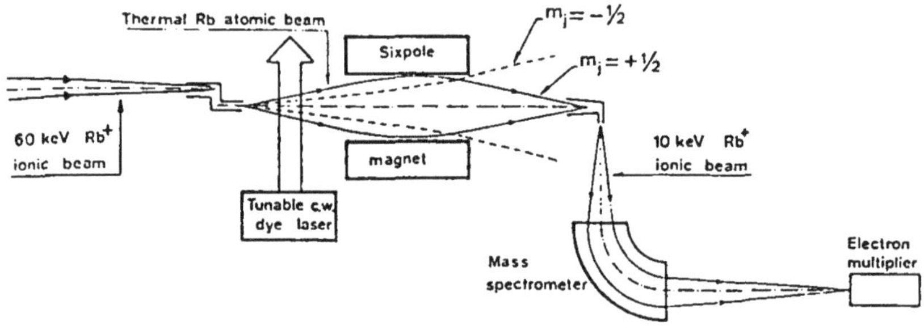

Figure 6: LIOP set-up for spectroscopy of unstable alkali isotopes.

Resonance ionization mass spectroscopy (RIMS): The next example makes also use of exciting an atomic beam at right angles, but this time by three different laser beams which raise the electron stepwise into the continuum. The resulting ions are mass separated and counted. This Resonance Ionization Mass Spectroscopy (RIMS) has first been applied by a Gatchina-Troitsk-Collaboration to a large number of neutron-deficient rare earth isotopes (see Fig. 7), establishing an impressive systematics of nuclear radii and moments in this region [40,41]. With some modifications the method has been further applied by a Mainz-ISOLDE-Collaboration [42] and a French-Canadian-Collaboration [43] at Orsay to refractory elements, namely gold and platinum. Quite recently an Orsay-Mainz-ISOLDE-Collaboration has taken up RIMS measurements in this region again and added chains of osmium and iridium to the picture besides extending the gold and platinum chains further off stability [44]. The results are shown in the synopsis of Fig. 8 which also contains data on heavier, neighbouring elements obtained by different methods [21].

The most dramatic feature in Fig. 8 is the sudden, large increase of $\langle r^2 \rangle$ of odd mercury isotopes far off stability [45,46]. Quite unexpectedly, they indicated that a strong deformation can occur even close to a magic proton number (Z = 82). These measurements date back before the age of laser spectroscopy and have triggered the field a lot. The even mercury isotopes, on the other hand, follow the trend of their heavier, spherical neighbours. It turns out that spherical and deformed shapes are almost degenerate in this region such that slight changes in the single particle structure may cross the nuclei to flop from one shape to the other. Also gold shows this sharp transition to deformed shapes.

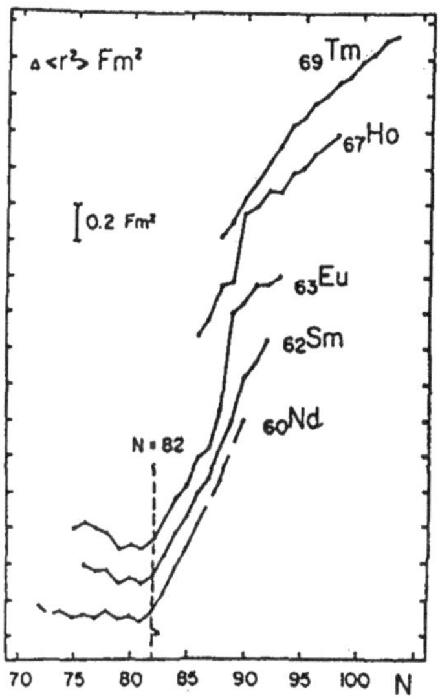

Figure 7: IS of unstable rare earth elements obtained by resonance ionization spectroscopy [42].

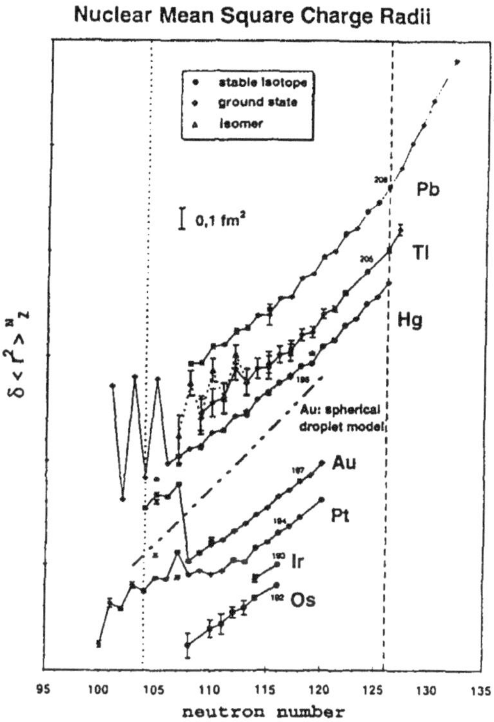

Figure 8: Systematics of IS from Os to Pb.

Collinear laser spectroscopy: The last method I am going to describe in this context is ideally adapted to on-line applications, since it works directly on the mass separated beam without stopping and re-evaporating (see Fig. 9) [47,48]. Still, the beam is manipulated in most cases, namely, it is neutralized in flight by charge transfer when passing an alkali vapour. The laser excites the beam collinearly and thus takes advantage of a strong contraction of the longitudinal velocity spread of the beam which is a result of a phase space transformation during the acceleration of the ions by the high voltage at the ion source. So, it is also a Doppler free spectroscopy. Resonant laser excitation is usually detected by the fluorescence from the beam. But when highest sensitivity is asked for at the wings of the isotopic yield curves, then several variants of collinear laser spectroscopy apply which transform the laser excitation into a signal of particle detection, e.g. by optical pumping and state selective re-ionization [49]. Collinear laser spectroscopy has proved to be the most versatile and fruitful method in the field. As an early example, Fig. 10 shows resonances in radium, the heaviest element which has been investigated systematically, so far [50,34].

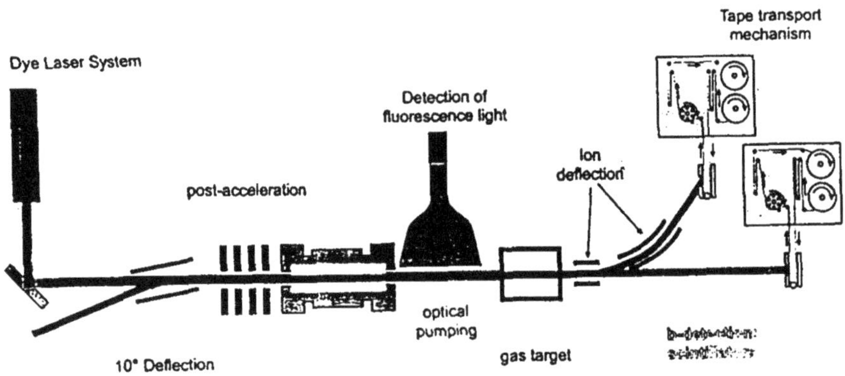

Figure 9: Collinear laser spectroscopy set-up.

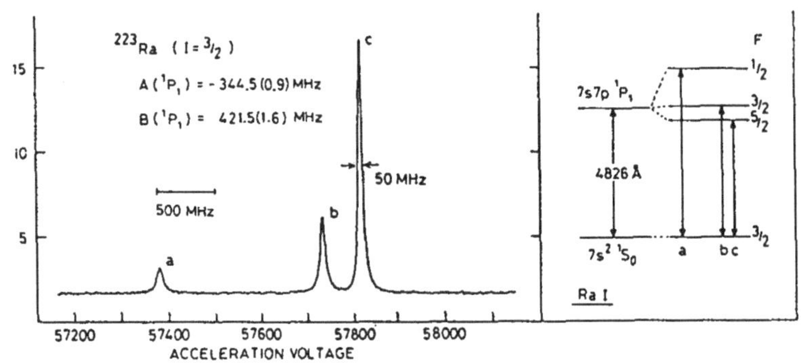

Figure 10: Scan of Ra hfs by collinear laser spectroscopy [51].

6 Discussion of IS systematics

Kink at shell closures and nuclear deformation: Fig.s 7 and 8 have given already an impression of the wealth and completeness of present data. Let us now turn to the discussion of specific examples. We start with the IS of the Eu chain which crosses the N = 82 magic number (Fig. 11) [41,51,52,21]. A very striking feature in this plot is the sharp kink of the IS at the magic number. It is repeated systematically in all isotones (see Fig. 7) and repeats again at N = 50 (see Fig. 13). A shell effect of equal clarity and persistence is only found in the systematics of nuclear binding energies. The standard phenomenological analysis of the kink is performed by the two parameter model (10), where the first, volume dependent term is inserted from the droplet model and the second, shape dependent one is restricted to β_2 only [20,21]. To both sides of N = 82 the course of the experimental $\delta\langle r^2\rangle$-values in Fig. 13 crosses lines of constant β_2-deformation parameters. Their slope is calculated from the droplet model and represents the first term in (10), i.e. the mere increase in nuclear volume. We see on the neutron-rich side the deformation to increase first steadily and then in a large final step between ^{151}Eu and ^{152}Eu. From that point on, nuclear shape has stabilized again and the volume expansion follows the droplet prediction along a line of constant deformation $\langle\beta_2^2\rangle^{1/2} = 0.32$. This value agrees perfectly with those derived from $BE2$-values as well as from spectroscopic quadrupole moments Q_s. Closer to the magic number this agreement is not so perfect anymore, neither on the neutron rich nor on the neutron deficient side. There the IS always points to larger deformation than the other observables, $BE2$ (compare Fig. 13) and Q_s, which, in contrast to $\langle r^2\rangle$, do not exhaust the full collective strength of the nuclear wave function. The same observations apply to all chains which are crossing magic neutron numbers N = 82 and N = 50 (compare Fig. 13).

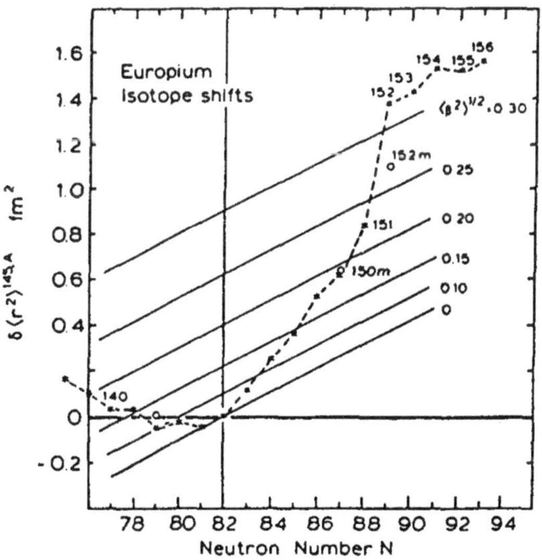

Figure 11: Europium IS presented in a plot with equideformation lines $\langle\beta^2\rangle^{1/2}$ = const.

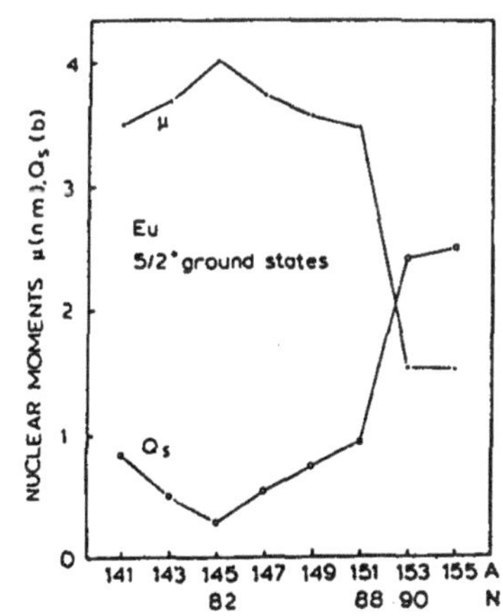

Figure 12: Magnetic (upper line) and quadrupole moments (lower line) of even N Eu isotopes.

Europium moments: Fig. 12 supplements the discussion on the IS of europium with the systematics of the nuclear moments of its even N isotopes[52]. The spin is always I = 5/2. It is ascribed to a $2d_{5/2}$ shell model state of the valence proton. Note, that ^{145}Eu can be regarded as a proton hole state to the double magic ^{146}Gd. There, the magnetic moment reaches a maximum of 80 % of the Schmidt-value, whereas the quadrupole moment acquires a minimum of two times the single particle value. At the shape transition the sharp drop of the magnetic moment proves the transition to a totally different single particle state, a $|\ 413\ |\ 5/2\rangle$ Nilsson state, actually, whereas the quadrupole moment gains a factor of 3 by the onset of a global deformation and the strong coupling of the spin to the intrinsic axis.

Relativistic mean field calculations: Coming back to kinks again a weaker but nevertheless distinct kink of the IS is observed at the shell closure N = 126 in the chain of Pb isotopes (see Fig. 8). Since here, at Z = 82, also the proton shell is closed we are definitely not allowed to interpret this residual kink in terms of the phenomenological model (10) by claiming a deviation from spherical shape responsible. Because contrary to the rare earth or Kr region (see below) the potential energy surface remains stiff with respect to collective coordinates in the neighbourhood of N = 126 [53]. This problem remained open for many years, until relativistic mean field (RMF) calculations entered the scene. The basic progress with respect to the earlier non-relativistic codes lies in the inherently proper treatment of the spin orbit interaction. This matters particularly at shell closures. Hence modern RMF calculations match perfectly the course of the IS in lead [53,54]. This applies also to the semi magic cases around N = 82 and N = 50. Fig. 13 shows a comparison of experimental [55] and RMF [56] values for the IS in Kr. The agreement is very good. Earlier calculations, using density-dependent forces, but in non-relativistic mechanics, reproduced these IS on the heavy side above the neutron shell closure reasonably well. But they did not show the kink at the shell closures [57]. The neighbouring chain of Rb runs all the way parallel to that of Kr, except at the heavy end where between N = 59 and N = 60 the nucleus jumps into a strong static deformation similar as shown in Fig. 11 for Eu. This discovery came as a surprise and was one of the early successes of the LIOP method [58].

Parabolic IS and core polarization models: For isotope chains close to proton shell closures one frequently observes a parabolic shape of the IS which spans over the whole neutron shell in question with its maximum curvature placed at midshell. Also binding energies show this feature. An early example of such an IS is the Ca chain which covers the $1f_{7/2}$ shell from N = 20 to N = 28 [59]. It has attracted the interest of many theoreticians. In this

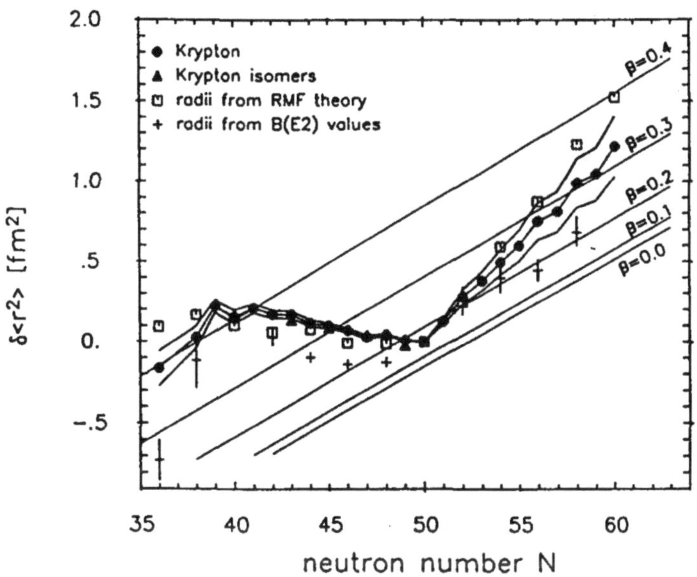

Figure 13: Chain of Kr IS (full symbols) together with calculations from RMF theory (squares) and from BE2 values through (10).

region one is tempted, of course, to try an interpretation in pure shell model terms. Talmi has interpreted a polarization of the ^{40}Ca core by the valence neutrons. Polarization by an even multipole of order 2 ℓ then yields a shift [38]

$$\delta\langle r^2\rangle^{(n)} \propto \frac{2 \mid \langle j||T_\nu^{(2\ell)}||j\rangle^2}{(2j+1)(2j-1)}\{n(2j+1-n)-\frac{1}{2}[1-(-1)^2](j+\frac{1}{2})\}, \quad (11)$$

where n is the number of valence neutrons. The first term in the wavy brackets represents the parabola, the second term the odd-even staggering with the correct sign. An expansion of the core volume according to the $A^{1/3}$ law is not considered in the model.

It is interesting that with some modification the model can be extended to the major shell $50 \leq N \leq 82$. The formula

$$\delta\langle r^2\rangle = a(N-50)+b(N-50)(82-N)+\frac{1}{2}c[1-(-1)^N] \quad (12)$$

perfectly fits the large data set of Cd [60], In [61] and Sn [62] isotopes (see Fig. 14). In contrast to (11) a linear term accounting for the volume increase has been added. Like in (11) the polarization term is also taken to be proportional to the number of particles times the number of holes in the shell. Here any influence of subshells is submerged in a global dependence on the two cornerstones N = 50 and N = 82 only.

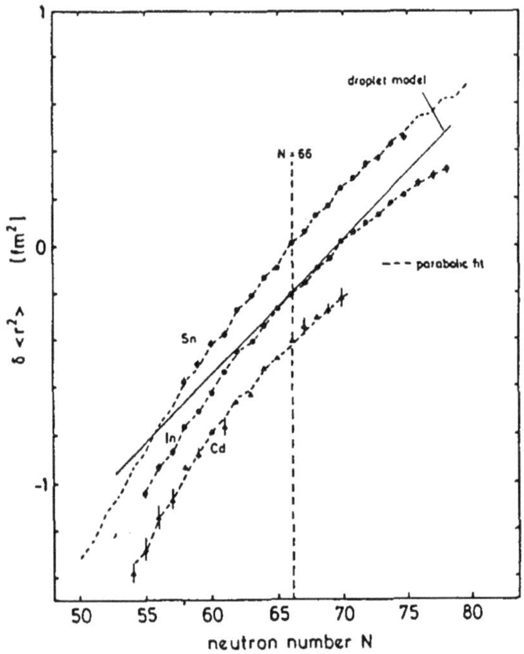

Figure 14: IS in Cd, In, Sn together fits by the parabolic model (12) (broken lines).

Isotone and isobar shift: Together with absolute $\langle r^2\rangle$ data in the stable valley obtained from electron scattering or muonic spectra the $\delta\langle r^2\rangle$ data from IS measurements on long isotopic chains have established nowadays a dense network of nuclear radii - similar to that of masses - which covers large areas of the nuclear chart reaching far out into the exotic regions. So we may also have a look to isotone shifts and isobar shifts even. Only little attention has been payed so far to these questions. Sprouse has analyzed isotonic shifts at the N = 82 shell closure [63]. Data in this interesting chain span from $_{54}$Xe to $_{66}$Dy. A kink at the double magic $^{146}_{64}\mathrm{Gd}^{82}$ is observed, but with opposite sign surprisingly. A kink of usual sign is observed when the N = 66 isotones cross the proton magic tin line [21].

An access to isobar shift we find for instance in the chain A = 118 where the data span from $^{118}_{48}\mathrm{Cd}^{70}$ to $^{118}_{55}\mathrm{Cs}^{63}$ with seven nucleons exchanged. Corrected for the different deformations of these nuclei according to (10) we find an increase of the mean square charge radius by about 0.8 fm^2 [21] in accordance with the parameterized droplet model prediction [20]. Note, that in a naive, homogenously charged liquid drop model the isobar shift would be identically zero since the charge radius then depends only on the total mass number A. A more flexible assumption would be to keep the mean squared mass radius $\langle r^2\rangle_m$ of a nucleus constant in an isobaric chain. This implies still imcompressibility of nuclear matter. Under this boundary condition a change of the mean squared radius of the proton distribution is compensated by that of the neutrons. To first order we find

$$\delta\langle r^2\rangle_n = -\frac{Z}{N}\delta\langle r^2\rangle_p . \tag{13}$$

Applying this formula one estimates that the neutron radius would expand from ^{118}Cs to ^{118}Cd by about 0.15 fm [21], a reasonable value which is also in fair agreement with that expected from a density-dependent Hartree-Fock-Calculation [18].

7 Present trends in isotope shift measurements

The last three decades have yielded an impressive systematics of nuclear charge radii determined from IS measurements in optical spectra. The discussion in the preceding section has also made clear that the main features are pretty well explained nowadays by gross models of nuclei and even better by microscopic calculations. What is still to do? First we notice still some large gaps in the systematics. They concern in particular unstable isotopes of refractory elements which are hard to deal with. But there is also a number of deticated problems which attract the interest of the researchers in the field. I like to mention three of them:

Hydrogen and few electron spectra: Precision spectroscopy of hydrogen and its isotopes as well as of hydrogen like spectra of heavy elements plays a major role in atomic physics. It involves fundamental constants of physics, test of QED to a high order, as well as nucleon and nuclear structure. A recent experiment by Hänsch and collaborators on hydrogen isotopes [64] have determined the H-D IS to the incredible precision of ten digits δ_ν = 670994334.64 (15) kHz. Although, largely dominated by the mass shift (1), one still derives from this number the most accurate value for the difference in mean squared charge radii

$$\delta\langle r^2\rangle_{ch}^{d,p} = 3.812(15) fm^2 , \tag{14}$$

which supersedes results from elastic electron scattering clearly.

It is also possible today to produce and store highly charged ions of the heaviest elements stripped down to the K-shell or even naked. Particularly suited for precision spectroscopy are $2p \to 2s$ transitions in lithium like spectra because of their quite narrow radiative width. A group at Livermore recently managed to measure this way the isotope shift in uranium and derived a value of $\delta\langle r^2\rangle^{238,235} = 0.250(32) fm^2$ [65]. The merit of this measurement lies in the fact that the result is free of systematic errors since this tightly bound few electron spectrum can be calculated and analyzed safely.

Light elements and trapping techniques: We cannot expect the IS of light elements to be governed by systematics in a way similar to that of heavy ones. The addition of a single nucleon there matters a lot for the nuclear structure. In our context it would be particularly interesting to approach the driplines and to study halo effects. Note, that also a neutron halo would increase the mean squared charge radius since it causes the core to move around the center of mass.

Experiments on the entire chains of He, Li and Be isotopes seem possible but require dedicated methods of highest accuracy and sensitivity. Several groups are preparing respective proposals. Its seems that laser cooling and storing the atoms in magneto-optical traps are a promising route after their feasibility has been demonstrated off stability for Na [66], Rb [67] and Fr [68] isotopes and in trace analysis for the case of ^{81}Kr [69].

Apart from the aforementioned hydrogen case, the lightest chain so far investigated is the one of neon isotopes. The experiment is being performed by R. Neugart and collaborators at the ISOLDE, using collinear laser spectroscopy [70]. The results cover a range from mass 17 to 28. They show a rather irregular structure. The lightest bound isotope ^{17}Ne seems to have a radius even larger than the deformed ^{20}Ne. Could this be interpreted as a sign of a proton halo, a phenomeon which has been searched for so much in recent years? If this turns out to be true, it would mark a great progress in our understanding of the proton drip line. Since the tiny volume shift in Ne is superposed by a huge mass shift, however, a sign of caution is still appropriate until the evaluation of the data is finalized.

Super deformation and the heaviest elements: Our last example points to the opposite extreme, where the volume shift reaches the fabulous value of 10 wave numbers and shadows out any other nuclear structure effect seen in atomic spectra. It concerns the IS of the super deformed ($\beta_2 = 0.66$) fission isomers ^{240f}Am and ^{242f}Am with respect to the nuclear ground state of neighbouring ^{241}Am and ^{243}Am. The short-lived isomers are produced at tiny rates from ^{242}Pu by (d,2n) and (p,3n) reactions, respectively. After extensive methodological development work, a Mainz-Heidelberg-Collaboration [71] managed to perform resonance ionization experiments on these isomers. The delayed spontaneous fission served as a highly selective and 100 % efficient detection signal after the laser ionized species had been extracted onto the fission fragment detector. Fig. 15 shows the RIS signals which are composed of only 192 events. The experiment constitutes so far the only direct proof of the extraordinary large extension of the charge distribution of fission isomers. Apart from this spectacular effect the field of hfs and IS investigations of the heaviest elements is virginal but scientifically quite promising. The main obstacle in accessing this region is formed by the extremely low production yields. On the other hand, the demand for ultra high sensitivity is eased a bit by the rather relaxed requirements for spectroscopic precision since the effects are getting quite large. Specific proposals to investigate this region are under way which also will make use of trapping techniques [72].

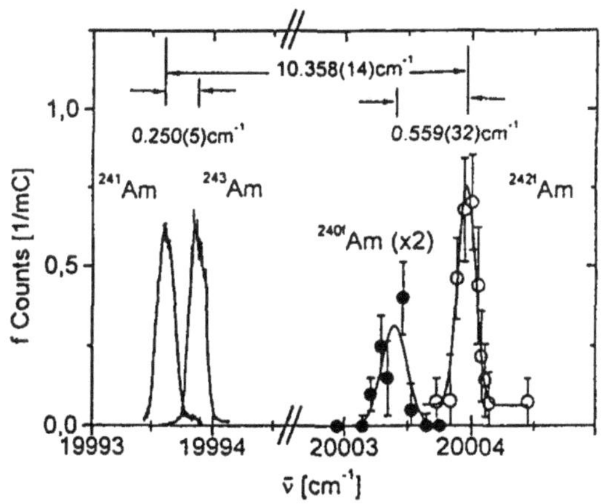

Figure 15: Optical IS of fission isomers ^{240,242f}Am with respect to neighbouring ^{241,243}Am.

Concluding remark: More generally speaking, we see now, at the turn of the century, a strong branch of nuclear physics focussing on the investigation of radioactive species with a great methodological variety. Here, modern techniques of atomic physics are still going to play key roles. This concerns in particular trapping of atoms and ions in optical and electromagnetic fields as a starting point for any kind of sensitive and precise spectroscopy. This will be optical as well as mass and beta spectroscopy.

References

1. J.E. Rosenthal and G. Breit, Phys. Rev. **41**, 459 (1932).
2. G. Breit, Revs. Modern Phys. **30**, 507 (1958).
3. H. Schüler, Naturwissenschaften **19**, 320 (1931).

4. H. Kopfermann, Naturwissenschaften **19**, 400 (1931).
5. H. Schüler and J.E. Keystone, Z. Physik **70**, 1 (1931).
6. D.A. Jackson, Z. Physik **75**, 223 (1932).
7. H. Kopfermann, Z. Physik **75**, 363 (1932).
8. E. Bretscher and E. Guth, Phys. Zeitschrift **32**, 670 (1931).
9. G. Racah, NATURE **129**, 408 (1932).
10. G. Breit, Phys. Rev. **42**, 348 (1932).
11. R.A. Bodmer, Nucl. Phys. **9**, 371 (1959).
12. E.P. Seltzer, Phys. Rev. **188**, 1916 (1969).
13. A.M. Mårtensson-Pendrill, in Proc. of 1. Euroconf. on Atomic Physics at Accelerators, Mainz, Germany, 1999 (eds. L. Schweikhard and H.-J. Kluge), Hyperfine Interactions, in press.
14. G. Fricke et al, At. Data Nucl. Data Tables **60**, 177 (1995).
15. G. Breit, Revs. Modern Phys. **30**, 507 (1958).
16. P. Brix and H. Kopfermann, Revs. Modern Phys. **30**, 517 (1958).
17. G. E. Brown in *Facets of Physics*, ed. D.A. Bromley and V. Hughes (Academic Press, New York and London, 1970), p. 141.
18. I. Angeli, M. Beiner, R.J. Lombard, D. Mas, J. Phys. **G6**, 303 (1980).
19. W.D. Myers and W.J. Swiatecki, Ann. Phys.(N.Y.) **55**, 395 (1969).
20. W.D. Myers and K.H. Schmidt, Nucl. Phys. **A410**, 61 (1983).
21. E.W. Otten in *Treatise on Heavy Ion Science*, (ed. D.A. Bromley, Plenum Press, NY and London, 1989), vol. 8, p. 517.
22. H. Schüler and Th.Schmidt, Z. Physik **92**, 148 (1934).
23. H. Schüler and Th. Schmidt, Z. Physik **94**, 457 (1935).
24. P. Brix and H. Kopfermann, Nachr. Akad. Wiss., Göttingen, Math. Phys. Kl, , 31 (1947).
25. P. Brix and H. Kopfermann, Z. Physik **126**, 344 (1949).
26. G. Breit et al, Phys. Rev. **78**, 390 (1950).
27. H.A. Jahn and E. Teller, Proc. R. Soc. London, Ser. **A161**, 220 (1937).
28. P.G. Reinhard and E.W. Otten, Nucl. Phys. **A420**, 173 (1984).
29. W.J. Tomlinson and H.H. Stroke, Nucl. Phys. **60**, 614 (1964).
30. B. Hoffmann et al, Z. Phys. **315**, 57 (1984).
31. B.S. Reehal and R.A. Sorensen, Nucl. Phys. A **161**, 385 (1971).
32. D. Zawischa, Phys. Lett **155B**, 309 (1985).
33. D. Zawischa et al, Phys. Lett. **185B**, 299 (1987).
34. S.A.Ahmad et al, Nucl. Phys. A **483**, 244 (1988).
35. A. Coc et al, Nucl. Phys. A **468**, 1 (1987).
36. W. Borchers et al, Hyperfine Interactions **34**, 25 (1987).
37. G.A. Leander and R.K. Sheline, Nucl. Phys. A **413**, 375 (1984).
38. I. Talmi, Nucl. Phys. A **423**, 189 (1984).

39. H.L. Ravn, Phys. Rep. **54**, 201 (1979).
40. G.D. Alkhazov et al, JETP Lett. **37**, 274 (1983).
41. G.D. Alkhazov et al, JETP Lett. (USA) **46**, 170 (1987).
42. K. Wallmeroth et al, Phys. Rev. Lett. **58**, 1516 (1987).
43. H.T. Duong et al, Phys. Lett. B **217**, 401 (1989).
44. F. Le Blanc et al, Phys. Rev. C **60**, 054310-1 (1999).
45. J. Bonn et al, Phys. Lett. **38B**, 308 (1972).
46. J. Bonn et al, Z. Phys. A **276**, 203 (1976).
47. K.R. Anton et al, Phys. Rev. Lett. **40**, 642 (1978).
48. reviewed e.g. in: R. Neugart, in *Progress in Atomic Spectroscopy* (eds. H.J. Beyer and H. Kleinpoppen, Plenum Press, New York, 1987), Part D, Ch. 2, p. 75.
49. R. Neugart et al, Nucl. Instr. Meth. B **17**, 354 (1986).
50. S.A. Ahmad et al, Phys. Lett. **133B**, 47 (1983).
51. K. Dörschel et al, Z. Phys. A **317**, 233 (1984).
52. S.A. Ahmad et al, Z. Phys. A **A321**, 35 (1985).
53. P.G. Reinhard et al, in Proc. of 1. Euroconf. on Atomic Physics at Accelerators, Mainz, Germany, 1999 (eds. L. Schweikhard and H.-J. Kluge), Hyperfine Interactions, in press, and references therein.
54. M.M. Sharma, G.A. Lalazissis, and P. Ring, Phys. Lett. B **317**, 9 (1993).
55. M. Keim et al, Nucl. Phys. A **586**, 219 (1995).
56. G.A. Lalazissis and M.M. Sharma, Nucl. Phys. A **586**, 201 (1995).
57. X. Campi and M. Epherre, Phys. Rev. C **22**, 2605 (1980).
58. C. Thibault et al, Phys. Rev. **C23**, 2270 (1981).
59. A. Andl et al, Phys. Rev. C **26**, 2194 (1982).
60. F. Buchinger et al, Nucl. Phys. A **462**, 305 (1987).
61. J. Eberz et al, Nucl. Phys. A **464**, 9 (1987).
62. J. Eberz et al, Z. Phys. A **326**, 121 (1987).
63. G.D. Sprouse et al, Hyperfine Interactions **59**, 91 (1990).
64. A. Huber et al, Phys. Rev. Lett. **80**, 468 (1998).
65. S.R. Elliot et al, Phys. Rev. C **57**, 583 (1998).
66. Z.-T. Lu et al, Phys. Rev. Lett. **72**, 3791 (1994).
67. G. Gwinner et al, Phys. Rev. Lett. **72**, 3795 (1994).
68. J.E. Simsarian et al, Phys. Rev. Lett. **76**, 3522 (1996).
69. C.Y. Chen et al, Science **286**, 1139 (1999)
70. R. Neugart, in Proc. of 1. Euroconf. on Atomic Physics at Accelerators, Mainz, Germany, 1999 (eds. L. Schweikhard and H.-J. Kluge), Hyperfine Interactions, in press.
71. H. Backe et al, Phys. Rev. Lett. **80**, 920 (1998).
72. H. Backe et al., in *SHIP-Trap* Proposal to the GSI (1999).

GREGORY BREIT AND THE FERMION-FERMION INTERACTION

G.E. BROWN
Department of Physics & Astronomy, SUNY at Stony Brook
Stony Brook, NY 11794, USA

Gregory Breit's most important conceptual research was in the development of the Breit Interaction between two electrons. Carried out long before Feynmann Graphs and the modern covariant formulation of quantum electrodynamics, it lacked the proper treatment of pair theory; i.e., the negative energy electron states were left unoccupied. This resulted in higher-order terms being incorrect. From comparison of results with the fine structure of Helium, Breit realized that these higher order terms were wrong. He formulated the rule that these e^4 terms should be neglected, or that his interaction should be used only as an expectation value. In thinking about effects of possible exchange of "heavy photons", vector mesons between nucleons much later ~ 1960, Breit saw that a large number of phenomena: short-range repulsion, strong short-range spin-orbit force and others required by the empirically known nucleon-nucleon interactions could be understood from introduction of heavy vector meson exchange into the nucleon-nucleon interaction.

1 The Breit Interaction

Gregory Breit's entry into the fermion-fermion interaction began with his treatment in quantum theory of the retardation in the interaction between two electrons. His previous publications involved electronic interactions in more applied situations and it is clear that he wanted to understand them from fundamentals. His 1929 paper "The Effect of Retardation on the Interaction of Two Electrons" [1] begins from the Dirac equation, and then notes that

$$\frac{dx_k}{c\ dt} = \frac{2\pi i}{k}\left[(E/c), x_k\right] = -\alpha_k, \tag{1}$$

further that this expression "has been contributed to the writer by letter by Dr. P.A.M. Dirac in the summer of 1928." He remarks that the approximate Lagrangian and Hamiltonian taking into account the retardation of potentials were derived on the classical theory by Darwin [2]. (When encountering difficulties in extending the interaction to higher orders, he always came back to this classical expression.) The form for the interaction Breit arrived at was

$$I = \frac{e^2}{r_{12}} - \frac{e^2}{2r_{12}}\left(\vec{\alpha}^{(1)} \cdot \vec{\alpha}^{(2)} + \frac{\vec{\alpha}^{(1)} \cdot \vec{r}_{12}\ \vec{\alpha}^{(2)} \cdot \vec{r}_{12}}{r_{12}^2}\right). \tag{2}$$

Normally, one would think of the current-current interaction being twice the $\vec{\alpha}^{(1)} \cdot \vec{\alpha}^{(2)}$ term in eq.(2). The last term includes also retardation of the Coulomb interactions.

In a second paper [3] Breit took up the fine structure of He as a test of his Hamiltonian. The terms of order e^4 were found to be in contradiction with experiment. One should note that Oppenheimer [a] [4] had already shown how to eliminate the longitudinal (and scalar) photons leaving an instantaneous Coulomb potential, so that terms of higher order than e^2 in the expansion were genuinely an expansion in α. Breit rederived his equation by field theory [5]. Still the terms of order $e^4(\alpha^2)$ were found to be in contradiction with the fine structure in He, and Breit came to the realization that that part of the interaction I involving transverse photons had to be limited in use to lowest order. Indeed Bethe, ignoring Breit's prescription, used the Breit interaction to second order in his 1933 Handbuch article [6], again obtaining results that were in conflict with the empirical He fine structure.

In my thesis work on the effect of the proton motion and size on the fine and hyperfine structure of hydrogen [7], we did not need to go beyond order α. Indeed, we could show that the effective mass approximation to the proton motion was corrected only in relative order $\alpha(m_e/m_N)$, not order α, and this important result still stands.

The situation was clearly unsatisfactory, although by comparison with classical theory Breit had arrived at rules limiting the use of his interaction to situations where it was correct. He speculated in his papers that the errors in higher order might result from negative energy states, as indeed was later shown to be the case (see below).

Following my Ph.D. I went to Birmingham in England to work with Rudolph Peierls. Freeman Dyson was also in Birmingham and it seemed natural to try to derive the electron-electron interaction from field theory, which I undertook with Geoff Ravenhall.

We early understood that because of negative-energy states the Breit interaction was "meaningless" [8]. This could be seen by first constructing the helium atom as two noninteracting electrons in the field of the nucleus, and then turning on the electron-electron interaction slowly.

As shown in Fig.1, the two-electron wave function will quickly dissolve, one electron, upon interaction with the other, dropping into a negative energy state and the other going off into a positive energy one so as to conserve energy. (Amusingly, this effect lay unnoticed in the literature until people

[a] Indeed, the current-current interactions would be $I = -e^2\vec{\alpha}^{(1)} \cdot \vec{\alpha}^{(2)}/r_{12}$. The difference between this and the term involving the α's in eq.(2) comes from projecting out the longitudinal photons in order to obtain the instantaneous Coulomb interaction. This projection involves $\vec{k} \cdot \vec{\alpha}$ terms in momentum space which become the $\vec{r}_{12} \cdot \vec{\alpha}$ terms in configuration space.

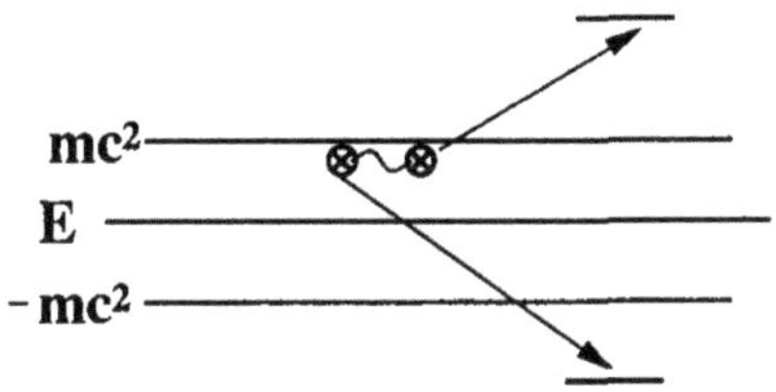

Figure 1: "Continuum dissolution" of the two-electron wave function. If the two electrons in helium interact one electron can scatter into a negative energy state, the other into a positive energy one, so as to conserve energy. Thus, the wave function will quickly dissolve, unless the negative energy states are filled with electrons so as to block this process.

began putting two-body correlations into many-body calculations, upon which strange things happened. In 1985 Joe Sucher clarified these matters[9], referring back to our paper.

Of course, the negative energy states must be filled with electrons so as to prevent the continuum dissolution. This had, however, not been done in Breit's derivation. He had let the electrons make transitions, in building up the correlations of order e^4, to negative energy states and back. Basically these transitions were effected by operators such as

$$O_{++} = \frac{\vec{\alpha}^{(1)} \cdot \vec{\alpha}^{(2)}}{(\vec{p}_1 - \vec{p}_2)^2 - (E_{\vec{p}_1} - E_{\vec{p}_2})^2} \tag{3}$$

where

$$E_{\vec{p}} = \sqrt{\vec{p}^2 + m_e^2} \tag{4}$$

whereas for transition from positive to negative energy states the operator

$$O_{+-} = \frac{\vec{\alpha}^{(1)} \cdot \vec{\alpha}^{(2)}}{(\vec{p}_1 - \vec{p}_2)^2 - (E_{\vec{p}_1} + E_{\vec{p}_2})^2} \tag{5}$$

should have been used. The change in sign of E_{p_2} comes because it is the energy of a negative energy state. Projection operators onto positive and negative energy states, in a plane wave basis, have to be introduced, and then it is clear which of the operators, O_{++} or O_{+-} should be used. Of course, the Brown-Ravenhall calculation was constructed properly in pair theory.

The net effect of using O_{+-} rather than O_{++} is to have a term $\sim 4m_e^2$ in the denominator, replacing the term $\sim (\vec{p}_1 - \vec{p}_2)^2$. This does away with the incorrect α^2 terms, the effects from negative energy states contributing only in order α^3. This is as expected since the operator (5) to negative energy states essentially has $4m_e^2$ in the denominator, whereas with $(E_{\vec{p}_1} - E_{\vec{p}_2})^2$ in the denominator, it would have been $\sim \vec{p}^2 \sim \alpha^2 m_e^2$.

The most important part of the Brown-Ravenhall paper was the continuum dissolution. This was first noticed only twenty-some years later when workers began trying to improve atomic Hartree-Fock calculation by adding two-body corrections. Then their wave functions suddenly "went down the drain". Joe Sucher explained this in his paper[9]. Thus Breit's prescription for how to use his interaction turned out to be completely right.

One might think that developments following the renewed activity in quantum electrodynamics in the late 1940's and 1950's, especially the formulation of the Bethe-Salpeter equation, would have replaced the Breit interaction. However, for reasons I don't have time to go into, such equations have to be reduced to equal time ones, for electrons essentially the Breit equation, before calculations can be carried out. I believe that the Breit interaction remains the most important contribution Gregory Breit made to theoretical Physics.

An unpleasant feature of the Brown-Ravenhall treatment separating positive and negative energy states, which was necessary to get the correct sign for $E_{\vec{p}}$ in the denominator of O_{+-}, is that projection operators

$$\Lambda^{\pm}(\vec{p}) = \frac{\sqrt{\vec{p}^2 + m^2} \pm (\vec{\alpha} \cdot \vec{p} + \beta m_e)}{2\sqrt{\vec{p}^2 + m_e^2}}, \tag{6}$$

which are nonlocal in configuration space, had to be employed; i.e., each $\vec{\alpha}$ in eq.(5) had to be replaced by

$$\Lambda^+(\vec{p}_1)\vec{\alpha}\Lambda^-(\vec{p}_2) + \Lambda^-(\vec{p}_1')\vec{\alpha}\Lambda^+(\vec{p}_2') \tag{7}$$

so that one would know the correct sum $E_{\vec{p}_1} + E_{\vec{p}_2}$ in the denominator. (For those parts of the interaction with $\Lambda^+(\vec{p}_1)\vec{\alpha}\Lambda^+(\vec{p}_2)$, the denominator would involve $E_{\vec{p}_1} - E_{\vec{p}_2}$.) Since the problems in atomic physics were most conveniently formulated in configuration space; i.e., correlations would be functions of $\vec{r}_{12}$, the $\Lambda^{\pm}(\vec{p})$ bring in ugly nonlocal operators. Thus far, to my knowledge, the correct treatment has, as in the case of the fine structure of He, shown that contributions from negative energy states are postponed until order α^3.

In his 1933 paper [5] Breit arrived, after eliminating the longitudinal and scalar photons which gave the instantaneous Coulomb interaction [4], at his equation (11) of that paper:

$$(\Delta E)_{n,n} = \left(-\frac{1}{4}\right) \sum_{i,j} e_i e_j \int\int \frac{1}{r_{p'p''}} \left[\vec{\alpha}^i \cdot \vec{\alpha}^j + \frac{\vec{\alpha}^i \cdot \vec{r}_{p'p''} \vec{\alpha}^j \cdot \vec{r}_{p'p''}}{r_{p'} r_{p''}}\right]$$
$$\times D_i(r_i - r')D_j(r_j - r'')dV_{p'}dV_{p''}. \tag{8}$$

Quantum numbers of the systems to which this interaction is applied to are denoted n_1, n_2, $\cdots$; note that Breit stuck to the diagonal part $(\Delta E)_{n,n}$, believing that it was the off-diagonal elements that gave the trouble. In fact, in application to the fine structure of He, this diagonal interaction already had problems, in that the e^4 terms in the wave function, once solved for by using this interaction, included terms which in a plane-wave representation mix positive and negative energy states. This is an ugly business, but Breit's prescription to ignore the e^4 terms was exactly right, the ugliness entering in order α^3. The arguments leading to eq.(8), which Breit derived over and over again, were very persuasive.

Breit conjectured [5] that eq.(8) "may have a wider range of applicability than the He fine structure". Two K electrons around a bare nucleus would be expected to obey this equation. The experimentally accessible K lines are unfortunately complicated by the screening due to other electrons."

Aumsingly, if one goes to heavy atoms with $Z \gg 1$, the situation simplifies [10]. In this case, expansion can be made in the ratio of electron-electron to (central) nuclear interaction giving

$$H_{int} = \frac{e^2}{4\pi} \frac{(1 - \vec{\alpha}_1 \cdot \vec{\alpha}_2) \exp\left[i|E_m - E_n|(r_{12})\right]}{r_{12}} \tag{9}$$

where E_m and E_n are the energies of the initial and final hydrogen-like states in the nuclear potential. For two K-electrons $E_m = E_n$ and

$$H_{int}(m = n) = \frac{e^2}{4\pi} \frac{(1 - \vec{\alpha}_1 \cdot \vec{\alpha}_2)}{r_{12}}; \tag{10}$$

i.e., the effect of retardation disappears. At first sight this is surprising: the faster the electrons move, the smaller the retardation. But then one realize that this is quantum mechanics; to a good approximation the electrons are in *stationary* states.

2 The Nucleon Nucleon Interaction

As Mac Hull writes in his Biographical Memories of Breit,[11] much of the work that occupied Breit during his remarkable career involved the scattering or reaction of charged particles – usually both positive (i.e., an interaction in which the repulsive Coulomb force opertes). Solutions for the radial wave function we called Coulomb wave functions, when only the Coulomb force acts. Breit published the first of a long series of papers on the subject with F.L. Yost and J.A. Wheeler in 1936, later a comprehensive Handbuch article with Mac Hull in 1959.

For a short time I was involved in the calculation of Coulomb wave functions, trying WKB methods with various projects. When actually asked to join the computational project I protested, saying that if this was physics, I was going to go back to Wisconsin and work with someone else. Having the G.I. bill (I had been in the navy 1943-6) gave me a certain freedom for all of my time in graduate school, and Breit saw that he would lose me unless he gave me problems that did not involve numerical work. He then gave me gorgeous problems involving the calculation of effects of nuclear motion (and of the finite size of the nucleons) on the fine and hyperfine structure of hydrogen. The published papers [7] are still classics.

When Breit had to make trips, he would invite me to walk with him to the train station in New Haven. The inducement was "We can talk Physics". We had wonderful far-ranging conversations, with him doing most of the talking, me asking questions. These talks influenced strongly my development in nuclear physics, as can be seen from my publications for many years after I left him. Most of all, I learned a persistence, keeping after a problem, until I felt satisfied that I really understood it.

In the 1950's the Berkeley cyclotron started churning out data on proton-proton scattering. Aside from Coulomb distortion, the differential cross section was roughly isotropic, right up to incident proton energies of $\sim$ 300 MeV. The cross sections were too large for the scattering to be S-wave. The question was why the cross section held up at the larger angles $\sim 90°$. As matters evolved, the connection between the short-range repulsion in the nucleon-nucleon interaction and the spin-orbit interaction, in the form of the Thomas term connected with the hard core proved to be attractive. At least three of us had this idea. Breit published a characteristically thorough paper "Nucleon-Nucleon Spin-Orbit Interaction and the Repulsive Core in 1960 [12], also S.N. Gupta in the same year [13]. I was again thinking along the same lines as Breit and had proposed this as a thesis problem to D.L. Sprung in Birmingham in 1958. His published paper [14] appeared in 1961 after he moved to Cornell.

At the time I began theoretical physicists, the experts in the U.S., could be numbered easily on the fingers of two hands. I have always considered myself to have been extremely lucky to have been able to interact so strongly – more intensely, I believe, than any of his other students – with one of these giants.

The Legacy of Gregory Breit has finally put his work into perspective, showing his persistence in pursuing a fundamental conceptual understanding of theoretical physics. No matter how much effort was required, he attacked important problems head on. In retrospect it seems absolutely amazing what a deep understanding of so many essential questions be achieved.

References

1. G. Breit, Phys. Rev. 34 (1929) 553.
2. Darwin, Phil. Mag. 39 (1920) 537.
3. G. Breit, Phys. Rev. 36 (1930) 383.
4. J.R. Oppenheimer, Phys. Rev. 35 (1930) 461.
5. G. Breit, Phys. Rev. 39 (1932) 616.
6. H.A. Bethe, *Handbuch der Physik* (1933) Geiger-Scheel, Bd. 24, I. Jeil, p. 375.
7. G. Breit and G.E. Brown, Phys. Rev. 74 (1948) 1278; G.E. Brown and G.B. Arfken, Phys. Rev. 76 (1949) 1299; G. Breit and G.E. Brown, Phys. Rev. 76 (1949) 1307; see, also, G.E. Brown, Proc. Nat. Acad. Sci. Wash. 36 (1950) 15.
8. G.E. Brown and G. Ravenhall, Proc. Roy. Soc. A 208 (1951) 552.
9. J. Sucher, Phys. Rev. Lett. 55 (1985) 1033.
10. G.E. Brown, Phil. Mag. 43 (1952) 467.
11. M.H. Hull, Biographical Memoirs, National Academy of Sciences, 74 (1998) 27.
12. G. Breit, Phys. Rev. 120 (1960) 287.
13. S.N. Gupta, Phys. Rev. 117 (1960) 1146.
14. D.W.L. Sprung, Proc. Poy. Soc. 267 (1962) 244.

Ultra-Relativistic Nuclei: A New Frontier

L. McLerran
Physics Dept. Bdg. 510A, BNL P. O. Box 5000
Upton NY, 11713 USA
E-mail: mclerran@bnl.gov

The collisions of ultra-relativistic nuclei provide a window on the behavior of strong interactions at asymptotically high energies. They also will allow us to study the bulk properties of hadronic matter at very high densities.

1 Two Fundamental Issues

There are two important central issues which will be studied through the use of ultrarelativistic nuclei.

The first is the gross structure of high energy hadronic interactions. How do things behave in the limit $E \to \infty$? Can one understand this simply from first principles in QCD? What is the correct space-time picture of high energy hadronic interactions? It is surprising to me that such simple issues are not really understood, even though we have been studying strong interaction processes for over half a century.[1]

The second issue arises from the use of high energy hadronic interactions as a tool to study the gross structure of hadronic matter in bulk. We use ultrarelativistic heavy ion collisions to produce matter at very high energy density. This matter which has either or both of high temperature and high baryon number density. Is there a phase transition to a deconfined or chirally symmetric quark gluon plasma? What is the equation of state of such matter? Is our understanding of mass generation and confinement correct? [2]

We should ask how these studies fit into our conception of nuclear physics. I think that traditional nuclear physics was the study of nuclei. It turned out that to understand nuclei, one had to posit a theory of strongly interacting matter, and that this theory developed a life and dimension of its own.

The currently accepted theory of strong interactions, QCD, not only incorporates nuclei but all of the interactions of hadrons at any energy. In fact, viewed within the framework of QCD, nuclei are an afterthought, a residual of strong interactions which barely binds nucleons together. Traditionally, nuclear physics was concerned only with the low energy aspects of QCD. In recent year, I think it is fair to say that all aspects of QCD are of interest now to practicing nuclear physicists, and is more a subject of nuclear physics than of particle physics.

Another extension of traditional nuclear physics was from the theory of nuclei to that of high density matter in general. This has had applications in the theory of neutron stars and the big bang early universe. The ideological limits of nuclear physics are a little fuzzy here. Many of us who have studied high energy density nuclear matter have used similar techniques to study the physics of matter at energy density scales many orders of magnitude higher than scales typical of nuclear matter, for example at temperatures of the order of the electroweak scale $T \sim 100\ GeV$ or even the Planck scale $T \sim 10^{19}\ GeV$. Although this has very little to do with nuclei, many elements of the theory of matter at these diverse scales are common, as is the mathematical language used.

Finally, another area which nuclear physics plays a strong role is in complicated dynamics in field theory. This comes simply from the fact that nuclei are complicated objects, and to study them one needs a powerful arsenal of theoretical methods. In QCD, for example, one is studying a strongly interacting field theory with all of its complications. The techniques one uses here generalize to any field theory.

The study of ultrarelativistic nuclei straddles all of the above areas except for one. It is not an attempt to understand nuclei! Ultrarelativistic nuclear physics is a world which has been generalized from traditional nuclear physics, but incorporates much of its soul:

- Strong Interactions

- Bulk Properties of Matter at High Energy Density

- Complicated Dynamical Problems

2 The Ultra-Relativistic Nucleus

2.1 The Un-Relativistic Nucleus

For our purposes, the nucleus is a simple object. I will treat it as an object (oftentimes a cube!) with a characteristic size of $A^{1/3}$ where A is its baryon number. This ranges from $R \sim 1 - 10\ Fm$. Time scales characteristic of nuclei will be in the range $1 - 100\ Fm/c$ which correspond to the time it takes light to travel a proton size to the characteristic time for a sound wave in nuclear matter to propagate across the nucleus. The characteristic baryon density of a nucleus is $\rho \sim .1 - .2\ Fm^{-3}$ and energy density $\epsilon \sim .1 - .2\ GeV/Fm^3$

2.2 The Ultra-Relativistic Nucleus

The ultra-relativistic nucleus is almost the simple Lorentz boost of the un-relativistic nucleus. Indeed, the baryon number is Lorentz contracted to a size of order $1/\gamma$ where $\gamma = E/M$ is the Lorentz gamma factor of the nucleus. An essential difference is that meson degrees of freedom are important for the ultra-relativistic nucleus. These can be created in a collisions, and are in fact an important part of the Fock space wavefunction for high energy processes. These meson degrees of freedom have a range of longitudinal energies which vary continuously from the lab energy of the nucleus down to a typical hadronic energy, 200 MeV. The size scale where these components are important is $z \sim 1/p_z$ where p_z is the longitudinal energy. These degrees of freedom therefore smear out the nucleus so it always has a size of at least 1 $Fm \sim 1/200\ MeV$.

The ultrarelativistic nucleus therefore has its baryon number in a size region of order 1 $Fm\ /\gamma$, but its energy density is spread out in a region of size 1 Fm.

This distribution of meson longitudinal energies ranges from $p_z \sim 200\ MeV$ to $p_z \sim \gamma \times 200\ MeV$ If we plot things in terms of rapidity,

$$y \sim ln(E/E_0) \tag{1}$$

where E is the meson energy and $E_0 \sim 200\ MeV$, then it turns out the meson distribution of mesons is a slowly varying function of energy.

In Figure 1, a cartoon for an ultrarelativistic nucleus is shown. The meson degrees of freedom as shown as the circles and extend a distance of order 1 Fm outside the region where the baryon number is concentrated. The baryon number is concentrated in the the thin sheet of matter shown.

The distribution of mesons as a function of rapidity y, dN/dy, has a range of $0 \leq y \leq ln(\gamma)$ for fixed energy of the nucleus. What happens when we increase the value of the energy of the nucleus? Then the range for y increases. Empirically, the fast mesons, those in the original range of rapidity with rapidities closest to that of the nucleus change very little. What happens is we add new degrees of freedom, soft mesons, in the rapidity range of $0 \leq y \leq ln(\gamma'/\gamma)$. The modification of going to higher energy is simply to add more soft mesons to the nuclear wavefunction. Those with momentum close to the nucleus are little changed.

In Figure 2, a rapidity density of partons is shown. In Fig 3, we see the result of going to higher energy. We simply add in more partons at small x. The distribution in in the region of large positive y is shifted, since it now has higher energy, but its shape is the same. The physics of the fast partons is therefore the same, and the new physics of high energy is associated with small x.

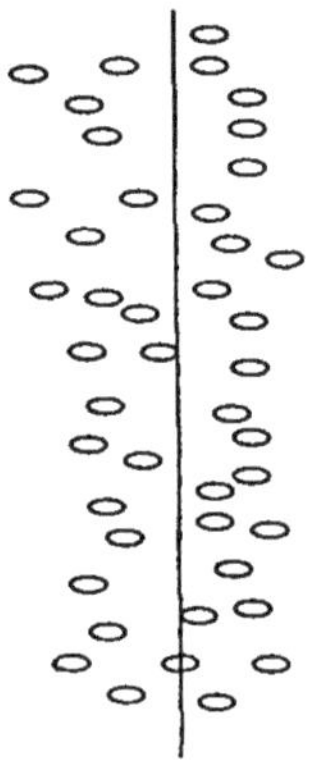

Figure 1: An Ultra-relativistic Nucleus

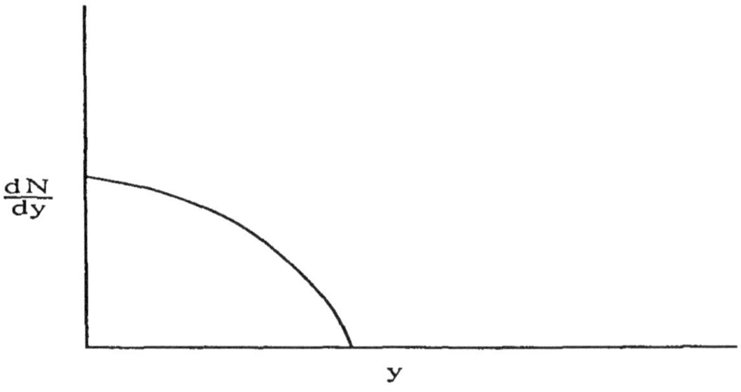

Figure 2: The rapidity distribution of partons

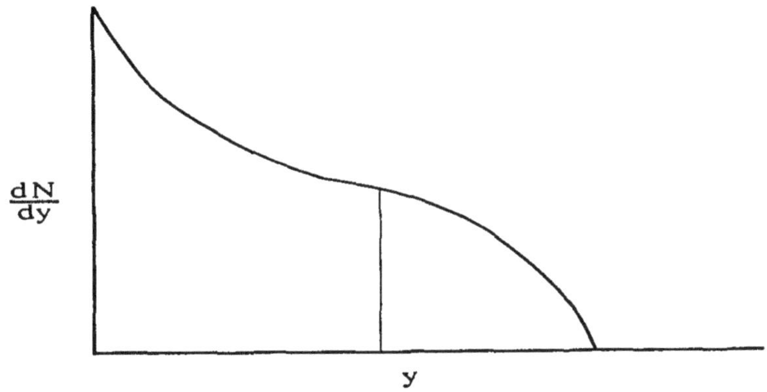

Figure 3: The changed rapidity distribution of partons for higher hadron energy

3 The High Density of Mesons

One can measure the density of partons (a generic word for mesons or gluons or whatever the fundamental degrees of freedom are) in deep inelastic scattering at HERA. If we let $x = \gamma_{parton}/\gamma_{nucleus}$, then it is found that

$$\frac{dN}{dy} \sim \frac{A^{1/3}}{x^\delta} \tag{2}$$

(A is the nuclear baryon number.) The number δ is about .2 − .3 at the x values typical of HERA.

We see that there are two possible ways to increase the density of partons. We can either decrease the minimum possible x which corresponds to increasing the maximal energy or we can increase A. An increase of A can change this density by about an order of magnitude. To get an order of magnitude increase through the beam energy would require a nucleus with a factor of $10^3 - 10^5$ greater energy. (Going to the center of mass frame would make this requirement a little less strong.) Getting such a higher energy costs a lot of money, since it would require a much larger accelerator.

Clearly, if we want to make the density of partons higher, it makes sense to use nuclei. This of course assumes that the physics is controlled by the density of partons per unit area. Another advantage is that there will be some conceptual simplifications if one can ignore the transverse size of the nucleus, and this is a better approximation if the size of the nucleus is larger.

It is clear that the physics of small x is the physics of high energy since going to higher energy simply adds on the small x component of the hadron wavefunction. One might also help in getting this asymptotic limit by using nuclei. We also see that the physics of high energy is the physics of high energy density matter.

Notice that the density of partons per unit area is dimensionful, $\Lambda^2 = dN/dy/\pi R^2$ has the dimensions of an inverse length squared or an energy squared. When

$$\Lambda >> \Lambda_{QCD} \tag{3}$$

we expect that by asymptotic freedom, the QCD interaction strength is weak, $\alpha_S(\Lambda) << 1$. Therefore, the high energy limit is controlled by the physics of weakly coupled high energy density matter.

4 Experimentally Probing High Density QCD

Traditionally, the study of the small x part of the hadron wavefunction was done using ep scattering. I am arguing here that it might be better for some purposes to do eA.

Experiments with ep have their correspondence in AA at RHIC and LHC. We can estimate the initial energy densities and see that roughly speaking HERA with ep corresponds to the initial energy densities expected at RHIC. HERA with eA corresponds roughly to AA at LHC.

There is a new proposal from Brookhaven National Laboratory to build a 10 *GeV* electron ring in the RHIC tunnel and scatter the electrons from the heavy ion beam. In addition one could do polarized ep scattering. The energy ranges are a little smaller than that of HERA but the slow variation of the gluon density with x means the physics is not much different. There are also some advantages of eRHIC verses HERA due to the possibility of considering a continuous range of collision energies and higher design luminosities. Such a program would directly complement the existing experimental AA program.

5 AA Scattering

The physics of AA scattering includes the study of the high energy limit of QCD. It will provide the environment where the highest energy densities will be achieved and will therefore correspond to much high energies than could be achieved in say a $p\bar{p}$ collider. In addition, ultrarelativistic nuclear collisions allow us to study matter as a high temperature or baryon number density gas.

There is a subtle difference between the matter which makes up the wavefunction of an ultra-relativistic nucleus and that of a produced quark-gluon plasma. For the wavefunction, the matter is highly coherent, in fact in a single quantum state. The constituents of this matter are virtual and are not propagating real quanta.

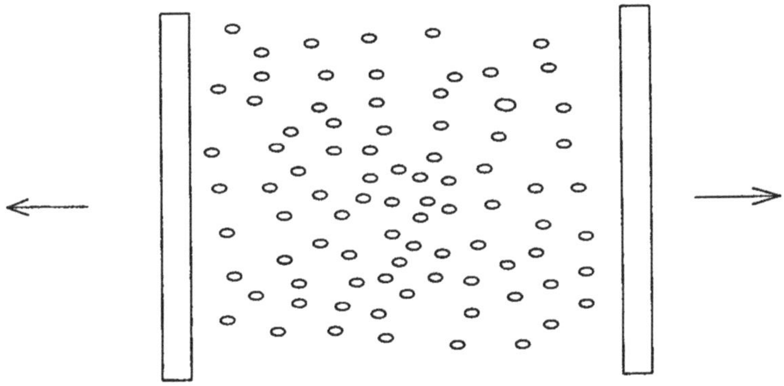

Figure 4: A Collision of Ultra-relativistic Nuclei

Only relatively late in a collision are particles actually produced. In AA collisions, the high density of such produced particles forces them to scatter with themselves many times, eventually producing thermalized matter.

In Figure 4, a collision of two nuclei is shown. After the collision, the very fast degrees of freedom of the two nuclei are unchanged for a time of order $\gamma \times 1\ Fm/c$. Here γ is the Lorentz gamma factor for the center of mass. The fast degrees of freedom change in their own rest frame on a time scale $t \sim 1\ Fm/c$, but in the center of mass frame, their reaction is time dilated. The parton degrees of freedom are produced quickly. This is because some of them are moving slowly in the center of mass frame. The reaction of these degrees of freedom is not Lorentz time dilated.

This thermalized quark-gluon plasma is produced as a rapidly expanding gas. If the nuclei are large enough, this plasma can exist as a thermalized system for times as large as 20 Fm/c. During this time, the matter which is initialy quarks ands gluons reassembles itself as hadrons, perhaps going through a phase transition which separates quark-gluon matter from ordinary hadronic matter. Eventually, the hadronic matter cools and stops interacting with itself and its remnants are fly off.

The typical energy density we expect when the matter is formed is of order $\epsilon \sim \Lambda^4$. This may be somewhere in the range of $\epsilon \sim 10^2 - 10^4\ GeV^2$ for the values of Λ discussed above. This happens in a relatively large region, for gold $R \sim 7\ Fm$, where the typical wavelength associated with a 1 GeV scale size is .2 Fm.

One has a fairly complete space-time scenario for the production of matter

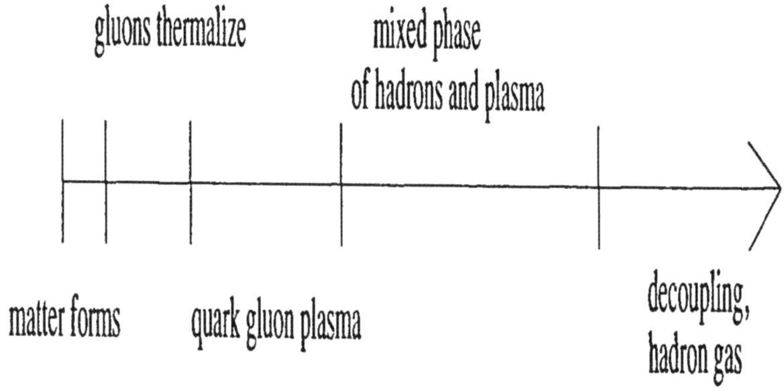

Figure 5: A Time Line for the Evolution of Matter in an Ultra-relativistic Nuclear Collision

in heavy ion collisions. I will go through this picture for the central region of heavy ion collisions in the center of mass frame.

At very early time, $t \leq 1/\Lambda$, the matter is coherent and quantum mechanical. For gold at RHIC energy, $1/\Lambda \sim .2\ Fm/c$. There are gluons in the form of coherent color fields. These fields arise from sources which are very close together, and their field strengths add together coherently.

At a time of the order of $t \sim 1/\Lambda$, the coherent gluon fields begin to radiate gluon quanta. These quanta begin to scatter from one another and thermalize. A typical time for thermalization is of order $t \sim 3 - 5/\Lambda$.

Once the gluons are thermalized, they form a gluonic plasma. Quarks are coming into equilibrium abundances even after the gluonic plasma has thermalized.

At a time of order $t \sim 2 - 3\ Fm/c$, the matter begins forming hadrons. This might take place at a first order phase transition. If so, then there is a coexistence of a plasma phase and a hadron gas phase. If there is no first order phase transition, but a continuous change in the properties of the system, then we should think of the dynamics as evolving from a system where the simple description is in terms of quarks and gluons into one where it is hadrons. This time of evolution can be quite long, perhaps as long as $t \sim 10\ Fm/c$.

A time line for the evolution of matter produced in the central region of the collision of two nuclei is shown in Figure 5.

In the late stages of the collision, the matter expands as a hadronic gas, eventually becoming a pion gas. At some late time, this pion gas decouples, and stops interacting with itself.

6 Basic Physics We Want to Study

There are several goals of the study of high density hadronic matter. If we think we understand confinement, then producing a system where deconfinement occurs provides an entirely non-trivial test of these ideas.

When the system become deconfined, quarks lose their mass. The origin of the masses of particles such as the proton is a mystery in QCD. If one were to take the naive QCD theory and predict the proton mass, it would be of order 100 MeV. It is in fact an order of magnitude larger. We say this happens because chiral symmetry is broken. Chiral symmetry would require that if QCD had massless quarks, all baryons would either be massless or appear in a doubled spectrum with partners of opposite parity. QCD to a good approximation has small mass quarks, but the baryons have big masses and do not appear in doublets. We therefore say that chiral symmetry is broken. How does this happen? Can we systematically study this and test our ideas?

Another very intersting possibility for QCD is that there is a real first order phase transition which separates a quark-gluon plasma from a hadronic gas. If so, this is the strongest first order phase transition we could ever study. The latent heat per particle is of the order of the rest mass energy per particle! In condensed matter physics, the latent heat per particle is 9-10 orders of magnitude smaller than the rest mass!

Such a strong first order phase transition generates large energy density fluctuations. We are all familiar with this phenomenon. When water boils, steam bubbles form which are orders of magnitudes larger than typical atomic sizes. Such density fluctuation might be a signal for a quark gluon plasma transition in heavy ion collisions.

Perhaps more important, such a transition provides an experimental realization of various phase transitions which occur in the early universe. For example, at the electroweak scale, a similar phase transition to that at the QCD scale might occur. Such a transition has been postulated to generate the baryon asymmetry of the universe and seed magnetic fields for galaxies. An essential ingredient in this formation is the formation of density inhomogeneities at the phase transition.

Acknowledgments

This manuscript has been authored under contract No. DE-AC02-98CH10886 with U. S. Department of Energy.

References

1. L. McLerran, Proceedings of 8'th Mexican School of Particles and Fields (and references therein), Oaxaca de Juarez, Mexico (1998).
2. L. McLerran, *Rev. Mod. Phys.*, **58**, 1021 (1986) (and references therein).

SUPERSYMMETRY IN NUCLEI

F. IACHELLO
Center for Theoretical Physics, Sloane Physics Laboratory, Yale University, New Haven, CT 06520-8120
E-mail: francesco.iachello@yale.edu

The recent confirmation of the occurrence of supersymmetry in nuclei is reviewed.

1 Supersymmetry

Supersymmetry is a complex type of symmetry that deals simultaneously with two different types of particles: bosons and fermions. In contrast with "normal" symmetries in which symmetry operations transform bosons into bosons *or* fermions into fermions, in supersymmetry the symmetry operations transform bosons into bosons, fermions into fermions *and* bosons into fermions (and viceversa). For a supersymmetry to occur, the following conditions should be met: (i) To each bosonic quantum state there should be an associated fermionic state with related energy. What states are there and their energy relation is given by the irreducible representations of the supergroup, $\mathcal{G}$, and by the supersymmetric Hamiltonian H (for non-relativistic situations), or Lagrangian $\mathcal{L}$ (for quantum fields) that describe the system. (ii) All observables in the bosonic and fermionic system are likewise related by supersymmetry transformations.

The early history of supersymmetry is rather confused. In 1966, Miyazawa[1] introduced a construction of supersymmetry by generalizing the flavor symmetry of Gell-Mann and Ne'eman and Gürsey and Radicati, but this paper went unnoticed. Ramond [2] and Neveu and Schwarz [3] reintroduced it in the context of dual models. However, supersymmetry did not acquire prominence in physics until 1973-74, following the work of Volkov and Akulov [4] and Wess and Zumino [5]. An example of supersymmetric field theory is provided by the Wess-Zumino Lagrangian

$$
\begin{aligned}
\mathcal{L}(\S) = & -\frac{1}{2}\left(\partial_\mu A(x)\right)^2 - \frac{1}{2}\left(\partial_\mu B(x)\right)^2 \\
& -\frac{1}{2} i\bar{\psi}(x)\gamma^\mu \partial_\mu \psi(x) \\
& -\frac{1}{2} m^2 A^2(x) - \frac{1}{2} m^2 B^2(x) - \frac{1}{2} i m \bar{\psi}(x)\psi(x)
\end{aligned}
$$

$$-gmA(x)\left[A^2(x)+B^2(x)\right]$$
$$-\frac{1}{2}g^2\left[A^2(x)+B^2(x)\right]$$
$$-ig\bar{\psi}(x)\left[A(x)-\gamma_5 B(x)\right]\psi(x)$$

The fundamental ingredients in this Lagrangian are two scalar fields $A(x), B(x)$ (bosonic) and a spinor field $\psi(x)$ (fermionic). The Lagrangian is invariant under supersymmetry transformations between the bosonic and fermionic fields. As a consequence, the coefficients in front of the various terms are not independent, but related by supersymmetry. There is thus only one mass m and one coupling constant g.

2 Supersymmetry in Nuclei

Although supersymmetry was originally introduced for applications in particle physics, soon afterwards it was applied to other fields of physics. In 1980, I suggested that the spectra of some heavy nuclei could be linked by a supersymmetry transformation and showed evidence for it [6].

Supersymmetry in nuclei is based on a model of nuclear structure, the Interacting Boson Model, that I had introduced earlier [7], [8], in which nucleons pair together into angular momentum $J=0$ and $J=2$ (s- and d-wave pairing). Pairs are treated as bosons. The building blocks of the model are six boson operators $b_\alpha(\alpha=1,...,6)\equiv s, d_\mu(\mu=0,\pm1,\pm2)$. These boson operators span a six-dimensional space with algebraic structure $U(6)$. The elements of $U(6)$ are the bilinear products of boson creation and annihilation operators $G_{\alpha\beta}=b_\alpha^\dagger b_\beta$. The Hamiltonian and transition operators are

$$H=E_0+\sum_{\alpha\beta}\varepsilon_{\alpha\beta}b_\alpha^\dagger b_\beta+\sum_{\alpha\alpha'\beta\beta'}v_{\alpha\alpha'\beta\beta'}b_\alpha^\dagger b_{\alpha'}^\dagger b_\beta b_{\beta'}+...,$$
$$T=t_0+\sum_{\alpha\beta}t_{\alpha\beta}b_\alpha^\dagger b_\beta+... \quad .$$

These operators can be rewritten in terms of elements of $U(6)$ by replacing the bilinear products of creation and annihilation operators by $G_{\alpha\beta}$.

The Interacting Boson Model has beeen used extensively to describe properties of nuclei with an even number of protons and neutrons (even-even nuclei). In this description, a major role is played by dynamic symmetries of the Hamiltonian. These are situations in which the Hamiltonian contains only certain operators (called invariant or Casimir operators, C_i) of a chain of subalgebras originating from $U(6)$ and terminating with the angular momentum algebra

$O(3)$ (the problem is rotational invariant). In other words, H is a functional of invariants only,

$$H = f(C_i).$$

For the Interacting Boson Model, there are three and only three such situations corresponding to the breakings:

$$\begin{array}{ll} U(6) \supset U(5) \supset O(5) \supset O(3) \supset O(2) & (I), \\ U(6) \supset SU(3) \supset O(3) \supset O(2) & (II), \\ U(6) \supset O(6) \supset O(5) \supset O(3) \supset O(2) & (III). \end{array}$$

When a dynamic symmetry occurs, all properties can be calculated in explicit analytic form. In particular, the energies of the states are given by

$$\begin{aligned} E^{(I)}(N, n_d, v, n_\Delta, L, M_L) &= E_0 + \varepsilon n_d + \alpha n_d(n_d + 4) \\ &+\beta v(v+3) + \gamma L(L+1) \\ E^{(II)}(N, \lambda, \mu, K, L, M_L) &= E_0 + \kappa(\lambda^2 + \mu^2 + \lambda\mu + \\ &+3\lambda + 3\mu) + \kappa' L(L+1) \\ E^{(III)}(N, \sigma, \tau, \nu_\Delta, L, M_L) &= E_0 + A\sigma(\sigma + 4) \\ &+B\tau(\tau+3) + CL(L+1) \end{aligned}$$

Here N is the total number of bosons, L their angular momentum and M_L its projection on a fixed axis. The other quantum numbers, $n_d, v, n_\Delta, \ldots$, label the representations of the groups in the chains (I),(II),(III).

[These formulas are similar to Bohr's formula for the hydrogen atom

$$E(n, \ell, m_\ell) = -\frac{A}{n^2}$$

(with dynamic symmetry $O(4)$), and to mass formulas of particle physics.]

Since their introduction in the years 1976-79, several examples of dynamic symmetries in nuclei have been found [8].

The Interacting Boson Model and its symmetries provide a classification of spectra of even-even nuclei. However, one has also odd-even and even-odd nuclei. For these nuclei, at least one particle must be unpaired. Furthermore, in even-even nuclei, at higher energies, some of the pairs may break and one must consider explicitly states with two unpaired fermions. A more general model is thus one in which there are at the same time pairs (bosons) and individual particles (fermions). This model, called Interacting Boson-Fermion Model, is used to describe more general properties of nuclei [9, 10].

The building blocks of the Interacting Boson-Fermion Model are boson, $b_\alpha(\alpha = 1, ..., 6) \equiv s(J = 0); d_\mu(J = 2; \mu = 0, \pm 1, \pm 2)$, and fermion, $a_i(i = 1, ..., \Omega) \equiv a_{jm}(j; m = \pm j, \pm(j-1), ..., \pm\frac{1}{2})$, operators. The model Hamiltonian is

$$H = H_B + H_F + V_{BF}$$

with

$$\begin{aligned} H_B &= E_0 + \sum_{\alpha\beta} \varepsilon_{\alpha\beta} b_\alpha^\dagger b_\beta + \sum_{\alpha\alpha'\beta\beta'} v_{\alpha\alpha'\beta\beta'} b_\alpha^\dagger b_{\alpha'}^\dagger b_\beta b_{\beta'} \\ H_F &= E_0' + \sum_{ik} \eta_{ik} a_i^\dagger a_k + \sum_{ii'kk'} u_{ii'kk'} a_i^\dagger a_{i'}^\dagger a_k a_{k'} \\ V_{BF} &= \sum_{\alpha\beta ik} w_{\alpha\beta ik} b_\alpha^\dagger b_\beta a_i^\dagger a_k \end{aligned}$$

Here H_B and H_F denote the boson and fermion Hamiltonian and V_{BF} their interaction.

Within the context of the Interacting Boson-Fermion Model, a supersymmetry occurs whenever: (i) the boson and fermion single particle states are related by a supersymmetry transformation; and (ii) the coefficients in the Hamiltonian H describing the bosonic and fermionic degrees of freedom are also related by supersymmetry transformations. Supersymmetries require a new type of mathematical construction for their description: Graded Lie Algebras (sometimes called superalgebras). The basic properties of these algebraic constructions are: there are two types of generators, called bosonic, G_α, and fermionic, F_i, respectively, satisfying the following commutation and anticommutation relations

$$\begin{aligned} [G_\alpha, G_\beta] &= \sum_\gamma c_{\alpha\beta}^\gamma G_\gamma \\ [G_\alpha, F_i] &= \sum_j d_{\alpha i}^j F_j \\ \{F_i, F_j\} &= \sum_\alpha g_{ij}^\alpha G_\alpha \end{aligned}$$

The presence of anticommutators makes the algebra not a Lie algebra.

The mathematical background needed to describe supersymmetries in nuclei was developed in the 1980's [11]. It makes use of unitary superalgebras

$U(n/m)$. These superalgebras can be constructed with bilinear products of creation and annihilation operators usually arranged in a matrix form

$$\begin{pmatrix} b^\dagger_\alpha b_\beta & b^\dagger_\alpha a_i \\ a^\dagger_i b_\alpha & a^\dagger_i a_j \end{pmatrix}$$

The bosonic and fermionic generators are:

$$\begin{aligned} G_{\alpha\beta} &= b^\dagger_\alpha b_\beta \\ G_{ij} &= a^\dagger_i a_j \\ F^\dagger_{\alpha i} &= b^\dagger_\alpha a_i \\ F_{i\alpha} &= a^\dagger_i b_\alpha \end{aligned}$$

In the Interacting Boson-Fermion Model, the dimension n of the bosonic space is $n = 6$ and the dimension m of the fermionic space is $m = \Omega = \sum_i (2j_i + 1)$. The Hamiltonian H and other operators can then be written in terms of the elements of the superalgebra $U(6/\Omega)$. (The case discussed in 1980 [6] had $j = \frac{3}{2}$, $\Omega = 4$. This case is particularly interesting since it has a structure similar to supersymmetric gravity with $J = 0, J = 2$ (graviton), and $J = \frac{3}{2}$ (gravitino)).

A consequence of supersymmetry is that if the bosonic states and their energies are known, one can predict the fermionic states and their energies. Both are given by the same formula. For example, for $U(6/4)$ supersymmetry, the energy formula is:

$$\begin{aligned} E(\mathcal{N}, (\sigma_\infty, \sigma_\in, \sigma_\ni), (\tau_\infty, \tau_\in), \nu_., \mathcal{J}, \mathcal{M}) = E_0 & \\ +A\left[\sigma_1(\sigma_1 + 4) + \sigma_2(\sigma_2 + 2) + \sigma_3^2\right] & \\ +B\left[\tau_2(\tau_2 + 3) + \tau_1(\tau_1 + 1)\right] & \\ +CJ(J+1). & \end{aligned}$$

Here $\mathcal{N}$ is the total number of bosons plus fermions and $\sigma_1, \sigma_2, \sigma_3, ...$ are the quantum numbers that label the states. [In particle physics, if one knows the mass of quarks, one can predict the mass of their supersymmetric partners, called squarks, or if one knows the properties of gluons, one can predict those of gluinos (not yet found).]

In the early 1980's, several examples of nuclear spectra with supersymmetric properties were found [12], relating spectra of even-even nuclei with those of odd-even nuclei (odd proton or odd neutron). An example is shown in Tables

Table 1: Comparison between experimental and calculated energies in the supersymmetric partners Os and Ir. Only 21 of the known 49 states are shown.

^{190}Os		
σ_1, τ_1, J^P	$E_{\exp}(keV)$	$E_{th}(keV)$
$9,0,0^+$	0	0
$9,1,2^+$	187	180
$9,2,2^+$	558	360
$9,2,4^+$	548	500
$9,3,0^+$	912	540
$9,3,3^+$	756	660
$9,3,4^+$	955	740
$9,3,6^+$	1050	960
^{191}Ir		
$\frac{17}{2},\frac{1}{2},\frac{3}{2}^+$	0	0
$\frac{17}{2},\frac{3}{2},\frac{1}{2}^+$	82	120
$\frac{17}{2},\frac{3}{2},\frac{5}{2}^+$	129	200
$\frac{17}{2},\frac{3}{2},\frac{7}{2}^+$	343	270
$\frac{17}{2},\frac{5}{2},\frac{3}{2}^+$	179	360
$\frac{17}{2},\frac{5}{2},\frac{5}{2}^+$	351	410
$\frac{17}{2},\frac{5}{2},\frac{7}{2}^+$	686	480
$\frac{17}{2},\frac{5}{2},\frac{9}{2}^+$	502	570
$\frac{17}{2},\frac{5}{2},\frac{11}{2}^+$	832	680
$\frac{15}{2},\frac{1}{2},\frac{3}{2}^+$	539	560
$\frac{15}{2},\frac{3}{2},\frac{1}{2}^+$	624	680
$\frac{15}{2},\frac{3}{2},\frac{5}{2}^+$	748	760

1 and 2, where the observed states of the supersymmetric partners ^{190}Os and ^{191}Ir related by $U(6/4)$ supersymmetry and their electromagnetic couplings are compared with the symmetry predictions. If one defines, as a measure of supersymmetry breaking the quantity

$$\phi = \frac{\sum_i \mid E_i^{th} - E_i^{\exp} \mid}{\sum_i E_i^{\exp}},$$

where the sum goes over the observed states, one finds $\phi = 14\%$.

In view of these results, by 1984-85 it was concluded that supersymmetry in nuclei had been found (with a breaking of the order of 20%) [13].

Table 2: Comparison between experimental and calculated B(E2) values in the supersymmetric partners Os and Ir.

^{190}Os			
$(\sigma_1, \tau_1, J)_i$	$(\sigma_1, \tau_1, J)_f$	$BE(2)_{\exp}(e^2b^2)$	$BE(2)_{th}(e^2b^2)$
$9,1,2$	$9,0,0$	0.478(12)	0.478
$9,2,2$	$9,0,0$	0.046(2)	0
$9,2,2$	$9,1,2$	0.259(15)	0.654
$9,2,4$	$9,1,2$	0.622(44)	0.654
$9,3,4$	$9,1,2$	0.010(2)	0
$9,3,4$	$9,2,2$	0.488(100)	0.375
$9,3,4$	$9,2,4$	0.362(72)	0.340
$9,3,6$	$9,2,4$	1.038(330)	0.715
^{191}Ir			
$\frac{17}{2},\frac{3}{2},\frac{1}{2}$	$\frac{17}{2},\frac{1}{2},\frac{3}{2}$	0.130(3)	0.425
$\frac{17}{2},\frac{3}{2},\frac{5}{2}$	$\frac{17}{2},\frac{1}{2},\frac{3}{2}$	0.640(30)	0.425
$\frac{17}{2},\frac{3}{2},\frac{7}{2}$	$\frac{17}{2},\frac{1}{2},\frac{3}{2}$	0.293(6)	0.425
$\frac{17}{2},\frac{5}{2},\frac{3}{2}$	$\frac{17}{2},\frac{1}{2},\frac{3}{2}$	0.073(13)	0
$\frac{17}{2},\frac{5}{2},\frac{5}{2}$	$\frac{17}{2},\frac{1}{2},\frac{3}{2}$	0.0111(4)	0
$\frac{17}{2},\frac{5}{2},\frac{7}{2}$	$\frac{17}{2},\frac{1}{2},\frac{3}{2}$	0.065(6)	0

3 Supersymmetry in nuclei confirmed

Soon after the introduction of the Interacting Boson Model, it was realized that an improvement in the description of nuclear properties could be obtained by distinguishing proton pairs from neutron pairs. This model is called Interacting Boson Model-2. The building blocks here are boson operators $b_{\alpha\pi}, b_{\alpha\nu} (\alpha = 1, ..., 6; \pi = \text{proton}, \nu = \text{neutron})$. The boson operators span a (six+six)-dimensional space with algebraic structure $U_\pi(6) \oplus U_\nu(6)$. Consequently, when going to nuclei with unpaired particles, one has a model with two types of bosons (proton and neutron) and two types of fermions (proton and neutron), called Interacting Boson-Fermion Model-2.

If supersymmetry occurs for this very complex systems one expects now to have supersymmetric partners composed of a quartet of nuclei, even-even, even-odd, odd-even and odd-odd, called a magic square [14]. An example is

$$\begin{array}{cc} ^{194}Pt & ^{195}Pt \\ ^{195}Au & ^{196}Au \end{array}$$

Spectra of even-even and even-odd nuclei have been known for some time. However, spectra of odd-odd nuclei are very difficult to measure, since the density of states in these nuclei is very high and the energy resolution of most detectors is not sufficiently good. In a major effort, led by J. Jolie, that has

Table 3: Comparison between experimental energies and those predicted by supersymmetry in Au.

^{196}Au			
$(\sigma_1,\sigma_2,\sigma_3),(\tau_1,\tau_2)$	J^P	$E_{\text{exp}}(keV)$	$E_{th}(keV)$
$(\frac{13}{2},\frac{1}{2},\frac{1}{2}),(\frac{1}{2},\frac{1}{2})$	1^-	6	0
	2^-	0	18
$(\frac{3}{2},\frac{1}{2})$	0^-	...	221
	1^-	253	230
	2^-	307	303
	3^-	355	329
	3^-	375	377
	4^-	258	413
$(\frac{11}{2},\frac{1}{2},\frac{1}{2}),(\frac{1}{2},\frac{1}{2})$	1^-	167	146
	2^-	162	164
$(\frac{3}{2},\frac{1}{2})$	0^-	388	367
	1^-	348	376
	2^-	456	449
	3^-	466	475
	3^-	491	523
	4^-	520	559
$(\frac{11}{2},\frac{3}{2},\frac{1}{2}),(\frac{3}{2},\frac{1}{2})$	0^-	42	138
	1^-	...	146
	2^-	198	218
	3^-	234	245
	3^-	287	293
	4^-	213	329

involved several laboratories for several years, it has very recently been possibile to measure spectra of heavy odd-odd nuclei with unprecedented accuracy and, most importantly assign spin and parities to the individual levels. In particular, the magnetic spectrometer at the Ludwig-Maximilians Universität in Münich, Germany, developed by G. Graw, can separate levels only a few keV apart. It has thus been possible to measure the spectrum of ^{196}Au, the missing supersymmetric partner of ^{194}Pt, ^{195}Pt and ^{195}Au [15]. A portion of this spectrum together with the supersymmetry predictions is shown in Table 3. Here only the lowest 20 of the measured 46 states are shown. The degree of agreement for the 26 states not shown here is comparable to that of the table. Only few of the states exptected by supersymmetry (in the table the 0^- state at 221 keV and the 1^- state at146 keV) have not been conclusevely separated.

The observed spectrum of ^{196}Au meets all the criteria for supersymmetry described above: (i) to each bosonic quantum state there are fermionic partners obtained from it by a supersymmetry transformation. Here the transformation is rather complex, $U_\pi(6/4) \oplus U_\nu(6/12)$, since it involves both protons and neutrons; (ii) the energies of the states are given by a single formula and are related by supersymmetry; (iii) the measured intensities (not reported here) follow the supersymmetry predictions. In view of this new evidence, one must conclude that supersymmetry in nuclei has been confirmed.

4 Implications to other fields

(a) Particle Physics

Supersymmetry has been sought in Particle Physics for decades. The confirmation of supersymmetry in nuclei indicates that this very complex type of symmetry can occur in Nature. It gives hope that, although badly broken, supersymmetry may occur in particle physics. However, supersymmetry in Nuclear Physics is a symmetry that relates composite objects (pairs) with fundamental objects (nucleons). Can it be the same in particle physics (Nambu, Gürsey [16],...)?

(b) Condensed matter physics

Some supersymmetric theories have been constructed in condensed matter physics (Parisi-Sourlas [17]). Nambu has suggested that supersymmetry may occur in Type II superconductors [18]. The occurrence of supersymmetry in Nuclear Physics may lead to other supersymmetric theories between composite objects and their constituents.

5 Conclusions

Supersymmetry, one of the most fundamental types of symmetry that one may encounter in Nature, has been found and confirmed in Nuclei.

6 Aknowledgements

This article is dedicated to the memory of Gregory Breit.

Acknowledgments

This work is supported in part by U.S. D.O.E. Contracts DE-FG02-91ER40608.

References

1. H. Miyazawa, *Progr. Theor. Phys.* **36**, 1266 (1966).
2. P. Ramond, *Phys. Rev. D* **3**, 2415 (1971).
3. A. Neveu and J. Schwarz, *Nucl. Phys. B* **31**, 86 (1971).
4. D.V. Volkov and V.P. Akulov, *Phys. Lett. B* **46**, 109 (1973).
5. J. Wess and B. Zumino, *Nucl. Phys. B* **70**, 39 (1974).

6. F. Iachello, *Phys. Rev. Lett.* **44**, 772 (1980).
7. F. Iachello, in *Nuclear Structure and Spectroscopy,* H.P. Blok and A.E.L. Dieperink, eds., (Scholar's Press, Amsterdam, 1974), p.163; A. Arima and F. Iachello, *Phys. Rev. Lett.* **35**, 1069 (1975).
8. F. Iachello and A. Arima, *The Interacting Boson Model* (Cambridge University Press, Cambridge, 1987).
9. F. Iachello and O. Scholten, *Phys. Rev. Lett.* **43**, 679 (1979).
10. F. Iachello and P. van Isacker, *The Interacting Boson-Fermion Model* (Cambridge University Press, Cambridge, 1991).
11. A.B. Balantekin, I. Bars and F. Iachello, *Nucl. Phys. A* **370**, 284 (1981); I. Bars, *Physica D* **15**, 42 (1985).
12. F. Iachello, *Physica D* **15**, 85 (1985); R.F. Casten, *Physica D* **15**, 99 (1985).
13. R.F. Casten and D.H. Feng, *Physics Today* **37**, 26 (1984).
14. P. van Isacker, J. Jolie, K. Heyde and A. Frank, *Phys. Rev. Lett.* **54**, 653 (1985).
15. A. Metz, J. Jolie, G. Graw, R. Hertenberger, J. Groger, Ch. Gunther, N. Warr and Y. Eisermann, *Phys. Rev. Lett.* **83**, 1542 (1999).
16. S. Catto and F. Gürsey, *Nuovo Cimento A* **86**, 201 (1985).
17. G. Parisi and N. Sourlas, *Phys. Rev. Lett.* **43**, 744 (1979).
18. Y. Nambu, *Physica D* **15**, 147 (1985).

100 YEARS OF THE QUANTUM – THE GLORY AND THE SHAME[†]

JOHN ARCHIBALD WHEELER

Physics Department, Princeton University, Princeton, NJ 08544 USA

E-mail:jawheeler@pupgg.princeton.edu

What is the greatest mystery that stands out on the books of physics today? Every one of us will have a different answer. Some will say it is the structure of the elementary particles. Others ask in what form is the mass or the attraction that has held the universe together thus far against perpetual expansion. Others will think of the November 11, 1999 dedication of America's first 6 kilometer by 6 kilometer gravity wave detector and will ask what information it will bring us about events going on in the deep and secret places of the universe. My own hope is that we'll see in time to come the answer to a question now almost a century old, "How come the quantum?" Gregory Breit would have sympathized with the investigations at each of these fronts, and could surely have filled us in on what the pioneers thought and said about the question, "How come the quantum?"

Glory? Yes, the centerpiece of the discoveries that physics has given us in the last hundred years. Shame? Yes, shame that after a hundred years, we still don't know what deeper idea lies at the root of the quantum. Yet a hundred years is a short enough time compared with the 229 years that elapsed from Isaac Newton's treatise on gravity and the miracles it worked to Albert Einstein's general relativity, with its lesson, "Mass tells spacetime how to curve, and spacetime tell mass how to move." Such was the time required to give substance to Newton's prophetic words, "That one body may act upon another at a distance through a vacuum, without the mediation of any thing else, by and through which their action and force may be conveyed from one to another, is to me so great an absurdity, that I believe no man, who has in philosophical matters a competent faculty of thinking, can ever fall into it."

Today we are closing in on the 100th anniversary of Max Planck's December 10, 1900 revolutionary note to the Prussian Academy in Berlin. James Franck, who was in Berlin at the time, and who later won the Iron Cross for heroism on the battlefield, has told me the circumstantial details of the great moment. On Sunday afternoon, October 7, 1900, the 35-year-old Heinrich Rubens and his wife came to tea at the Plancks' home (Planck was then 42). Rubens brought with him the latest results of measurements that he and Ferdinand Kurlbaum had made on the intensity of heat radiation (black-body radiation) at temperatures up to 1,800 K and infrared

[†] I thank Kenneth Ford for his help in formulating and preparing this written version of my talk.

wavelengths up to more than 50,000 nm. Recently developed techniques of precision intensity measurements and precision wavelength measurements into the far infrared made it possible for Rubens to present the results to startling accuracy, accuracy sufficient to show clearly the invalidity of earlier efforts to develop formulas for the intensity of black-body radiation.

After the visitors had been thanked and gone, Planck sat down to find a formula that would fit the curve. He wrote out the formula—not *de novo,* for it was one he had already been toying with—and calculated points with it. As I have learned from the wonderful book *Subtle is the Lord,* by Abraham Pais, Planck wrote his formula on a postcard and mailed it to Rubens that very night. It is Franck's recollection that his method of mailing it may have been to take it to the nearest pneumatic tube of the postal system and send it flying to the physics institute.

Curve fitting is one thing, but explaining "how come?" is another. Planck had the advantage of enough reading of Boltzmann to know how to break up a continuum into a finite number of elements. Though this was in the beginning only a mathematical algorithm, it had the payoff of reproducing exactly Planck's formula. Within two months, Planck advanced the quantum interpretation of his formula. Even sooner, on October 25, 1900, Rubens and Kurlbaum published a paper showing that their measurements agreed exactly with the Planck formula and disagreed with formulas attributed to Wien, Thiessen, Rayleigh, and Lummer and Pringsheim.[1] So, with his quantum-based formula, Planck could accurately calculate his new constant *h,* the Boltzmann constant *k,* Avogadro's number, the mass of an atom, and the charge of the electron! (He could also introduce the fundamental combinations of constants that we have later called the Planck length and the Planck time.)

Nowadays one is accustomed to seeing the radiation intensity at this, that, or the other temperature plotted against frequency, rising to a maximum at a certain frequency and then falling, with the frequency of the maximum shifting to the right as one turns attention from a particular temperature to a higher temperature. If, however, one does not look at the usual textbook plots of radiation as a function of frequency but looks instead at a curve of the kind Rubens presented to Planck on that famous Sunday afternoon, one sees the truly impressive feature of the radiation: The intensity at a given wavelength varies linearly with temperature at higher temperatures, curving toward the horizontal only at low temperature (Figure 1). The extrapolated linear portion intercepts the horizontal temperature axis at the point $kT = \frac{1}{2}h\nu$, showing visually the soon-to-become famous half quantum of "zero-point" energy.

Looking back on those days through today's lenses, we see Planck's discovery as the climax of a quartet of discoveries: Wilhelm Roentgen discovering X rays in 1895, Henri Becquerel discovering radioactivity in 1896, J. J. Thomson discovering the electron in 1997, and Planck discovering the quantum in 1900. (Not until 1905 did Einstein arrive at the revolutionary insight that light arrives at its target in quanta of energy $E = h\nu$.)

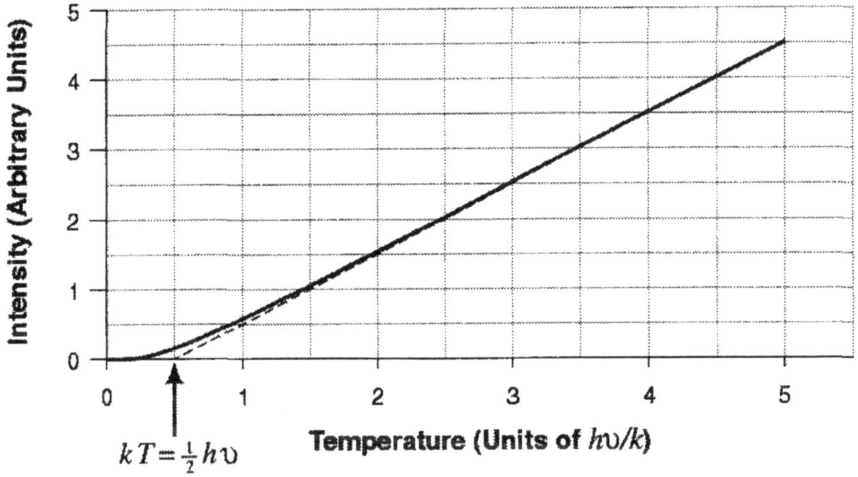

Figure 1: Intensity of black-body radiation vs. temperature for fixed frequency, according to the Planck formula. The 1900 data of Rubens and Kurlbaum (Reference 1), reaching to about 6 on this temperature scale, lent strong support to Planck's formula, even before Planck provided the quantum interpretation.

I must mention one American who played an important role in the discovery of the quantum, one not often mentioned. Three experimental advances were essential for obtaining the Rubens-Kurlbaum results that definitively supported Planck's formula: precision of intensity measurement, precision of wavelength measurement, and the ability to form and use narrow bands of wavelength in the far infrared. The gifted American experimenter Ernest Nichols (1869-1924) contributed to all of these refinements. In 1894-1896, still in his twenties, he took a leave of absence from Colgate University to work in Emil Warburg's lab in Berlin. There he invented a radiometer far more sensitive to infrared radiation than any previously available. Then he and Rubens developed a method of successive reflections from quartz and other crystals to obtain as *Reststrahlen* (residual rays) nearly monochromatic infrared radiation at far greater wavelengths than had been possible before. These achievements were published in 1897.

Everyone in the world of physics knows the glorious unwinding of the quantum story in the 99 years since Planck's discovery. It might not teach us "how come the electron?" but it did explain to us "how come X rays?" and "how come radioactivity?" It is not necessary to describe each victory if one wants to describe Napoleon's campaign; in the same spirit we can list only some of the conquests of the quantum idea without detailing them. It's a good idea to give it credit before punching it in the nose and saying, "how come you are here?"

It wasn't long after Max Planck's account of blackbody radiation that Peter Debye successfully applied the same kind of reasoning to the oscillatory degrees of freedom of crystalline and other solids. Walther Nernst had enough chemical background to know and care about entropy as much as energy. It followed from his analysis that the entropy of the solid—and of most substances—does not go to zero as the temperature falls, but reaches a so-called, "zero-point entropy" at the absolute zero of temperature. The Belgian industrialist Ernest Solvay in bequeathing his fortune to science leaned heavily on Nernst for advice, and Nernst helped conceive the wonderful series of Solvay Congresses that did so much to advance physics in the earlier days of this century, bringing together Planck, Einstein, Lorentz, Ehrenfest, Franck, and others. I understand from Franck that Nernst wanted more attention given to his zero-point entropy than the others felt warranted. In spite of the difficulties that arose from that score, Nernst remained an honored member of the community. He died in Berlin in 1941. When the war ended his body was transferred to a site in West Berlin and still later, in 1951 he was re-buried again in Göttingen. "How come you are always at these burials?", Franck found himself asked by a colleague. "Oh, he cannot be buried too often for me," was the reply.

Then came the famous work of Niels Bohr on the hydrogen atom. Following his student research, Bohr went to Cambridge University to work with J. J. Thomson, the discoverer of the electron. The shy young Dane, not yet fully at home in English, had brought along with him Thomson's book on the electron and, opening it up, pointed to passages that were doubtful, or even wrong, and solicited Thomson's response. The young Bohr, seeking only to show by his questions how thoroughly he had read Thomson's book and how much he appreciated it, got off in this way to an awkward start for his intended year with Thomson. Fortunately fate brought to Cambridge from time to time a young visitor from Manchester, in the shape of Ernest Rutherford, with whom the young Bohr struck up a warm friendship and from whom he got the stimulus that he had sought by coming to England. Rutherford soon had Bohr installed at Manchester. There he could see what it meant to observe the scattering of alpha particles by atomic nuclei and become convinced of the concentration of the electric charge of the nucleus in a very small region. Soon Bohr was calculating the motion of electrons in the electric field of force of the atomic nucleus. Then came quantum states and quantum jumps from one state to another and the fruitful idea of the correspondence principle. But Karl Popper tells us that a new outlook wins acceptance by predicting something totally unexpected, which thereupon is observed. Helium, deprived of one electron, offers a hydrogen-like atomic system. The spectrum Bohr predicted, showing spectral lines like those of hydrogen with almost four times the frequency—the "almost" referring to the correction of the reduced mass for the four-times heavier mass of the helium nucleus—showed up clear and strong. Back in Copenhagen Bohr got a letter from

Rutherford telling him he had to publish his results. Bohr wrote back that nobody would believe him unless he could explain the spectra of all the atoms. Rutherford replied, in effect: Bohr, you explain hydrogen and you explain helium and everyone will believe all the rest.

That business of explaining all the rest had then become the business of the group that rose around Bohr at Copenhagen.

If the quantum of Planck became so soon so central in explaining the structure of an atom, it revealed itself in a new light in Arthur Compton's study of the scattering of X-rays by electrons. The photon of the incident X-ray hit the electron with one energy but came off at this, that, or the other angle with this, that, or the other reduced energy. Compton found that the effect he had discovered could only be understood by applying the laws of conservation of energy and momentum to the collision between the photon and the electron—one more triumph for the quantum!

If light, so clearly a wave phenomenon could thus show a particle-like behavior, it was inspiring to find that electrons, so clearly particles, could show a wave-like character as in the experiments of Davisson and Germer on the scattering of electrons by crystals. This discovery, made in 1927, and that of Compton, made in 1923, showed physics live and doing well in this country. Around this time in France a descendent of a long line of distinguished French admirals resolved to show himself worthy of his ancestors. Louis de Broglie submitted a Ph.D. thesis at the University of Paris arguing for the Bohr quantum conditions for the electron in the hydrogen atom as the consequence of fitting electron waves together around the circuit of the orbit. The idea was so new that the examining committee went outside its circle for advice on the acceptability of the thesis. Einstein gave a favorable opinion and the thesis was accepted. The thesis also came to Debye in Zurich who gave it to one of his young graduate students to report on at the weekly seminar. When the student, Schrödinger, finished his report, Debye said in effect, "You speak about waves. But where is the wave equation?" Schrödinger went on to produce and publish the soon-famous wave equation, the master key for so much of modern physics.

At this time others in Europe were seeking a totally different way to describe the mechanics of the atom, putting emphasis on quantities considered to have a tangible meaning, like the electric dipole moment of the system, the total momentum of the electrons. This work, inspired not least by the initial efforts of Hans Kramers, and developed by Max Born and Pascual Jordan and the young Werner Heisenberg, led to the birth of the so-called matrix mechanics, with its famous non-commutative algebra. Among those who contributed to showing the equivalence of the Schrödinger wave mechanics and the Born-Jordan matrix mechanics was the young Paul Dirac. He had come to Copenhagen in the 1920s at the recommendation of Rutherford. A few months later, when Bohr happened to be in London, Rutherford asked him how the young Dirac was doing. "He doesn't say very much," Bohr replied. Where upon Rutherford told him the story of the parrot. The purchaser found that the parrot he had bought was not able to speak and took it back to the dealer from whom he had bought it. The dealer, after looking at the parrot said: "Oh, how terrible, you wanted the parrot that speaks and I sold you the parrot that thinks!"

But if the mathematical equivalence of the new matrix mechanics and the wave mechanics soon became clear, it took a little longer to see the physical consequences of the authoritative mathematics. The young Heisenberg, who had lingered after a lecture Bohr gave in Munich, and had been drawn off into a long walk with him, gave up thoughts about a possible career in music and came to Copenhagen. Then came the summer of the historic discussions between Bohr and Heisenberg about the uncertainty principle. Hendrik Casimir tells us how widely spread among scientists in Copenhagen was news of the disagreements and agreements between Bohr and Heisenberg during that summer. Among all the professors at the University of Copenhagen the one whom Bohr as a student had admired the most and whose lectures he found the most stimulating was Harald Høffding, Professor of Philosophy, and now occupant of the House of Honor, which Niels Bohr was destined later to receive as home. Høffding invited the two men to come around one evening and explain to him what they agreed on and what they disagreed on and what should be done to settle the disagreement. Casimir came along with them and tells us how the double-slit experiment came up in the discussion of the uncertainty principle. At one point Høffding put his finger on the diagram and inquired: "Where can the electron be said to be?" Whereupon Bohr replied: "To be? To be? To be? What does it mean, to be?"

Heisenberg went off to the island of Bornholm for the rest of the summer to write up his famous paper and send it in for publication. When he brought the proof to Bohr, he found himself having to make important corrections in the proof. Bohr's role in a paper so great and so important is not well known. Those of us who want to follow further the relations between the two men will probably be seeing the play "Copenhagen," opening in New York after its success in London.

Dirac, with his passion for clear theory and precise results, had given us a relativistic theory of the electron, in which the proper treatment of the electron spin by way of the Dirac matrixes brought in negative energy states. The history of those Dirac matrixes reminds us of how chancy it is for strange ideas to get passed on. I had spent a happy month at the Dublin Institute of Advanced Study and went to thank the founder of that institute for his vision, as some years before I had thanked Abraham Flexner for his vision in setting up the Princeton Institute for Advanced Study. The Dublin founder, Eamon de Valera, when I went to call on him at the end of my stay, turned out to my surprise still to be president of Ireland at his advanced age. He told me how as a young man he had taken part in the unsuccessful Easter Day revolt against the British power. He and his companions were locked up in prison and sentenced to be shot at dawn the next morning. To quiet his mind during that long night, he kept writing down over and over again the key formula that

William Rowan Hamilton had conceived one Sunday afternoon when walking with his wife in Dublin, and had scratched into the stonework in the bridge over the river Liffy:

$$i^2 = j^2 = k^2 = ijk = -1.$$

De Valera, however, had an American strain in his ancestry and was well known to Irish circles in America. And this was a time when Britain hoped the United States would join it in the war against Germany. Therefore when morning came de Valera was the only who was not shot.

In the subsequent years Hamilton quaternions and their magic in describing rotations passed out of currency. When Dirac learned about them in a course at Cambridge there was only one other student in the course. Thanks to Dirac these objects found their rightful place in the relativistic quantum mechanics of the electron and the understanding of many of the mysteries of the cosmic rays, culminating in the discovery of the positron and the mesons. In the field of elementary particle physics, these early victories gave the community the confidence needed to go on to the fascinating developments in the realm of quarks and heavy particle transformations that we all follow in the literature with such interest from week to week.

To Gregory Breit I owe a wonderful introduction to the glory of quantum mechanics. About the shame of the subject I can never remember his speaking. What shame? That we don't know where the quantum comes from. Still less do we know how come existence itself. No, not to question quantum mechanics, but to learn from a master of the subject how to put it to use: That was the great push for a 22-year-old to come from Baltimore to New York in 1933 to work with Breit. And thorough he was. "Wouldn't you help me check these calculations on the Dirac electron?" he asked me one day. Naturally I was honored to, even though I had to start out as a beginner. Breit explained to me the algebra of the Dirac matrices, of which only later was I to learn the romantic Irish history. And I had to become acquainted with Dirac's bold idea to take all the negative energy states of the electron to be filled. Breit realized right from the beginning that the states of the electron in the vacuum are so free-moving that no single photon could raise an electron to a positive energy state. It would take two photons, which meant for us a second-order calculation. Breit had done it and I checked it out. So we came to a formula for the effective cross-section of the collision of two photons to produce a positron-electron pair. Looking for an application, we found ourselves driven to what seemed astrophysical conditions, hard to observe. Would that we could have foreseen the fantastic photon flux obtainable more then half a century later at the Stanford Linear Accelerator. With it my Princeton colleague Kirk McDonald, working in 1997 with partners from Stanford University, the University of Rochester, the University of Tennessee, and Princeton, directly observed numerous instances of the production of an electron pair out of the vacuum.

During my 1933-34 postdoctoral year with Breit, I ventured to apply for a second year of my National Research Council fellowship to work with Niels Bohr in Copenhagen. Why? Because, I wrote in my application, "He sees further ahead than anyone alive." Breit supported my application. The summer of 1934 saw me saying "Good-bye, Professor Breit," "Hello, Professor Bohr," good-bye to learning how to answer questions, hello to judging what to ask, good-bye to regarding the quantum as handed down from heaven and hello to the question how come the quantum.

So quantum mechanics served year after year as a faithful hunting dog and companion of the explorer of elementary-particle physics, as well as serving as the indispensable partner of colleagues exploring the nature of the chemical bond and unraveling the mysteries of the superconductor. But unanswered is the fundamental question about quantum mechanics itself: Where does it come from? As I. I. Rabi would have said, "Who ordered this?"

That was the subject of the on-again, off-again debates between Bohr and Einstein that extended over more than a dozen years. Once, probably in 1942, I went around to Einstein's house to tell him about the path-integral or "sum-over-histories" formulation of quantum mechanics that Richard Feynman had just developed. He listened to me patiently for 20 minutes but when I finished he said, "I still cannot believe God plays dice. But maybe I have earned the right to make my mistakes."

Not long after World War II, when I thought that David Bohm might have important new insights on the quantum, I induced him to move from Berkeley to Princeton. But I could not follow the path that he and his disciples took. Following up an idea of Louis de Broglie, he argued that the electron in the double-slit experiment acquired its instruction where to go on the photographic plate from microscopically fine adjustment in the initial path of the electron entering the apparatus. I remember struggling to make sense of his picture of the double-slit interference experiment. In his picture the electron or photon was like a missile travelling through a mountainous potential barrier, with valleys leading to one or other of familiar outcomes. Given the initial velocity and direction, the missile had no escape from ending up at this, that, or the other interference fringe. The uncertainty all arose from the screwdriver that positioned the original missile launch. But nowhere in the physics could one discern any microscopic screw that a screwdriver could adjust to give always one fringe or another.

Niels Bohr never ceased to trouble about the quantum. Some thoughts of his and of von Weizsäcker on the idea of delayed choice got through to me. So in 1978 I wrote a paper proposing more than one conceivable "delayed choice" experiment. Carroll Alley, with Oleg Jakubowicz and William Wickes, took up the idea at the University of Maryland and found that indeed the results came out as predicted. One could at last say, "Look here and see how you can decide, after an event has already happened, whether it shall have happened or not." That is delayed choice with a vengeance. Confirming evidence of delayed choice came from colleagues at

Berkeley. One can even visualize a delayed-choice experiment at the cosmological level.

As Niels Bohr liked to say, a quantum event is brought to a close by an irreversible act of amplification. Contrast this with Wigner's argument that the elementary quantum phenomenon is only then brought to a close when the results enter the consciousness—a position the Wigner later gave up.

On the windowsill of my home on an island in Maine I keep a rock from the garden of Academe, a rock that heard the words of Plato and Aristotle as they walked and talked. Some day there will arise the equivalent of that garden where a few thoughtful colleagues will see how to put it all together and save us from the shame of not knowing "how come the quantum." How could the proposed academy best function, have the great questions brought to mind over and over day after day, bring Bohr and Einstein and the others back to life to stimulate and inspire their endeavors? There are surely as many ideas on how to do this as members of my Yale audience. But copies of two books I would surely provide for each member of the little academy for fireside reading every evening. One is the recently published Volume 7 of Niels Bohr's Collected Works, *Foundations of Quantum Physics II (1933-1958),* edited by Jørgen Kalckar, full of those questions and proposals and insights that were at the heart of Niels Bohr's thinking. The other book for nighttime reading is a collection of papers on quantum theory and measurement put together in 1983 by Wojciech Zurek and me under the title *Quantum Theory and Measurement* and published by Princeton University Press. There are many threads, any one of which the members of our little academy could take up and pursue. For example, there is Bohr's statement about "the elementary quantum phenomenon," in his words, "brought to a close by an irreversible act of amplification," to which my wonderful colleague Eugene Wigner objected in his inimitable way, "What means irreversible?" And there is the whole series of issues to which this question gave rise. At the end of it all let our friends in their little academy save physics from its shame. Here is the science that considers itself the intellectual foundation for all chemistry, for geology, for astronomy and cosmology. And, proud foundation for so much, it does not yet know the foundation for its own teachings, how come the quantum? One can believe, and I do believe, that the answer to that question will prove to be also the answer to another question, "how come existence?"

Are these questions too philosophical? Are we entitled to reply that maybe philosophy is too important to be left to the philosophers? When will the shame of physics be redeemed and washed away in a cloud of glory? When our successors can confidently proclaim how come existence and how come the quantum and at last rightfully presume to practice the science that is the legitimate foundation of so many other sciences.

References

1. H. Rubens and F. Kurlbaum, *Preuss. Akad. Wiss. Berlin, Sitzungsberichte,* **41**, 929 *(October 25, 1900)*

ON GREGORY BREIT

MCALLISTER HULL
Department of Physics and Astronomy University of New Mexico, Albuquerque, NM, 87131
USA
E-mail: machull@unm.edu

It is a distinct pleasure to join old and new colleagues to recall the remarkable career of Gregory Breit in the year of the centenary of his birth. For nearly fifty years his work led the way in a number of significant fields of twentieth century physics. This symposium has demonstrated the staying power of the vision that inspired his pioneering work.

I met Gregory in the late summer of 1946 at Los Alamos, where Oppenheimer had organized a kind of special meeting of the American Physical Society to celebrate the end of the war and the contributions physics had made to the successful outcome. Several of we youngsters had persuaded senior members of the Los Alamos staff to give courses at what we called "Los Alamos University" before they dispersed to their own universities. We assigned ourselves to produce notes on the topics, and I got nuclear physics. Thus I knew of the "Breit-Wigner" resonance theory before Joe Hirschfelder, with whom I was working on phenomenology of the Bikini tests, introduced me to Breit at the meeting (Joe was a theoretical chemist at Wisconsin). Despite the fact that I would return to university as a junior physics major, Gregory invited me to come to Wisconsin to work with him. It was an association that lasted twenty years with day to day interactions, and to the end of his life as frequently as possible thereafter (I was actually chair of his department at SUNY Buffalo).

By a totally unscientific count, Gregory worked with nearly 90 collaborators: many of them were students and post-doctoral associates, and many are leading physicists of the century (not mutually exclusive sets!). He published over 300 papers between 1920 and 1970. The topics of this symposium attest to the diversity of his interests--and by no means are all of them represented. His teaching was meticulous and clear if somewhat plodding, and a book or books based on some of his courses would have been a major contribution. But there was always another problem to work out, and time for writing books never appeared. I tried to help by agreeing to collaborate on a book on nuclear physics, but could never pin him down enough to look at what I had started. When Blatt and Weisskopf appeared, he sighed with relief: the obligation I had urged had been satisfied! The review articles for the *Handbuch der Physik* were as near as he got to a book, and of course these were not general.

In both my 1948 and 1993 editions of *Encyclopedia Britannica*, Gregory appears as part of the Breit and Tuve experiment or the development of radar. Since this is not a topic of the symposium, I can, perhaps usefully reprise briefly his contribution to the device that played a major part in winning the Battle of Britain (for you youngsters, I can discuss WWII privately!). He wished, as a heritage of his electrical engineering background, to see whether the Kennely-Heaviside layer existed, and if so, how high it sat. Bouncing radio waves off the layer would reveal its existence, and if the time of the return signal could be measured, the height of the layer could be determined. Eventually, he and Merle Tuve pulsed their signal in order to leave time for the return signal to be recognized, and thus they invented the principle of radar. Their wave length was too long to allow aircraft ranging, but when they used the Naval Research Laboratory transmitter, planes rising from Washington airport produced diffuse signals they ignored once they had identified them. Of course, "long wave" radio depends for distance on reflection from the ionized layer, and while commercial radio has essentially abandoned this frequency range, the Navy still depends on it for a back up to satellite communication--or it did the last time I looked. It was typical of Gregory that he had no interest in possible applications of his ideas. Once the fundamental physics questions were answered, there was the next question to be addressed. In 1945, the Institute of Electrical and Electronic Engineering gave Gregory its Fellow Award for this "pioneering" work.

It was also typical that he participated in experiments in physics. His direct involvement extended to working with Tuve on accelerating nuclear particles as probes for studying nuclear properties, and Cockroft and Walton beat them by only a few months to the first published study of nuclear reactions with "artificial" beams. Gregory wanted 1 Mev, while his British counterparts were satisfied with a half. He continued to work with experimentalists as closely as possible at Wisconsin and at Yale when he came here. Allan Bromley can attest to his close interest in experiments with the Tandem van de Graaf: perhaps too close sometimes! He and Tuve demonstrated the betatron principle, and he continued to suggest not only experiments, but also accelerators with which to do them. The Heavy Ion Linear Accelerator proposal was led by Bob Behringer with scientific justification contributed by us. We noted that heavy ions could be accelerated also by cyclotrons, since for N=Z nuclei, totally stripped ions had the same charge to mass ratio as deuterons or alpha particles. The first heavy ion studies were done with a cyclotron at Oak Ridge in experiments by a former Yale student, Al Zucker. When the HILAC was awarded to Yale, Bob Gluckstern contributed practical design input to the conceptual designs we had worked out. Talks in this symposium by Larry McLerran and John Harris tell of heavy ion work well beyond any ideas Gregory and colleagues had, but he would be delighted to know how far the work has gone beyond his own. Gregory's enthusiastic support of experimental work continued with contributions to the proposal for the "meson factory" that Vernon headed. When the review panel awarded the machine to Los Alamos (instead of the circular machine they were actually proposing), Yale's design was built on the mesa--again with design input from Bob Gluckstern. Vernon got to do his mu meson work on the machine without having to be director of a lab! We usually suggested uses for proposed machines that we thought would be attractive to funding agencies even if we were not directly interested ourselves. Heavy ions to make transuranic elements, for example, or pions to treat cancer. A second HILAC was built at Berkeley for Glen Seaborg, and transuranic atoms have been made with heavy ions ever since, in Russia and Germany as well as Berkeley: one for Z=118 has been recently announced. A member of the Yale Medical School faculty went to New Mexico to try the pion cancer therapy, and demonstrated the value of the method quite successfully. It succumbed to funding cuts a few years ago. The Cancer Research and Treatment Center in the Medical School at my home university was founded to house the pion therapy research, and survives the funding cuts brilliantly as a broad based cancer research effort. As a cured cancer patient myself, I get my physicals there.

For at least thirty years, Gregory studied nucleon-nucleon interactions with a number of collaborators. John Wheeler, one of Gregory's earliest post doctoral associates, and a luminary of this symposium, was an early participant in this effort, and I was one of the longest running and last. He viewed these interactions as fundamental to a study of nuclear reactions and structure. Up to 300 or so Mev, that is still so, and we labored long and hard to define the phenomenological interaction accurately enough to characterize experiments in the field. We used a dual approach: to find the phase shifts in the scattering matrix (which Gregory had shown to be the fundamental characterization of the interaction), and by finding an interaction potential that would give the phase shifts. As the energy of the experiments went up, the scattering matrix and form of the potential got more complex. We used forms for spin and isospin that conformed to the available symmetries, and introduced the effects of the Breit interaction when needed. We used a Yukawa form for the potential with, finally, the experimental mass. The repulsive core that the increasingly numerous phase shifts suggested was intended to model the pion exchanges in higher order diagrams; today we'd claim it modeled the gluon-quark plasma. A number of nuclear structure calculations have been made by others using this potential. The persistence over so many years, and the insight that characterized the approach, are signatures of Gregory's way of doing physics.

During the war, Gregory initiated the volunteer withholding of papers in neutron and uranium physics: he had been one of the first to recognize the weapons potential of fission, and wanted to give no comfort to the enemy. He chaired the fast neutron project at Chicago, and his reports from that program contributed to future work in New Mexico. When he left for Washington to work on problems "less glamorous than fission" as he once told me, Oppenheimer took over, and the work eventually led to Los Alamos. It was typical of Gregory that he would work on important wartime problems that would be less attractive to others. His Washington work, on degaussing, the proximity fuse, and exterior ballistics, brought him citations from both the Army and Navy. The proximity fuse, that was also developed elsewhere, radically improved the efficiency of anti-aircraft fire during WWII. In 1967, President Johnson awarded him the National Medal of Science, in part for this work.

In retrospect, we see that Gregory contributed to the three major technological developments of WWII on the Allied side that made a significant difference in our war fighting capability: radar, the proximity fuse, and nuclear weapons. His work on degaussing and exterior ballistics was important, of course, but those problems were left over from WWI. The award of the National Medal of Science was well deserved!

Gregory's reputation for careful and penetrating examination of any problem he undertook to study was cited by Edward Teller in suggesting the he be asked to look at the possibility of a hydrogen bomb explosion setting off uncontrolled reactions in the atmosphere. A similar question had been raised before the Trinity test of the fission bomb, and had been answered in the negative by Hans Bethe (and others). In any case, Gregory accepted the assignment, and we set up a "secret" room in Sloane Lab to house the effort. Bob Gluckstern and I supervised some of the work under Gregory's direction. The final report indicated no problem pending the outcome of some new experiments, which were successful. This was, again, a typical response by Gregory to a need from his adopted country.

We were early users of computers to do theoretical physics. Gregory had known about analogue differential analyzers, and had used them in his war work for the Navy. The magnetic extrapolator he devised with their aid reduced the time for degaussing merchant ships by a factor of thirty. But he had no experience with digital computation, except with desk calculators. The tables of Coulomb wave functions we published were calculated by *persons* we called "computers" who were "programmed" by some of us. The first calculations we did with more powerful machines were done in the Yale business office with IBM accounting machines programmed by wiring plug boards--as Nick Metropolis had shown me how to do at Los Alamos. Eventually, we got time on much more powerful general purpose computers capable of using stored programs of arbitrary complexity--true Von Neumann/Turing machines. These were IBM computers in centers at NYU and at the IBM research laboratories at Poughkeepsie: we sometimes worked on "bread board" versions of the next generation of IBM computer at the Poughkeepsie lab. Of course, the desktop computer I'm writing on is more powerful than these room-sized behemoths. When Yale developed a Computer Center, our graduate students were the first consultants. Among the problems done with digital computers was the nucleon-nucleon interaction already mentioned. Gregory never learned programming, but it was typical of his vision that he encouraged the rest of us to use the new technology, and found the access we needed before Yale set up its center.

It is not news that Gregory was difficult at times on a personal basis (the old students in the audience will be remarking at my understatement). One occasion when I was an undergraduate, Gregory lost his temper over a misunderstanding about the way I had edited a joint paper, and I left his office saying that I would return "when he felt better". His usual (and genuine) apology became even more heartfelt later as I explained the misunderstanding and he never raised his voice to me again. We had our differences over the years, but they were always resolved without temper (albeit with some tension at times). Not everyone was so lucky, but there was never any doubt that Gregory had a deep concern about his associates on both an intellectual and personal basis. For example, when I went to Wisconsin, I was told to enroll in a writing class. I commented to Gregory that I thought it was an unwelcome requirement caused by my checkered academic career. Next day, Miles Hanley, the dean of American linguisticians, called and invited me into his graduate linguistics class as a substitute for junior writing. A lifelong interest was initiated. I never told Gregory that Miles had tried to recruit me after the course! Gregory's concern for his associates meant that no student or post doc ever left the group without a suitable position. He knew everyone of his generation, and his judgment of talent was respected. Gregory's politeness (except for the temper outbursts) was legendary. When I had decided to go to Wisconsin at his invitation, I was told never to get behind him and Eugene Wigner as they approached a door. Both were trained in European politesse, and it was impossible for them to get through : each would defer to the other until someone broke the impasse in desperation. He always stood when one entered his office (and expected you to do the same if he visited yours!). My wife, Mary, had an example of his politeness before I met him. She worked in the travel office at Los Alamos, and there were strict times set for lunch. Gregory went in to get some travel business done just at noon. "You'll have to come back," she said, not recognizing the great man! He bowed politely, and when she returned, there was Gregory, patiently waiting for her reappearance. He was, as usual with women, gravely courteous despite the delay in finishing his travel arrangements.

The phrase "*idee fixe*" could characterize some of Gregory's responses outside of physics. Once early in our relationship, he asked me to work on a calculation on a Saturday afternoon. I explained that I couldn't do the job when he asked, but would do it Sunday. Why the delay? Yale plays football Saturday afternoon, and I usually attend, I said (I don't know that he ever knew that I'd played football in high school--or that I played center field for the physics department softball team for 20 years, usually with Bob Gluckstern at third base!). For the rest of our association, he assumed I went to football games on Saturday afternoon--spring, summer, fall, winter, year after year! I never corrected his impression: it made my life easier at times.

Another fixation of Gregory's had to do with cigarettes. He was the classic hard case. He quit regularly, and when desperate, he would get a smoke off one of us (not me: I smoked a pipe and cigars), always keeping track and bringing around a pack to replace his borrowings. But he was absolutely convinced that cigarette manufacturers put in an additive to produce addiction! I explained about pure food and drug laws, but he didn't believe they worked with cigarettes. How pleased he would be to learn that his theory was half-right: nicotine is addictive, but it is not an additive (there are other noxious additives, however).

Despite his own unhealthy practice of smoking, Gregory was an advocate and practitioner of his version of *mens sana in corpore sano.* He did what we would call aerobic exercises himself, and was very concerned at any health problem of group members. He regularly called the group out for long walks--usually up the hill from Sloane Lab. Larry Biedenharn hated these outings, and tried to put Gregory off by engaging his attention just at a curb--hoping that he would jar himself stepping off and tire of the expedition. I don't think Larry was ever successful.

John Wheeler has said that Gregory was one of the least appreciated physicists of the century. In my obviously biased opinion, I agree, but I believe this splendid gathering makes some amends for the neglect, and I wish to express my thanks to my *alma mater* and my Yale colleagues for the recognition it affords my friend and mentor, Gregory Breit.

REMARKS FROM A FAMILY MEMBER

RALPH WYCKOFF
E-mail: rwyckoff@oregonvos.net

Unfortunately because of tragic circumstances within my immediate family I have been unable to hear what the prior speakers have said about my stepfather. In case they may have implied that he did not have a sense of humor I should like to dispel that notion and cite the following.

One evening when I was about 12 years old we went to the lab together for him to check on something. We found one of his students busily sawing on the leg of a chair that was usually used by another of his students.

GB: What are you doing with that chair?

Student: I am sawing on its legs.

GB: Why?

Student: This is So and So's (giving student's name which I have forgotten) chair and we are working together on project A (which GB obviously already knew) and he has such a one track mind and limited approach that we have been coming back here every night after he has left to saw 1/8th of an inch off each leg of his chair to see how long it would take before he would realize that something unusual had happened.

I expected an explosion at this point but what happened was that after about ten seconds of reflection and with no change of expression,

GB: That is quite interesting. Please let me know the result of your experiment.

Later outside the building as we were walking home, a sudden giggle and then the sawing student's name.

It was after World War II and I had come home from the War when he told me that at the time of the closure of the work on what became the Manhattan Project at the Chicago Football Stadium he had to take written material pertaining to their work to the Bethesda Naval Yard for ultimate movement to Los Alamos in our 1938 Chevrolet. Somewhere in Indiana he picked up a hitchhiker. I asked him if he did not consider it dangerous to pickup a hitchhiker under the circumstances. His response was, of course not, the man said he was a dog trainer. When I asked "What if he weren't a dog trainer?" His reply was "Nonsense, that would be speculation".

Speculation in our family was a dirty word.

When he got to Bethesda, he and his cargo were denied entry for want of proper security clearance. When he told me this story he thought it very humorous. I really don't know what he thought of his denial of entry at the time. I am pretty sure what it really was but that too would be speculation.

My stepfather's like of the water was not limited to canoes on a lake. From his experience on the Black Sea when young he became an excellent swimmer and seaman. In view of that we decided to scatter his ashes over the Pacific Ocean. To carry this out I approached a commercial fisherman in Newport here in Oregon who was getting ready to go to sea. I asked if he would take Professor Breit's ashes and scatter them while he and his crew were out. The skipper seemed hesitant so I told him that when my mother and stepfather were married he owned an oyster boat he kept on the Chesapeake that was a sailing vessel and the hull was made of three huge trees. The captain told me that he had been on the Chesapeake when he was young and seen such an oyster boat on display and that he would therefore be very pleased to take Professor Breit's ashes. A sever Pacific storm came up very soon after the ship put out. When I next saw the captain he told me of the storm saying that my stepfather's last ride had been a very wild one and probably was very appropriate. He then went on to say that he had entered the time the ashes had been scattered in the ship's log but that he had not entered the bearings of the scattering because they might not be correct due to the severity of the storm. He hoped that the omission would be all right. I assured him that it was certainly all right with all concerned.

I so much regret not hearing the prior speakers, but I am sure that they are in agreement that my stepfather was a superb teacher. In matters more varied than found in a university class room or laboratory I can confirm that opinion. As an undergraduate I was a physics-math major, but when he let me know that I was memorizing my classroom notes too much to be a good physicist, I decided to become a lawyer. My life with him prepared me excellently. I have been a War Crimes prosecutor, a federal compliance officer and litigation attorney, an Assistant Attorney General of Oregon, a defender in the 60's of the New York Times in Alabama when sued for libel and a Circuit Judge. I knew that strong family bonds could be based upon the blood, but I learned thanks to him that they could also be based upon the heart and mind.

COLOR, SPIN, AND FLAVOR-DEPENDENT FORCES IN QUANTUM CHROMODYNAMICS

R.L. JAFFE
Center for Theoretical Physics and Department of Physics
Laboratory for Nuclear Physics Massachusetts Institute of Technology
Cambridge, Massachusetts 02139
E-mail: jaffe@mit.edu

In Memoriam Kenneth A. Johnson 1931–1999

A simple generalization of the Breit Interaction explains many qualitative features of the spectrum of hadrons.

1 Introduction

I did not know Gregory Breit, but two of his closest associates influenced me deeply. I did research in Gerry Brown's group as an undergraduate at Princeton. Gerry welcomed young students into his research family and gave us wonderful problems to work on. He taught us that complicated problems often have simple answers – something that has proved true in particle physics over the past quarter century. Vernon Hughes invented deep inelastic spin physics in the early 1970's and has championed it relentlessly ever since. He and his collaborators have literally rewritten the book on the quark and gluon structure of the nucleon. It is a pleasure to speak at a symposium celebrating their teacher, Gregory Breit.

My talk will be largely pedagogical. I would like to show how a generalization of the "Breit Interaction" can account for some of the regularities of hadron physics. Although the basic ideas described here date back to the 1970's, they have not been presented quite this way before, and some developments are still at the forefront of modern research in QCD. This subject was a special favorite of my friend and collaborator, Ken Johnson, who died this past winter. Ken had many friends at Yale. He would have liked to hear this story, so I dedicate my talk to his memory.

It is hard to make definite statements about hadrons made of the light u, d, and s quarks. The nonperturbative regime is too complicated. It may never be solved to our satisfaction except on a computer. Nevertheless, the spectrum and interactions of baryons and mesons display remarkable regularities, which correlate with simple symmetry properties of the fundamental quark/gluon interactions. The role of models in QCD is to build simple physical pictures that connect the phenomenological regularities with the underlying structure.

The QCD "Breit Interaction" is the spin-dependent part of one-gluon exchange between light quarks in the lowest state of some unspecified mean field. It is summarized by an effective Hamiltonian acting on the quarks' spin and color indices,

$$\mathcal{H}_{\text{eff}} \propto -\sum_{i \neq j} \underset{\sim}{\lambda}_i \cdot \underset{\sim}{\lambda}_j \vec{\sigma}_i \cdot \vec{\sigma}_j \tag{1}$$

where $\vec{\sigma}_i$ and $\underset{\sim}{\lambda}_i$ are the spin and color operators of the ith quark. The spin operators are represented by the three 2×2 Pauli matrices, normalized to $\text{Tr}(\sigma_i^k)^2 = 2$ for $k = 1, 2, 3$, and the color operators are represented by the eight 3×3 Gell-Mann matrices, normalized the same way, $\text{Tr}(\lambda_i^a)^2 = 2$ for $a = 1, \ldots 8$. The sum over i and j extends over all quark pairs. For the moment, I ignore antiquarks. I also ignore quark mass differences. In reality the u and d masses are small enough relative to the natural scale of QCD that we can neglect them. The s quark is heavier. It is a reasonable first approximation to ignore its mass as well. The c, b, and t quarks are too heavy, and cannot be treated this way. The space-time dependence of $\mathcal{H}_{\text{eff}}$ is not well understood, but in this approximation it is universal, and need not concern us much.

$\mathcal{H}_{\text{eff}}$ can be read off the Feynman diagram for one-gluon exchange between quarks; see Fig. (1). At short distances where QCD is weakly coupled, we can trust perturbation theory, and one-gluon exchange should dominate. However, at typical hadronic distance scales QCD is strongly coupled. Still, there is reason to take the qualitative predictions which follow from $\mathcal{H}_{\text{eff}}$ seriously. Once the long range, spin-average, confining interactions in QCD have been integrated out, the resulting confining, bag-like mean field acts as an infrared cutoff, reducing the strength of the remaining QCD effects. Most phenomenological models of QCD – the Bag Model, the nonrelativistic quark model, and other quark models in particular – use this picture successfully. The absence of strong renormalization (higher twist effects) in deep inelastic scattering offers phenomenological support for a picture of hadrons where perturbation theory is qualitatively reliable once confinement has been implemented.

A couple of further notes on the form of eq. (1): First, the spin-averaged piece of one-gluon exchange has been set aside. It figures in the dynamics of confinement but not in the spectroscopy considered here. Second, the tensor and spin-orbit interactions generated by one-gluon exchange average to zero in the lowest quark state. Third, the appearance of $\vec{\sigma}$ matrices in eq. (1) does not mean the analysis is nonrelativistic. For light quarks this would be an unacceptable restriction. The Dirac $\vec{\alpha}$ matrices which appear in the relativistic quark currents reduce to $\vec{\sigma}$ matrices in the lowest orbital.

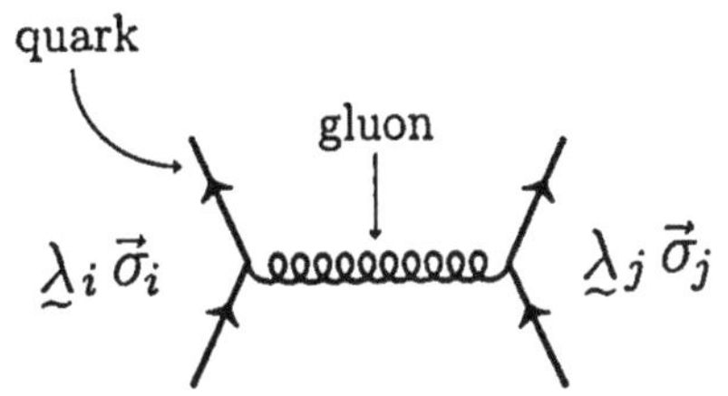

Figure 1. One-gluon exchange between quarks.

For obvious reasons the interaction of eq. (1) is known as the "colorspin" or "color-magnetic" interaction of QCD.

$\mathcal{H}_{\rm eff}$ was first introduced by De Rujula, Georgi, and Glashow in their pioneering paper on hadron spectroscopy in QCD.[1] Many of the spectroscopic results I will discuss were developed by them or by our group at MIT in the mid-1970's.[2,3,4] The subject of the scalar mesons has been revitalized recently by Schechter and his collaborators.[5] The possible role of $\mathcal{H}_{\rm eff}$ in quark matter has been the subject of much recent activity starting with the fundamental work of Alford, Rajagopal, and Wilczek.[6]

My talk is organized as follows: In Section 2 I review the basic symmetry structure of $\mathcal{H}_{\rm eff}$. In Section 3 I look at some properties of the baryons: the octet-decuplet splitting, the Λ–Σ splitting, and the pattern of excitations. Section 4 is devoted to mesons: first the pseudoscalar-vector splittings, next a remark on the absence of exotics, and finally a reevaluation of the $J^{PC} = 0^{++}$ mesons. In Section 5 I return to symmetry and extract a simple rule for the ground state of the Q^N configuration – the rule of flavor antisymmetry. In Section 6, I apply it to baryons (Q^3) and dibaryons (Q^6). Finally, in Section 7 I give a very brief introduction to the effects of $\mathcal{H}_{\rm eff}$ in quark matter – condensates, superconductivity, and unusual patterns of symmetry breaking.

2 Basics: Regularities of the Q–Q interaction

It is not necessary to use much mathematics to understand the implications of eq. (1) for the simplest case of two quarks. Most of what I need can be done merely by rewriting it in terms of color, spin, and flavor "exchange operators". The spin exchange operator, P^S_{12}, is defined by

$$P^S_{12} = \tfrac{1}{2} + \tfrac{1}{2}\vec{\sigma}_1 \cdot \vec{\sigma}_2 \ . \tag{2}$$

Two spin-$\frac{1}{2}$ particles may be coupled to a triplet of spin-1 states, $|3_s\rangle \equiv \{|\uparrow\uparrow\rangle, \frac{1}{\sqrt{2}}|\uparrow\downarrow + \downarrow\uparrow\rangle, |\downarrow\downarrow\rangle\}$, which are symmetric under spin exchange, or to a singlet spin-0 state, $|1_s\rangle \equiv \frac{1}{\sqrt{2}}|\uparrow\downarrow - \downarrow\uparrow\rangle$, which is antisymmetric under spin exchange.* The eigenvalues of $\vec{\sigma}_1 \cdot \vec{\sigma}_2$ are $+1$ and -3 in the spin-1 and spin-0 states respectively, so $P^S|3_s\rangle = +|3_s\rangle$, $P^S|1_s\rangle = -|1_s\rangle$. Thus P^S has the desired property that it gives ± 1 on states which are symmetric/antisymmetric under spin exchange.

Color and flavor states are both classified by $SU(3)$, so color and flavor exchange operators have the same form. To be specific, consider flavor. The quark labels are u, d, and s. I need a notation for the set of Gell-Mann matrices associated with flavor – denote them by $\{\underset{\sim}{\beta}\}$ to distinguish them from the $\{\underset{\sim}{\lambda}\}$ used for color.

There are nine flavor states of two quarks. Six are symmetric: $|6_f\rangle \equiv \{|uu\rangle, |dd\rangle, |ss\rangle, \frac{1}{\sqrt{2}}|ud + du\rangle, \frac{1}{\sqrt{2}}|ds + sd\rangle, \frac{1}{\sqrt{2}}|su + us\rangle\}$. Three are antisymmetric: $|\bar{3}_f\rangle \equiv \{\frac{1}{\sqrt{2}}|ud - du\rangle, \frac{1}{\sqrt{2}}|ds - sd\rangle, \frac{1}{\sqrt{2}}|su - us\rangle\}$. These states can be visualized most easily in the usual $SU(3)_f$ weight diagram in which the third component of isospin ($I_3 = \frac{1}{2}(n_u - n_d)$) is the x-axis and hypercharge ($Y = \frac{1}{3}(n_u + n_d - 2n_s)$) is the y-axis. The original quark triplet forms a triangle, point down, the 3_f. The antisymmetric set of two quark states forms a triangle, point up, hence the notation $\bar{3}_f$. The symmetric set of two quark states form a larger triangle, point down, the 6_f. All three representations are shown in Fig. (2).

A simple exercise with the Gell-Mann matrices lead to the exchange operator for flavor,

$$P_{12}^F = \tfrac{1}{3} + \tfrac{1}{2}\underset{\sim}{\beta}_1 \cdot \underset{\sim}{\beta}_2 \tag{3}$$

with the desired property that $P^F|6_f\rangle = +|6_f\rangle$ and $P^F|\bar{3}_f\rangle = -|\bar{3}_f\rangle$. Since color has the same structure as flavor:

$$P_{12}^C = \tfrac{1}{3} + \tfrac{1}{2}\underset{\sim}{\lambda}_1 \cdot \underset{\sim}{\lambda}_2 \tag{4}$$

and $P^C|6_c\rangle = +|6_c\rangle$, $P^C|\bar{3}_c\rangle = -|\bar{3}_c\rangle$.

Armed with eqs. (2), (3), and (4), we can rewrite the interaction between two quarks in terms of exchange operators,

$$\mathcal{H}_{\text{eff}} \propto -\underset{\sim}{\lambda}_1 \cdot \underset{\sim}{\lambda}_2 \vec{\sigma}_1 \cdot \vec{\sigma}_2 = -4P_{12}^C P_{12}^S + \tfrac{4}{3}P_{12}^S + 2P_{12}^C - \tfrac{2}{3} \,. \tag{5}$$

The next step exposes the reason for all this algebra: two quarks in the same orbital are *symmetric* under space exchange. Since the quarks are

*To avoid confusion, we label all states by their degeneracy. Thus the spin-1 states are $|3_s\rangle$ and the spin-0 states are $|1_s\rangle$.

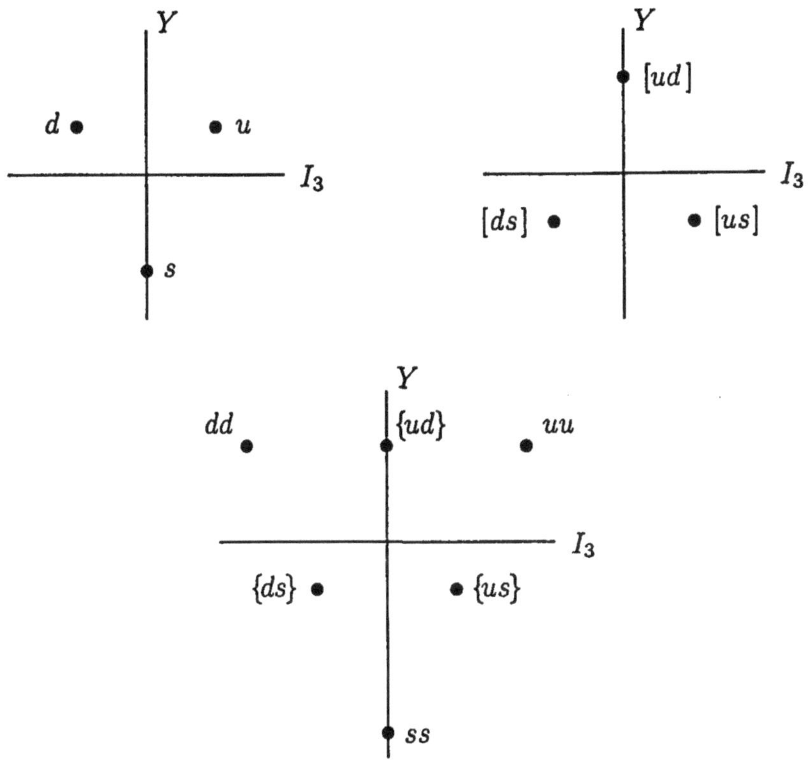

Figure 2. $SU(3)_f$ weight diagram for the fundamental 3_f, antisymmetric diquark, $\bar{3}_f$, and symmetric diquark, 6_f representations.

fermions, the wavefunction must be *antisymmetric* under the simultaneous exchange of the remaining labels: spin, color, and flavor. In terms of exchange operators,

$$\begin{aligned} P_{12}^C P_{12}^S P_{12}^F &= -1 \qquad \text{or} \\ P_{12}^C P_{12}^S &= -P_{12}^F \,. \end{aligned} \tag{6}$$

So eq. (5) can be rewritten as

$$\mathcal{H}_{\text{eff}} = 4P_{12}^F + \tfrac{4}{3}P_{12}^S + 2P_{12}^C - \tfrac{2}{3} \,. \tag{7}$$

Lo and behold: The dominant piece of the colorspin force (weighted by 4) is a *flavor*-exchange interaction. *This is why colorspin has significant implications for flavor-dependent mass splittings.*

Table 1. Flavor, spin, and color states of two quarks, and the eigenvalue of the QCD Breit Interaction of eq. (1).

Flavor	Spin	Color	ΔE
$\bar{3}(A)$	$1(A)$	$\bar{3}(A)$	-8
$\bar{3}(A)$	$3(S)$	$6(S)$	$-4/3$
$6(S)$	$3(S)$	$\bar{3}(A)$	$8/3$
$6(S)$	$1(A)$	$6(S)$	4

There are four totally antisymmetric configurations of flavor×spin×color. They and their associated colorspin interaction strengths are listed in Table 1. From eq. (7) it is clear that the configuration separately antisymmetric in flavor, spin, and color, is the most attractive channel. The other three configurations are symmetric in at least two indices, leading to dramatically less binding. The dominance of a single qq configuration, $|\bar{3}_f 1_s \bar{3}_c\rangle$, is the basic observation behind colorspin phenomenology. Colorspin favors antisymmetric (and therefore low-dimension) representations of the flavor and spin symmetry groups. Several applications will bear this out.

3 Baryons

3.1 The Octet-Decuplet Mass Difference

The lightest baryons composed predominantly of u, d, and s quarks can be grouped into an $SU(3)_f$ octet with spin-$\frac{1}{2}$ and an $SU(3)_f$ decuplet with spin-$\frac{3}{2}$. The $SU(3)_f$ weight diagrams in terms of I_3 and Y are shown in Fig. (3). States with the same I_3 and Y contain the same valence quarks. Thus the Δ^+ and p must both contain uud in addition to any $\bar{q}q$ pairs and gluons. For all pairs like the $p - \Delta^+$, the state in the decuplet is roughly 200–250 MeV heavier than the equivalent state in the octet. The *sign* of this effect is easily explained from the symmetry properties of the colorspin force: Since baryons are color singlets, each quark pair must be coupled to a color $\bar{3}_c$.† Because the decuplet has spin-$\frac{3}{2}$, each quark pair must be coupled to a spin triplet. Referring to Table 1, it appears that the $\bar{3}_c 3_s$ colorspin configuration has positive interaction energy. The spin coupling of the quarks in the octet is more complicated, but one thing is clear: at least the coupling of quark pairs

†So a color singlet can be made when the third quark is included: $\bar{3} \otimes 3 = 1 \oplus 8$, whereas $6 \otimes 3 = 8 \oplus 10$.

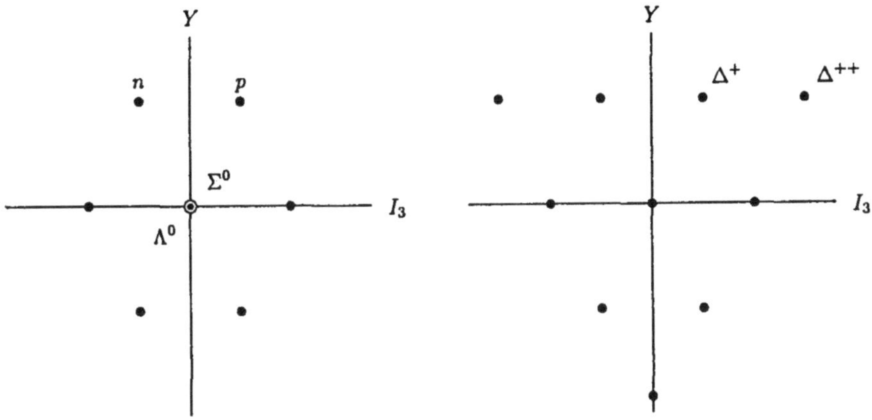

Figure 3. $SU(3)_f$ weight diagram for the baryon octet and decuplet. Some states discussed in the text are labeled.

to spin-0 is allowed. Therefore a state in the octet must be lighter than state with the same quark content in the decuplet.

3.2 The Λ–Σ Splitting

The lightest neutral strange baryons, the $\Lambda^0(1116)$ and the $\Sigma^0(1192)$ have the same quark content, *uds*. They differ in mass by 76 MeV. Before QCD this mass difference puzzled theorists who tried to understand the quark structure of hadrons. The puzzle was resolved by DeRujula, Georgi, and Glashow and led to much work on QCD inspired quark models.[1] The Λ^0 and the Σ^0 have the same quarks and the same spin. They differ in isospin:

$$|\Sigma^0\rangle \sim |(ud)^{I=1}s\rangle \qquad |\Lambda^0\rangle \sim |(ud)^{I=0}s\rangle \, . \tag{8}$$

Colorspin, like all spin-dependent effects in gauge theories, is a relativistic effect. For heavy quarks the coefficient we have suppressed in front of $\mathcal{H}_{\rm eff}$ goes like $1/m_q^2$. It is reasonable to assume that colorspin forces weaken with increasing quark mass. The u and d quarks are much lighter than the s quark, so the colorspin force between a ud pair is stronger than that between a us or ds pair. The ud pair in the Λ^0 is in the $I = 0$ or $\bar{3}_f$ state. In the Σ^0, the ud pair is in the $I = 1$ or 6_f state. The ud colorspin contribution to the mass is therefore attractive in the Λ^0, repulsive in the Σ^0. Since this dominates over ds and us, the Λ^0 must be lighter than the Σ^0. The magnitude of the $\Sigma^0 - \Lambda^0$ mass difference depends on the strength of the colorspin interaction (fixed by the decuplet-octet splitting) and the quark mass differences (fixed by $SU(3)_f$

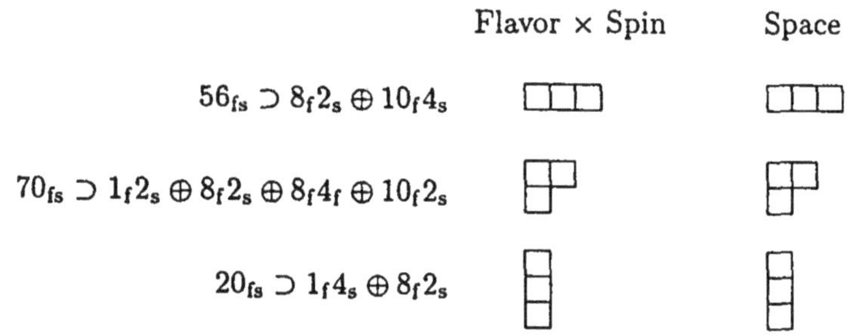

Figure 4. States of three quarks. $SU(6)_{fs}$ multiplets are shown along with their $SU(3)_f \otimes SU(2)_s$ content. The multiplets are represented by Young Diagrams. Boxes stacked vertically represent antisymmetrized quarks, those stacked horizontally are symmetrized. To make states symmetric under flavor×spin×space, quarks must have the same permutation symmetry under flavor×spin and space. Hence the identical Young Diagrams.

violating mass differences). The numerical value is roughly consistent with the 76 MeV observed.

3.3 Excited Baryon Spectroscopy

Tables of the masses and properties of excited baryons fill books.[7] Long ago, Dalitz and collaborators[8] showed that the fundamental structure of this spectrum could be understood in a "Symmetric Quark Model": Populate spin×flavor×space symmetric states of three quarks (color antisymmetry takes care of Fermi statistics) in the modes of some simple central potential. The lightest (positive parity) baryons have all quarks in the lowest orbital. To obtain negative parity excitations, promote one quark to the first ($L = 1$) excited orbital. To make positive parity excitations, promote two quarks to the first level or one quark to the second.

Dalitz's model contains the Gürsey's famous $SU(6)_{fs}$ symmetry[‡] built out of $SU(3)_f \otimes SU(2)_s$.[9] Baryon multiplets form representations of $SU(6)_{fs}$, which in turn contain $SU(3)_f$ multiplets of definite spin. The three $SU(6)_{fs}$ representations of three quarks are shown in Fig. (4) in terms of Young Diagrams. The 56_{fs} is totally symmetric in exchange of spin and flavor, the 70_{fs} has mixed symmetry, and the 20_{fs} is totally antisymmetric. The 56_{fs} and 70_{fs} are prominent in the baryon spectrum. For example, the lightest baryons form a 56_{fs} containing the decuplet with spin-$\frac{3}{2}$ and the octet with spin-$\frac{1}{2}$. In fact *all known baryons can be assigned to either* 56_{fs} *or* 70_{fs} *representations.*

[‡]I use the subscript fs in order not to mix it up with another $SU(6)_{cs}$ built of color×spin, which will be useful later.

What happened to the 20_{fs}? This representation goes with a totally antisymmetric space wavefunction, and could occur among the positive parity baryon resonances, which are quite well studied. The $SU(3)_f\times$spin content of the 20_{fs} is $|8_f 2_s\rangle$ and $|1_f 4_s\rangle$. This is the only occurrence of a spin-$\frac{3}{2}$, $SU(3)_f$ singlet. So observation of a spin $\frac{3}{2} - \Lambda^{*0}$ without the partners required to form an octet would be the signature of the 20_{fs}. There is no serious candidate for a $1_f 4_s$ state in the baryon spectrum. The QCD Breit interaction predicts this. Two quarks in the 20_{fs} must be antisymmetric in simultaneous exchange of flavor and spin. In contrast, quark pairs in the 56_{fs} must, and in the 70_{fs} may, be symmetric in flavor$\times$spin. So the 20_{fs} alone decouples from the most attractive configuration, $|\bar{3}_f 1_s \bar{3}_c\rangle$. Because of this, the 20_{fs} is promoted to higher energy and is less stable.

4 Mesons

The generalization of the Breit interaction to $\bar{q}q$ is simple. Once $\bar{q}q$ form a color singlet the residual interaction is attractive for spin-0 and repulsive for spin-1, much like positronium. The consequences are well known: The pseudoscalar mesons like the π are lighter than their vector partners like the ρ.

The fact that colorspin has other significant consequences for meson spectroscopy is less well known.

4.1 The Absence of Exotics

The absence of $SU(3)_f$ representations other than 1_f and 8_f for mesons, and 1_f, 8_f, and 10_f for baryons was one of the original inspirations for Gell-Mann's and Zweig's quark model. Later, QCD and confinement explained the absence of representations of with fractional charge, but quark model exotics like a meson 27_f-plet ($\bar{q}^2 q^2$) or a baryon $\overline{10}_f$-plet ($q^4\bar{q}$) are allowed by confinement. Gell-Mann's original paper on the quark model begs the question with a breathtakingly appropriate typographical error:

> Baryons can now be constructed from quarks by using the combinations (qqq), $(qqqq\bar{q})$, etc., while mesons are made out of $(q\bar{q})$, $(qq\bar{q}\bar{q})$, etc. It is assuming [*sic*] that the lowest baryon configuration (qqq) gives just the representations **1**, **8**, and **10** that have been observed, while the lowest meson configuration $(q\bar{q})$ similarly gives just **1** and **8**.[10]

and almost everyone has been assuming it ever since.

The experimental situation has changed little since Gell-Mann's day. Consider, for example, states of two pions at low energies. The $I = 2$ state, exemplified by $\pi^+\pi^+$ is exotic. It contains at least $\bar{d}^2u^2$. Data on $\pi^+\pi^+$ scattering shows no resonance at low energies, only a weak repulsion. The $I = 1$ and $I = 0$ states couple to $\bar{q}q$. The $I = 1$ phase shift resonates at the ρ. The $I = 0$ phase shift shows a strong, broad attraction from threshold up through $m_{\pi\pi} \approx 1$ GeV. The same absence of striking resonances occurs in every exotic channel.

QCD is consistent with the absence of strongly bound mesons made of more than $\bar{q}q$. Confinement saturates at q^3 or $\bar{q}q$. There are no van der Waals forces in QCD because there are no massless hadrons. So the mechanism that forms molecules from atoms in QED is absent in QCD. However, this does not explain why no $\bar{q}^2q^2$ states have been seen.

The QCD Breit interaction, $\mathcal{H}_{\text{eff}}$, gives a simple explanation for the absence of exotics. A complete treatment would require me to include $\bar{q}q$ interactions as well as qq. However, the gist of the argument can be read off Table 1. Exotic $SU(3)_f$ representations require either the quarks to be in the 6_f or the antiquarks to be in the $\bar{6}_f$ representation or both. If the quarks are in the $\bar{3}_f$ and the antiquarks are in the 3_f, then the resulting $\bar{q}^2q^2$ state transforms as an $8_f \oplus 1_f$, which is not exotic and could mistakenly be interpreted as a $\bar{q}q$ multiplet. Table 1 shows that colorspin interactions are repulsive in the 6_f and $\bar{6}_f$ representations. So colorspin pushes exotic flavor states of $\bar{q}^2q^2$ up above thresholds where they fall apart into the light pseudoscalar mesons from which they are built. A more sophisticated treatment including $\bar{q}q$ colorspin forces confirms this result.[3]

So the QCD Breit interaction explains the absence of prominent exotic meson resonances and also suggests that we look carefully at ordinary mesons to see if any $\bar{q}^2q^2$ states have been mistakenly cataloged as $\bar{q}q$.

4.2 The Scalar Mesons

The lightest (negative parity) light quark mesons have been known for forty years. The first excited states (with positive parity) are also well known. Of these, only the scalar mesons remain controversial. The rest were discovered and classified during the sixties and seventies. The identification and nature of the scalar mesons continues to be one of the most controversial subjects in hadron spectroscopy. The special role of the $\pi\pi$ s-wave in chiral dynamics and in nuclear physics makes this an important issue.

To make $J^{PC} = 0^{++}$ in the naive quark model it is necessary to add a unit of orbital angular momentum to a $\bar{q}q$ pair (to get positive parity).

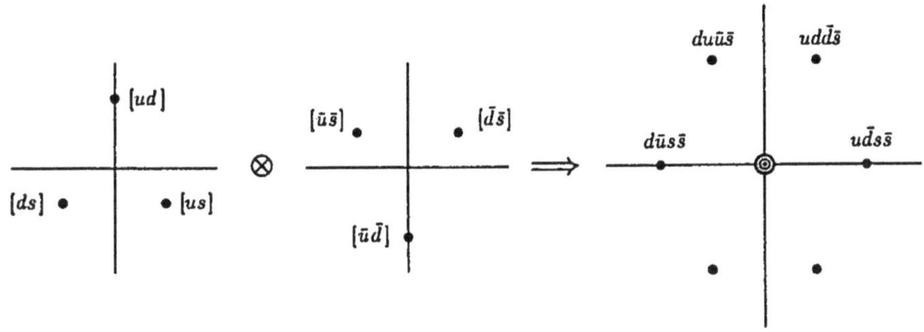

Figure 5. $SU(3)_f$ **weight diagrams for two antisymmetrized quarks, two antisymmetrized antiquarks, and for the flavor nonet contructed from them.**

Apparently this costs more than half a GeV in mass, since well established mesons with this makeup ($1^{++}, 1^{+-}$, and 2^{++}) are all known to lie between 1.2 and 1.6 MeV. In contrast $\bar{q}^2q^2$ can couple to 0^{++} without excitation. Surprisingly there are striking experimental effects in all flavor nonet 0^{++} channels at masses below 1 GeV, and they do not fit naive $\bar{q}q$ quark model expectations. Recently Black et al. reviewed the situation and concluded that the light 0^{++} mesons look more like $\bar{q}^2q^2$ than $\bar{q}q$, an idea originally suggested on the basis of the QCD colorspin interactions back in the 1970's.[3,5]

The QCD Breit interaction between quarks strongly favors the $|\bar{3}_f 1_s \bar{3}_c\rangle$ configuration for quark pairs. The $\bar{q}q$ interaction, which we have not discussed, mixes in a significant amount of the second most attractive configuration, $|\bar{3}_f 3_s 6_c\rangle$, which, notice, is also a flavor $\bar{3}_f$. Thus the lightest $\bar{q}^2q^2$ mesons have the flavor structure $(\bar{q}^2)^{3_f}(q^2)^{\bar{3}_f}$, which leads to an octet plus a singlet ($\equiv$ a nonet). Diagonalization of the $\bar{q}^2q^2$ generalization of $\mathcal{H}_{\rm eff}$ in the $\bar{q}^2q^2$ sector drives a single 0^{++} flavor nonet to remarkably low mass (600–1000 MeV in the Bag Model). Thus the *only* light meson channel that has defied classification for the past forty years is the one in which colorspin considerations leads one to expect $\bar{q}\bar{q}qq$ to be important.

Figure (5) shows how a nonet can be constructed from two quarks in a $\bar{3}_f$ and two antiquarks in a 3_f. The quark flavor content of some of the states are shown in the figure. The most striking feature of a $\bar{q}^2q^2$ nonet is an *inverted mass spectrum.* A standard, magically-mixed $\bar{q}q$ nonet has a degenerate isosinglet and isovector at the bottom, a strange isodoublet in the middle, and a lone isosinglet at the top. This ordering is easily understood by counting the number of strange quarks in the meson. A $\bar{q}^2q^2$ nonet is opposite: the $\bar{q}^2q^2$ isotriplet and one of the isosinglets contain hidden strange quarks, $\{u\bar{d}s\bar{s}, \frac{1}{\sqrt{2}}(u\bar{u} - d\bar{d})s\bar{s}, d\bar{u}s\bar{s}\}$ and $\frac{1}{\sqrt{2}}(u\bar{u} + d\bar{d})s\bar{s}$, and therefore lie at the top

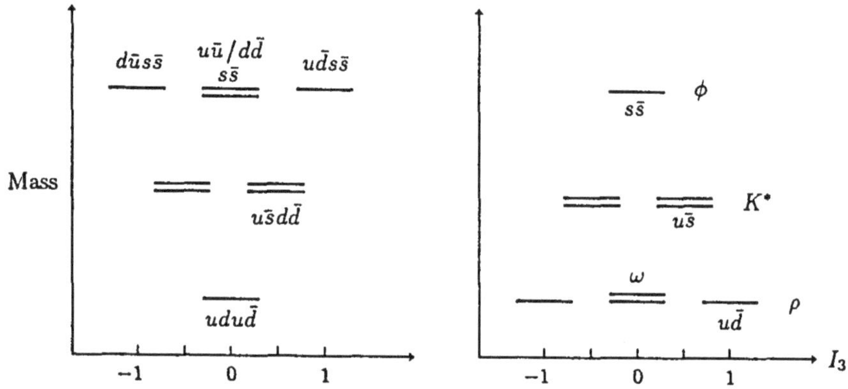

Figure 6. A cartoon representation of the masses of a $\bar{q}\bar{q}qq$ nonet compared with a $\bar{q}q$ nonet.

of the multiplet. The other isosinglet, $u\bar{d}d\bar{u}$ is the only state without strange quarks and therefore lies alone at the bottom of the multiplet. The strange isodoublets should lie in between. In summary, one expects a degenerate isosinglet and isotriplet at the top of the multiplet and strongly coupled $\bar{K}K$, an isosinglet at the bottom, strongly coupled to $\pi\pi$, and a strange isodoublet coupling to $K\pi$ in between. Fig. (6) shows the mass spectrum and quark content of a $\bar{q}^2q^2$ nonet and, for comparison, a $\bar{q}q$ nonet like the vector mesons (ρ, ω, and ϕ K^*).

The most well established 0^{++} mesons are the $I = 0$ $f_0(980)$ and the $I = 1$ $a_0(980)$. The f_0 couples to $\pi\pi$ and $\bar{K}K$. The a_0 couples to $\pi\eta$ and $\bar{K}K$. Both are so close to the $\bar{K}K$ threshold at 990 MeV that their shapes are strongly distorted by threshold effects, but it is clear that they couple strongly to $\bar{K}K$. This has always troubled those who would like to identify them as $\bar{q}q$ states. It is a natural consequence of the hidden $\bar{s}s$ component if they are predominantly $\bar{q}\bar{q}qq$.[§]

The contribution of Black et al. centers on the other isosinglet and strange isodoublet needed to fill up a nonet. If they are predominantly $\bar{q}^2q^2$ states, they should couple strongly to $\pi\pi$ and $K\pi$ respectively. Since the $\pi\pi$ and $K\pi$ thresholds are very low, one expects these states to "fall apart" into $\pi\pi$ or $K\pi$ with very large width. Enhancements in $\pi\pi$ and $K\pi$ s-wave scattering have been known for decades. Black et al. make the case that these enhancements correspond to broad states at approximately 560 MeV in $\pi\pi$ (known

[§]Another interpretation of the f_0 and a_0 as "$\bar{K}K$ molecules" is closely related to the $\bar{q}\bar{q}qq$ interpretation.[11]

as the $\sigma(560)$) and at 900 MeV in $K\pi$ (known as the $\kappa(900)$). Together the $f_0(970)$, $a_0(980)$, $\sigma(560)$, and $\kappa(900)$ make a nonet with mass spectrum, decay couplings and widths that look qualitatively like a $\bar{q}\bar{q}qq$ system. Should this assignment hold up, it will be striking confirmation of the role of the QCD Breit interaction in hadron spectroscopy.

5 Back to Basics: Spectroscopic Rules for Q^N

In Section 2 I looked at the colorspin force between pairs of quarks. Here I will generalize that analysis to the case of N quarks all in the same spacial orbital. This allows us to look at systems of up to 18 quarks (3 colors × 3 flavors × 2 spins) at short distances. It will allow us to draw interesting qualitative conclusions about baryons and "dibaryons" (i.e., 6-quark systems).

For N quarks,

$$\mathcal{H}^N_{\text{eff}} \propto -\sum_{i\neq j}^{N} \{\underset{\sim}{\lambda}\vec{\sigma}\}_i \cdot \{\underset{\sim}{\lambda}\vec{\sigma}\}_j \,. \tag{9}$$

Here I have grouped the color and spin operators of the ith quark together. The 24 matrices $\{\underset{\sim}{\lambda}\vec{\sigma}\}$ together with the 3 Pauli matrices $\{\vec{\sigma}\}$ and the 8 Gell-Mann matrices $\{\underset{\sim}{\lambda}\}$ together form the 35 generators of an $SU(6)_{\text{cs}}$ symmetry that we can call "colorspin". Let us denote the colorspin generators as a 35-dimensional vector of 6×6 matrices (in the fundamental representation), $\{\mu^r, r = 1, \ldots, 35\}$. The quadratic Casimir operator of $SU(6)_{\text{cs}}$ is defined as the sum of the squares of the colorspin generators,

$$C_6^N = \sum_{r=1}^{35}\Big(\sum_{i=1}^{N} \mu_i^r\Big)^2 \tag{10}$$

in analogy to the total spin

$$C_2^N = 4S_N(S_N+1) = \sum_{k=1}^{3}\Big(\sum_{i=1}^{N} \sigma_i^k\Big)^2 \tag{11}$$

and total color,

$$C_3^N = \sum_{a=1}^{8}\Big(\sum_{i=1}^{N} \lambda_i^a\Big)^2 . \tag{12}$$

Of course the color Casimir, C_3^N, is zero for any physical, i.e., color singlet, hadron.

It is an elementary exercise in matrix algebra to rewrite $\mathcal{H}^N_{\text{eff}}$ in terms of N, S_N, and C^N_6,

$$\mathcal{H}^N_{\text{eff}} \propto 8N - \tfrac{1}{2}C^N_6 + \tfrac{4}{3}S_N(S_N + 1)\,. \tag{13}$$

The Casimir of $SU(6)_{cs}$ is the sum of 35 normed generators. S_N is the sum of only 3 identically normed generators. Therefore C^N_6 dominates over S_N in eq. (13). So the spectroscopic consequences of $\mathcal{H}^N_{\text{eff}}$ for Q^N states are very simple. It is important to remember that we are assuming that all quarks are in the same orbital of some mean field and that their masses can be ignored.¶

- In this approximation, the Casimirs of spin, flavor, color and colorspin all commute with $\mathcal{H}^N_{\text{eff}}$, so mass eigenstates are (color singlet) states in irreducible representations of $SU(3)_f$, $SU(6)_{cs}$, and spin.
- For a given number of quarks, the largest C^N_6 is associated with the most symmetric representation – the most "horizontal" Young Diagram. Since quarks in the same orbital must be antisymmetrized in color×spin×flavor, the most symmetric colorspin goes with the most antisymmetric flavor. Thus the lightest state of N quarks is the one *most antisymmetric in flavor* subject to the constraint that the associated (conjugate) colorspin representation must contain a color singlet.
- The factor of $S_N(S_N + 1)$ is generally too small to affect the order of states.

In short: Nature prefers flavor antisymmetry. It is illuminating to apply this principle to the well-known case of $N = 3$ and then to the more speculative case of $N = 6$.

6 Baryons and Dibaryons

The simple rule for the spectrum of N quarks has interesting consequences for $N = 6$. First, however, let us see how it works for the baryons, $N = 3$.

6.1 Baryons Reconsidered

The allowed colorspin and flavor-irreducible representations are tabulated in terms of Young Diagrams in Fig. (7). The flavor $SU(3)_f$ representations, 1_f,

¶Ignoring the s-quark mass is not a good approximation. Fortunately s-quark mass effects can be calculated perturbatively.

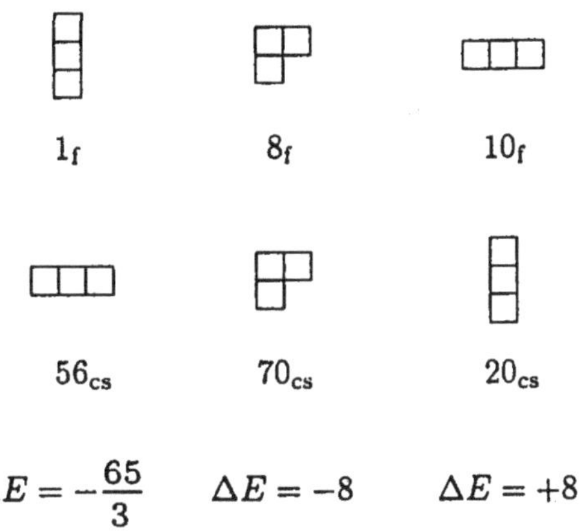

Figure 7. The states of three quarks, classified in $SU(6)_{cs} \otimes SU(3)_f$. For each multiplet the colorspin and flavor Young Diagrams are shown. To make states antisymmetric under flavor×color×spin, quarks must have the conjugate permutation symmetry under flavor and colorspin, obtained by interchanging rows and columns in the Young Diagram. The Breit Interaction favors the most antisymmetric flavor configuration, as evidenced by the eigenvalues displayed with the diagrams.

8_f, and 10_f are paired with the colorspin $SU(6)_{cs}$ representations 56_{cs}, 70_{cs}, and 20_{cs}, respectively. The "flavor antisymmetry" rule dictates

$$M_1 < M_8 < M_{10} \, . \tag{14}$$

In fact the expectation values of $\mathcal{H}_{\text{eff}}^{N=3}$ in the three states are $\langle \mathcal{H}_{\text{eff}} \rangle_1 = -65/3$, $\langle \mathcal{H}_{\text{eff}} \rangle_8 = -8$, and $\langle \mathcal{H}_{\text{eff}} \rangle_{10} = +8$. To complete the classification we must find out which of these colorspin multiplets contain color singlets. The $SU(3)_c \times SU(2)_s$ content of $SU(6)_{cs}$ representations is best known in the context of Gursey's old $SU(6)_{fs}$. We can copy those results over for color×spin:

$$\begin{aligned} 56_{cs} &\supset 8_c 2_s \oplus 10_c 4_s \\ 70_{cs} &\supset 1_c 2_s \oplus 8_c 2_s \oplus 8_c 4_s \oplus 10_c 2_s \\ 20_{cs} &\supset 1_c 4_s \oplus 8_c 2_s \end{aligned} \tag{15}$$

A novel perspective on the QCD picture of baryons emerges. The most attractive colorspin channel, which would be a candidate for the lightest baryon, the flavor singlet 56_{cs}, is not a physical state because it does not contain a color singlet. So the colorspin rule to antisymmetrize flavor is, in this case, frustrated by Fermi statistics: a color singlet, flavor singlet three quark state would have to be antisymmetric in spin, which is not possible for spin-$\frac{1}{2}$. Otherwise, the baryon ground state would be a flavor singlet, Λ-like uds baryon.

The flavor-antisymmetry rule works fine for the rest of the three quark states; the mixed symmetry multiplet, the 70_{cs}, is lightest. It contains a color singlet with spin-$\frac{1}{2}$ – the familiar flavor octet baryons. The antisymmetric colorspin multiplet, the 20_{cs}, is heaviest. It contains a color singlet with spin-$\frac{3}{2}$, the familiar decuplet baryons.

One final comment: It is tempting to suggest that the excluded multiplet, the 56_{cs}, could neutralize its color with a constituent gluon. The result would be a $qqqg$ $SU(3)_f$ singlet "hybrid" baryon. There is a rather light negative-parity Λ at 1405 MeV that has often caused problems for quark modelers. However current wisdom favors a q^3 or $q^4\bar{q}$ interpretation.[12]

6.2 Dibaryons – Known and Unknown

Now consider $N = 6$. The flavor Young Diagrams for six quarks are shown in Fig. (8). The colorspin Young Diagrams are conjugate (rows and columns interchanged). Only a few of the multiplets are shown. For the multiplets shown, the colorspin interaction energies are $\langle \mathcal{H}_{\rm eff} \rangle_1 = -24$, $\langle \mathcal{H}_{\rm eff} \rangle_8 = -28/3$, $\langle \mathcal{H}_{\rm eff} \rangle_{\overline{10}} = +8/3$, and $\langle \mathcal{H}_{\rm eff} \rangle_{27} = +3$.

The most attractive channel is the flavor singlet "dihyperon" (known as the H) with quark content $u^2d^2s^2$.[4] If colorspin were the only consideration, its interaction energy, -24, would bind it with respect to decay into two octet baryons (colorspin energy 2×-8). It would decay into the would-be flavor singlet baryon (colorspin energy $2 \times -65/3$), if that state were not excluded by color confinement. Models of confined light quarks are not sufficiently accurate to decide unequivocally whether the H is bound. Lattice calculations are improving and may answer the question before experimenters are able either to discover it or establish that it is not bound.[13]

The next most attractive channel is the flavor octet. The hypercharge $+1$ members have the quantum numbers of ΣN or ΛN. A strong resonance with these quantum numbers has long been known in ΛN scattering at 2129 MeV. Unfortunately this cannot be unambiguously interpreted as evidence for colorspin attraction because a loosely bound state with the same quantum numbers is expected as an $SU(3)_f$ analog of the deuteron.

The $\overline{10}_f$ and 27_f channels are interesting because the colorspin interaction energy is repulsive. These flavor representations are the smallest ones that contain states with the quantum numbers of the deuteron (the $\overline{10}_f$) and the $I = 1$ two nucleon system (the 27_f). Both appear at $Y = 2$ (which is strangeness zero) in the $SU(3)_f$ weight diagrams shown in Fig. (8). I interpret the positive colorspin interaction energy as evidence that the quark Breit interaction is repulsive at short distances in these channels. Once the three

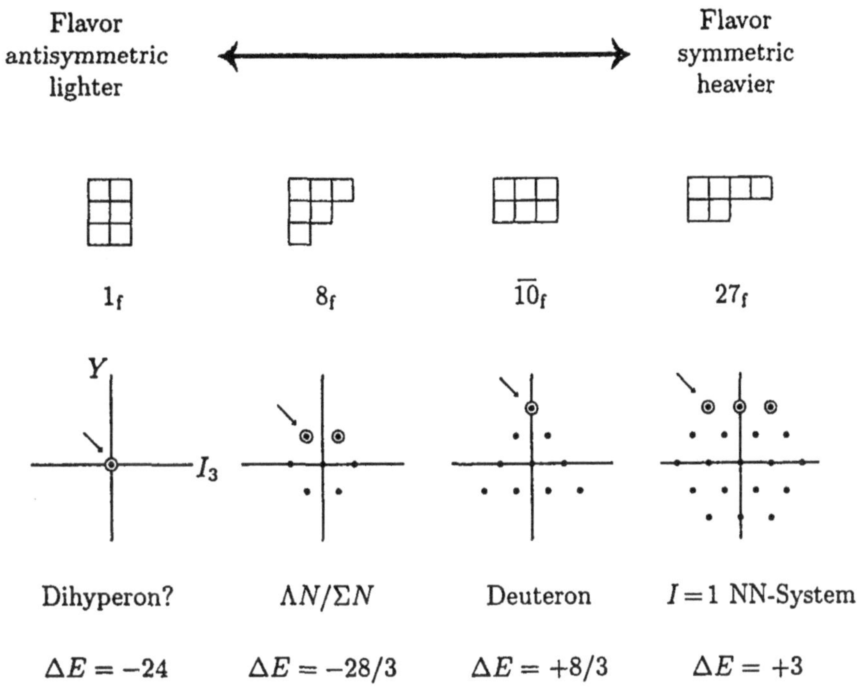

Figure 8. Some of the states of six quarks, classified in $SU(6)_{cs} \otimes SU(3)_f$. Only the flavor Young Diagrams are shown. The colorspin diagrams are conjugate. Again, the most antisymmetric possible flavor state has the most attractive colorspin interaction. The multiplets shown are interesting because the correspond to physically interesting channels: the H, the $\Lambda N/\Sigma N$ system, the deuteron, and the isotriplet dinucleon, respectively.

quark baryons have separated, apparently other effects generate an intermediate range attraction, which binds the deuteron and almost binds the $I = 0$ NN system.

7 Colorspin in Quark Matter

The most interesting recent application of colorspin systematics in QCD is the study of condensates and symmetries in cold, dense matter.[6] The attraction between quark-antiquark pairs is so strong in QCD that $\bar{q}q$ pairs condense in the vacuum, breaking chiral symmetry. Calculations show that the color singlet, spin-zero $(1_c 1_s)$ configuration is the most attractive channel for $\mathcal{H}_{\text{eff}}$. So condensation in this channel is consistent with colorspin considerations.

The $\bar{q}q$ condensate breaks the flavor $SU(3)_{fL} \times SU(3)_{fR}$ chiral symmetry spontaneously down to vector $SU(3)_f$, the standard flavor symmetry of the hadron spectrum. Interestingly, the colorspin interaction energy in the color antitriplet, spin-zero ($\bar{3}_c 1_s$) qq channel is half as attractive. But no condensate develops in this channel in vacuo. It is forbidden by the Vafa-Witten Theorem, which says that vector symmetries cannot break spontaneously in vector gauge theories like QCD.[14]

Things change in the presence of matter.[15,6] At zero temperature and finite density, degeneracy effects become important. At some point close to where the chemical potential for baryon number (μ_B) becomes comparable to Λ_{QCD}, nuclear matter passes over to quark matter. Studies show that the attraction that drives $\bar{q}q$ condensation weakens as μ_B increases. They also show that the qq colorspin attraction in the $\bar{3}_c 1_s$ channel can generate a condensate in the vicinity of the Fermi surface similar to BCS pairing in superconductivity. Alford, Rajagopal and Wilczek argue that this "color superconductivity" is robust – it occurs both in colorspin models of the quark force and in instanton models as well. Of course, QCD at moderate densities is still strongly interacting, and these predictions must be regarded as qualitative. Authoritative statements await a lattice treatment of quark matter at finite chemical potential.

If qq condenses in the $\bar{3}_c 1_s$ channel in quark matter, color symmetry is broken from $SU(3)_c$ to $SU(2)_c$. Of the eight gluons, the three generators of color-$SU(2)_c$ remain massless. The four gluons that are doublets under $SU(2)_c$ get mass a la the Meissner effect. One linear combination of the eighth (color hypercharge) gluon and the photon stays massless (much as the standard model photon survives weak $SU(2)$ symmetry breakdown). $SU(2)_c$ nonsinglet quark or gluon excitations still experience long-range confining forces, but $SU(2)_c$ singlet states do not. Thus, for example, two colors of each quark flavor (the ones that form the $SU(2)_c$ doublet) remain confined, but the third color, say "blue", is a freely propagating fundamental excitation in the medium.

For flavor symmetries, qq condensation has interesting consequences. The two-flavor and three-flavor cases differ dramatically. In two-flavor QCD (obtained by imagining the strange quark mass to be much larger than μ_B) the $\bar{3}_c 1_s$ condensate is a $SU(2)_{fL} \times SU(2)_{fR}$ singlet so it has no effect on the flavor symmetries (though it does violate baryon number). In particular, chiral symmetry is unbroken. Since chiral symmetry is broken spontaneously in the QCD vacuum, there must be a phase transition between $\mu_B = 0$ and the high density phase where superconductivity sets in.

In three-flavor QCD (imagine the strange quark mass to be negligible compared to μ_B) the qq condensate is in the antisymmetric $\bar{3}_c 1_s \bar{3}_f$ channel. Thus the condensate breaks both color and flavor $SU(3)$ symmetries. However, the condensate is invariant under *simultaneous* rotations in color and flavor, which means that a "diagonal" vector $SU(3)$ symmetry remains unbroken. This is the same flavor structure as the low density phase (only a vector $SU(3)_f$ remains unbroken in vacuo with three massless quarks), so it is possible that there is no phase transition from the low density to high density phase. Dubbed "color-flavor locking",[16] this phenomenon leads to several interesting speculations on the phase structure of QCD at finite densities. Implications for observations at RHIC and for the behavior of neutron stars are the subject of much recent interest.[17,18]

With this I've come to the end of this tour of the phenomenological implications of the Breit interaction in QCD. Insights into QCD in the confining domain are precious. The underlying Lagrangian is so simple, the phenomena are so rich, and the theoretical structure is so complex. It is remarkable that so much can be understood qualitatively in terms of the simple effective interaction,

$$\mathcal{H}_{\text{eff}} \propto -\sum_{i \neq j} \underset{\sim}{\lambda}_i \cdot \underset{\sim}{\lambda}_j \vec{\sigma}_i \cdot \vec{\sigma}_j \tag{16}$$

whose origins go back to Gregory Breit in the first half of the twentieth century.

Acknowledgments

This work is supported in part by funds provided by the U.S. Department of Energy (D.O.E.) under cooperative research agreement #DF-FC02-94ER40818.

References

1. A. De Rujula, H. Georgi and S.L. Glashow, *Phys. Rev.* D **12**, 147 (1975).
2. T. DeGrand, R.L. Jaffe, K. Johnson and J. Kiskis, *Phys. Rev.* D **12**, 2060 (1975).
3. R. Jaffe, *Phys. Rev.* D **15**, 267 (1977); *Phys. Rev.* D **15**, 281 (1977).
4. R. Jaffe, *Phys. Rev. Lett.* **38**, 195 (1977).
5. D. Black, A.H. Fariborz, F. Sannino, and J. Schechter, *Phys. Rev.* D **59**, 074026 (1999); hep-ph/9808415.
6. M. Alford, K. Rajagopal, and F. Wilczek, *Phys. Lett.* B **422**, 247 (1998); hep-ph/9711395.

7. C. Caso et al., *Eur. Phys. J.* C **3**, 1 (1998).
8. R.H. Dalitz, "Symmetries and the Strong Interactions", in *Proceedings of the XIIIth International Conference on High Energy Physics* (University of California Press, Berkeley, 1967); reprinted in J.J.J. Kokkedee, *The Quark Model* (W.A. Benjamin, New York, 1969).
9. F. Gürsey and L.A. Radicati, *Phys. Rev. Lett.* **13**, 173 (1964).
10. M. Gell-Mann, *Phys. Lett.* **8**, 214 (1964).
11. N. Isgur and J. Weinstein, *Phys. Rev.* D **41**, 2236 (1990).
12. S. Pakvasa and S.F. Tuan, *Phys. Lett.* B **459**, 301 (1999); hep-ph/9903551.
13. A. Pochinsky, J.W. Negele, and B. Scarlet, *Nucl. Phys. Proc. Suppl.* **73**, 255 (1999); hep-lat/9809077.
14. C. Vafa and E. Witten, *Nucl. Phys.* B **234**, 173 (1984).
15. D. Bailin and A. Love, *Phys. Rept.* **107**, 325 (1984), and references therein.
16. M. Alford, K. Rajagopal, and F. Wilczek, *Nucl. Phys.* B **537**, 443 (1999); hep-ph/9804403.
17. M. Stephanov, K. Rajagopal, and E. Shuryak, *Phys. Rev. Lett.* **81**, 4816 (1998); hep-ph/9806219; *Phys. Rev.* D **60**, 114028 (1999); hep-ph/9903292.
18. M. Alford, J. Berges, and K. Rajagopal, hep-ph/9910254.

NUCLEAR STRUCTURE WITH STABLE AND RADIOACTIVE BEAMS

R.F. CASTEN

Wright Nuclear Structure Laboratory, Yale University, New Haven, CT 06520

Selected frontier areas of nuclear structure physics, both near the valley of stability and in exotic nuclei near the edges of nuclear existence, are discussed. The emphasis is on pervasive themes of current work and on the structural evolution of nuclear systems with nucleon number.

1 Introduction

Nuclear structure physics is enjoying a broad-ranging renaissance. The driving forces behind this are simple emerging ideas about structure itself and its evolution with neutron and proton number, and the new technological capability to study exotic nuclei beyond the valley of stability. This capability extends in many cases to the boundaries of nuclear existence – the proton and neutron drip lines and the heaviest nuclei that can exist.

Studies in nuclei near the valley of stability focus on basic issues of the interplay of the single particle nature of nucleonic motions, the Pauli Principle, and the collective nature of many-body nuclear excitation modes. The development of extensive data on stable nuclei fosters the ability to pursue a "horizontal" approach to nuclear structure that focuses on the *evolution* of structure with changing numbers of constituents.

In exotic nuclei, new nuclear physics, relating to the properties of spatially extended, loosely bound, quantal systems, is rapidly emerging. With the advent, in the next decade, of advanced next-generation facilities for accelerating radioactive nuclear beams (RNBs), the study of exotic nuclei will enjoy an even more dramatic expansion. The study of evolutionary trends will proliferate and will benefit from ideas currently being developed for nuclei nearer stability.

A common way to study any physical system is to stress it and study its response. Stress in the general sense refers to any action that causes the system to deviate from its equilibrium condition. In elementary physics, we do this with springs, oscillators, and pendulums.

Nuclei can be stressed in many ways, as illustrated in Fig. 1. The three axes denote different "stress dimensions". The origin refers to the ground state of any nucleus. The circle (more accurately a hemisphere) centered at the origin denotes the modest stresses induced at low excitation energy and angular

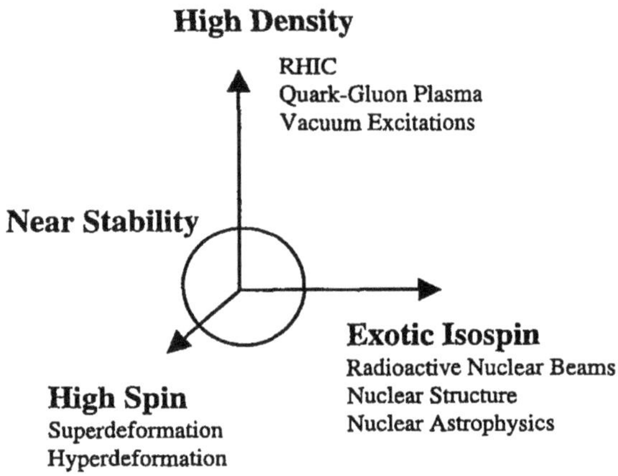

Figure 1. A schematic drawing of the different stress dimensions for nuclei.

momentum for nuclei that are stable or near stability. (In practical terms, these are nuclei that can be studied with reactions or interactions of stable accelerated beams and targets.) Nuclear states in this regime can be rather thoroughly studied with modern accelerators, reactors (neutron beams), and detectors.

This arena has provided the locus for much of nuclear physics to date, complemented by forays along each axis: Heavy ion projectile fragmentation experiments allow us to study the nucleus as a function of temperature, with modest density excursions. Studies of the liquid-gas phase transition, as discussed at this symposium by Gelbke [1], belong in this area. Nuclear excitations at high angular momentum have been a dominant focus of nuclear structure studies over the last quarter century, with spectacular discoveries of backbending and rotation alignment, superdeformation, identical bands, single particle rotation (not really rotation, but high angular momentum states composed of aligned particle orbits), and, recently, magnetic, chiral, and parity doublet rotational bands, The same heavy ion fusion reactions that have led us along the angular momentum axis have also given us one route out of the valley of stability since, by virtue of the curvature of the valley toward neutron rich species, the fusion of lighter nuclei inevitably leads to the left ("northwest") of stability. The other means of access beyond the valley is by fission of heavier nuclei (e.g., the actinides) which gives pockets of neutron rich nuclei.

In the present perspective on nuclear structure we will start with some general comments on approaches to modeling nuclei. The region of nuclei accessible with stable beams and targets remains at the core of much recent work in nuclear structure and is the focus of Section 3 of this overview. In Section 4 we will discuss the exciting types of studies far out along the "isospin" (exotic N/Z composition) axis currently being made possible by technological breakthroughs that allow us to produce and accelerate a wide variety of exotic, β-unstable nuclei far from the valley of stability.

2 Nuclear Structure

2.1 Approaches to Nuclear Structure

Historically, there have been three basic approaches to nuclear structure, illustrated in Fig. 2. All represent approximations and simplifications of the nuclear many body problem, at differing levels and with somewhat different goals, scope, and character.

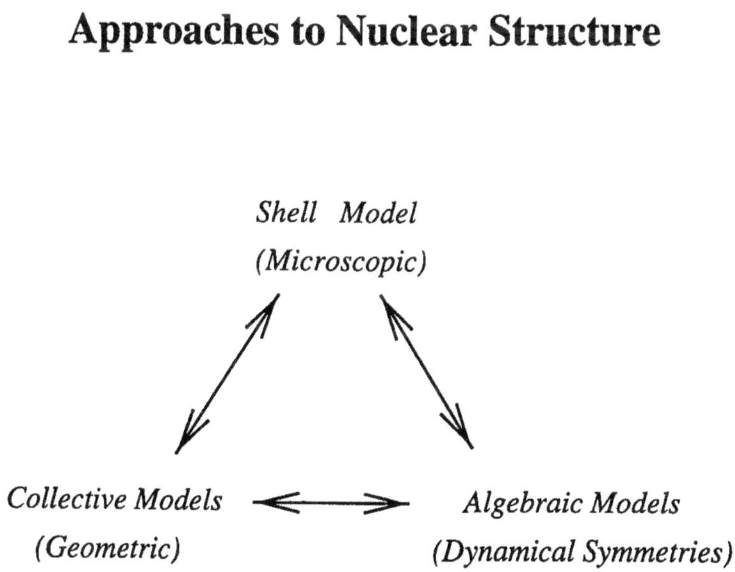

Figure 2. Theoretical approaches to nuclear structure.

In principle, we would like to understand the nucleus in terms of forces resulting from the interactions amongst quarks and gluons. We are far from such a goal. At the next level, we could take free nucleon-nucleon interactions and deduce nuclear structure from a Hamiltonian representing the sum of all nucleonic interactions. Except with great effort, and in the lightest nuclei, this too is yet impractical. Hence we are left with effective nucleon-nucleon interactions in the nuclear medium, which has a density far different from that of free nucleons. Even with such effective interactions the nuclear A-body problem is too complex for most nuclei.

Hence, we resort to simplifications. In the shell model (or better, the independent particle model) the set of 2-body effective interactions amongst all pairs of nucleons is replaced by a 1-body potential. The 3-dimensional Schrodinger equation is solved by separating its radial and angular coordinates. Today, in most applications, this is the standard microscopic model of the nucleus. The one body potential (usually taken to have Woods-Saxon form plus a non-central spin-orbit term) leads to a shell structure of levels as illustrated in Fig. 3, where each orbit is characterized by the quantum numbers n, ℓ, j, which are the principal quantum number (related to the number of nodes in the wave function in the radial degree of freedom), the orbital angular momentum ℓ (in units of $\hbar$) and the total nucleon angular momentum j=$\ell \pm {}^1/_2\hbar$.

The level groupings have a simple origin. In any plausible 1-body potential the orbit energy will increase with both n and ℓ. Higher n values mean more nodes and hence a shorter effective wave length: higher ℓ values imply a higher centrifugal force, a hence a narrower effective potential squeezed to larger radii. Hence two orbits, one with larger n and smaller ℓ – that is (n, ℓ) and $(n + \Delta n,\ \ell - \Delta\ell)$ can have similar energies. For most potentials of use for nuclei the relation $\Delta\ell = 2\Delta n$ characterizes close lying doublets, such as 2s-1d, 2p-1f, 1g-2d-3s. Since each orbit can contain at most (2j+1) nucleons of a given type, nuclei with successively larger numbers of nucleons fill this single particle sequence to higher and higher levels. Nucleon numbers (magic numbers) where large gaps occur in the single particle levels correspond to magic nuclei which have very special properties, rigidity to excitations and spherical structure. They are analogous to the inert gases of atomic physics.

Of course, the replacement of the set of all nucleon-nucleon interactions by a 1-body potential cannot be exact and one invariably needs to introduce "residual" interactions. Because of the stability of closed shells, however, the single particle model allows an important paradigm shift from the zero of particle energy to that of the last closed shell and thus, in most cases, one can treat the residual interactions amongst just the valence nucleons.

These residual interactions lead to correlations in the nucleonic motions and coherence in the nuclear wave functions, that is, to collectivity. Collective correlations can be described in terms of macroscopic oscillations of the nucleus and this has spawned the development of geometric collective models, the second leg of the structural triumvirate in Fig. 2. These two pictures, microscopic and geometric, are depicted schematically in Fig. 4. Typically, geometric models focus on an equilibrium nuclear shape along with vibrations and rotations (if the equilibrium shape is non-spherical) about it.

Large amplitude vibrations involving all the nucleons constitute the giant resonances located at 8-30 MeV of excitation. In shell model terms, they involve particle-hole excitations across major shell gaps. Smaller amplitude

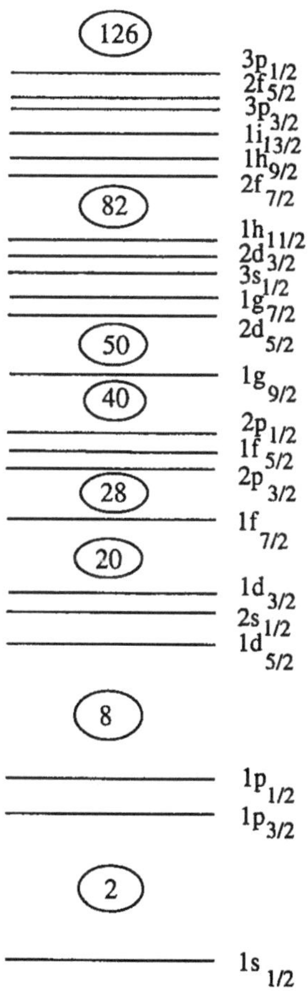

Figure 3. Shell structure of nucleonic orbits.

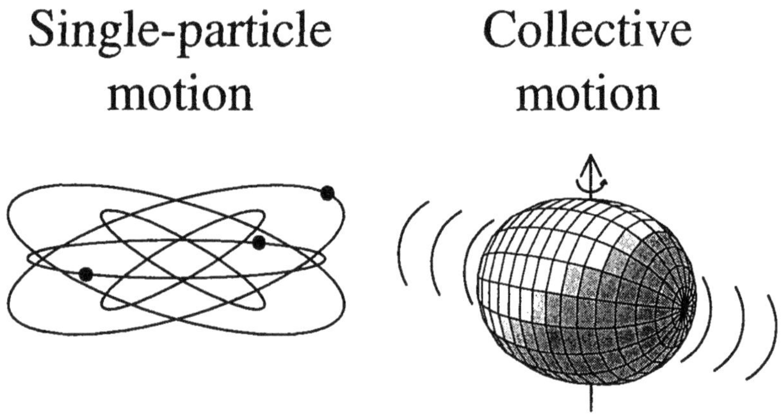

Figure 4. Complementary microscopic and geometric visualizations of nuclei. I am grateful M.A. Caprio for this figure.

vibrations (primarily involving in-shell excitations) constitute the lower lying collective modes such as (for the quadrupole deformed case) the so-called β and γ vibrations. Rotation can be superposed on such vibrational excitations, leading to the appearance, in deformed nuclei, of sequences of band structures, often extending to quite high spin. Such a sequence of rotational states is indicated pictorially in Fig. 5, along with the lowest two intrinsic excitations of the ground state band. Note the characteristic signature of rotational behavior, an $R_{4/2} = E(4_1^+)/E(2_1^+)$ ratio of 3.33. The lower part of the figure depicts vibrational or phonon modes of a spherical nucleus. We will discuss multi-phonon modes (in deformed nuclei) below.

The third approach to structure is an algebraic or group theoretic one which focuses on the concept of dynamical symmetries, as noted in Fig. 2. The idea of dynamical symmetries, resulting from group chain decompositions, gives powerful paradigms where energy eigenvalues and transition rates can be calculated analytically. The best known and most successful example of this approach is the Interacting Boson Model (IBM) proposed by Iachello and Arima [2] in 1974 and successfully tested, refined and expanded ever since.

The IBM assumes that low lying collective structure is determined by the valence nucleons but it then proceeds by an enormous truncation of the shell model space. It assumes that the key ingredient in collectivity is pairs of valence fermions that act as bosons and, further, it restricts these pairs to two

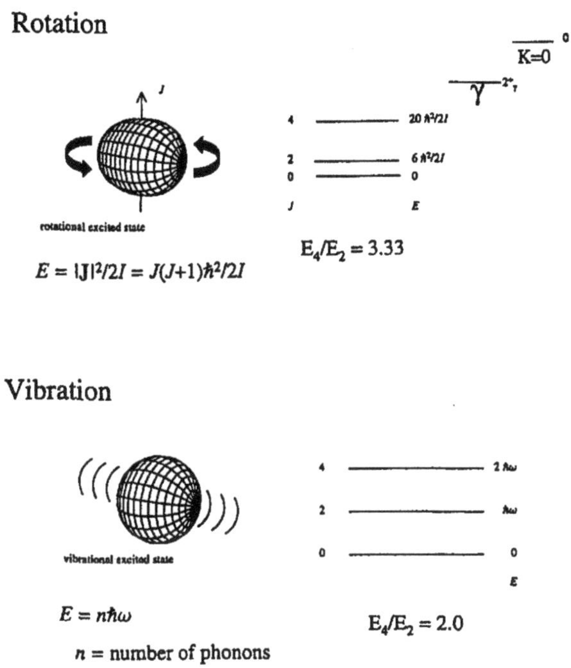

Figure 5. Rotational and vibrational excitations of nuclei.

angular momentum states L=0(s) and L=2(d). [Higher angular momentum bosons, as well as negative party bosons (p, f), have been introduced in extensions to the original IBM, as has the distinction between proton and neutron bosons in the so-called IBM-2 model.] The IBM Hamiltonian involves boson energy terms and interactions up to 2-body between the bosons.

In group theory terms, the s and d bosons span a six-dimensional space corresponding to the group U(6). U(6) and its physically viable sub-groups leads to three dynamical symmetries U(5), O(6), and SU(3). These correspond, respectively, to the geometric concepts of a harmonic vibrator, an axially asymmetric (γ soft) rotor and an axially symmetric rotor with particular degeneracies of vibrational modes. In each symmetry, each step in the sub-group decomposition introduces a new quantum number label, and splits some of the degeneracy of the higher group. In these dynamical symmetries the Hamiltonian is written in terms of Casimir operators of each sub-group. Hence the energies (as well as the wave functions and transition rates) can all be written down analytically. The group decomposition is illustrated schematically for O(6) on Fig. 6. The levels of the resultant spectrum are labeled by σ, τ, and J quantum numbers for the O(6), O(5), and O(3) sub-groups. All states of a given nucleus (boson number N) belong to a single representation of the parent group U(6).

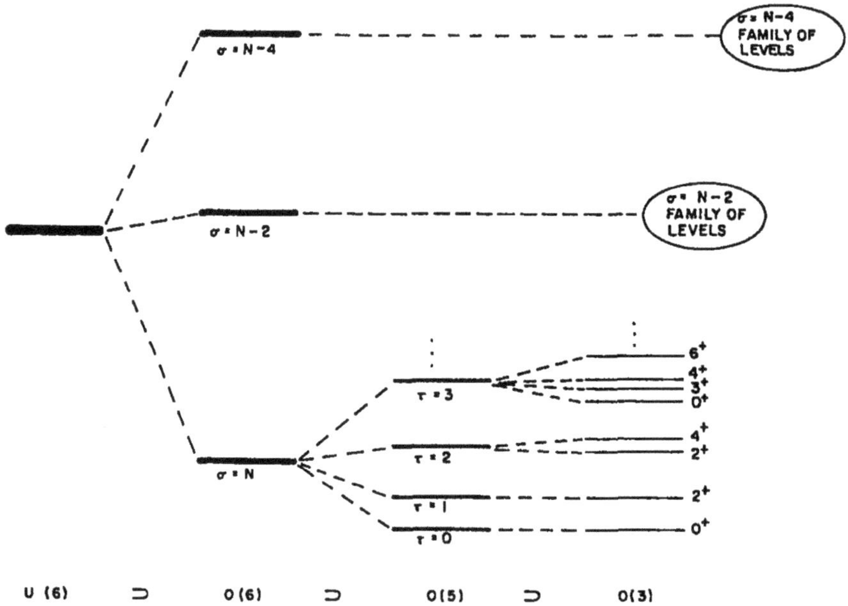

Figure 6. Group decomposition, level families, and quantum numbers, for the dynamical symmetry O(6).

All three of these dynamical symmetries have been discovered experimentally. Moreover, the evolution of structure in complex transition regions [e.g., O(6)→SU(3) or U(5)→SU(3)] is trivial to calculate in the IBM, often in terms of the variations of a single parameter, and nuclear transition regions have been successfully interpreted with this approach [3].

The shell model is microscopic. Geometric models are macroscopic. The IBM has elements of both, namely, its excitations have a close correspondence to the vibrations and rotations of geometric models. However, it also represents a very specific truncation of the shell model and its predictions have specific and characteristic variations with valence nucleon number (even with constant parameters in the Hamiltonian).

A number of years ago Iachello and collaborators extended the idea of dynamical symmetries (for even-even nuclei) to odd-A nuclei and proposed that a kind of Bose-Fermi dynamical supersymmetry (SUSY) [4] might actually be found in atomic nuclei. Early experimental efforts in the 80s and early 90s showed some evidence for nuclear SUSY but the clinching experiment has just been reported by a Swiss-German collaboration [5]. This spectacular work is discussed extensively by Iachello elsewhere in this volume. [6]

2.2 Structural Benchmarks – Transition Regions

Throughout the history of nuclear physics, structural interpretation has relied on a series of "benchmarks" in the zoological classification of nuclei. Some, such as the magic numbers, were known quite early and framed the interpretation of many pioneering studies (e.g., of single particle reactions) in the 1960s and 1970s.

Only recently has the fragility of magicity far from stability emerged as a new feature of the nuclear landscape – a feature that is, at the same time, both exciting in its implications for the evolution of structure and the study of nucleonic interactions and daunting in that key milestones or guideposts may be lacking for early third millennium explorers in these new lands.

Another class of benchmarks is equally important although much less predictable in terms of N,Z location, namely nuclei with structures close to the idealized paradigms of simple models such as the harmonic vibrator [7] and symmetric rotor [8] that we have discussed. As with doubly magic nuclei, such benchmarks facilitate the interpretation of nuclei close to them in structure. In the context of geometrical models (shape descriptions) of nuclei, the excellent rotors of the rare earth and actinide regions provide beautiful examples. In the context of the IBA, perhaps the best known examples are the so-called γ-soft [O(6)] nuclei in the Pt-Os and Xe-Ba regions. Any ideal paradigm plays the role of highlighting or magnifying deviations from it. In the rare earth nuclei, for example, the rotor paradigm has allowed recognition and study of subtle but important interactions (rotor-vibrator coupling or non-adiabaticity effects in the separation of microscopic and macroscopic degrees of freedom) that would otherwise have been invisible. Application of the vibrational paradigm in nuclei such as Cd brings to light evidence for anharmonicities and intruder states.

There is a third set of benchmarks that could in principle be of great value but has not really been developed, namely the recognition of characteristic patterns in different classes of spherical-deformed transition regions, that is, regions of nuclei where the structure evolves from one paradigm to the other. A classification of transition regions, and their distinguishing empirical manifestations, is an important current goal, as we enter new regions of nuclei, since a far greater number of nuclei will be transitional or intermediate in structure than those attaining the idealized structural symmetries. We now briefly discuss a start in this direction.

We want to ask if there are characteristic defining traits of typical shape transition regions of nuclei, such as, for example, a vibrator to symmetric rotor [U(5)→SU(3) in the IBA] transition or a vibrator to γ soft rotor to symmetric rotor [U(5)→O(6)→SU(3) in the IBA] transition. A related question is whether these shape transitions are gradual or rapid as a function of A, N, or Z and, if rapid, whether they can be thought of as true (albeit muted) phase transitions in finite nuclear systems. Our discussion is based on that in Ref. 9.

We start by considering Fig. 7 which shows a generic type of transition from a near-closed shell nucleus to a typical deformed rotor. As the number of valence nucleons, and in particular as the valence nucleon product N_pN_n, grows, the structure evolves from left to right. Configuration mixing amongst shell model configurations grows and coherence and collectivity develop. The collective states descend in energy, and transition rates increase. The value of $R_{4/2}$ is typically $\sim$ 1.5 for singly magic nuclei with simple configurations of the type $|j^nJ>$. As valence nucleons of both types are added, vibrational structure develops. For a set of phonon and multi-phonon states the typical vibrational pattern in the third panel of Fig. 7 emerges with 2-phonon states typically a little above twice the 1-phonon energy and comprising a fairly compact $0^+, 2^+, 4^+$ triplet. With further addition of valence protons and neutrons the equilibrium shape develops static deformation, and rotational spectra appear built both on the ground state and on excited collective vibrational modes, such as the classic β and γ vibrations. In the deformed region, $R_{4/2} \to 3.33$.

However, this generic scheme encompasses quite different modes of structural evolution. For example, an $R_{4/2}$ value of 2.5 can represent an anharmonic vibrator or an axially asymmetric rotor. The question arises whether we can identify distinct evolutionary morphologies.

Consider Fig. 8 which shows $E(2_1^+)$ and $R_{4/2}$ for isotopes of Gd, Th, and Ba (N<82) as a function of the the number of valence nucleons. Both observables change rapidly in Gd: $E(2_1^+)$ drops suddenly between N_n= 6 and 8 while $R_{4/2}$ rises from a value below 2.5 towards 3.33 in short order. This is consistent with our understanding of the rapid onset of deformation at N=90 in this region and of the underlying mechanism in which there are two competing configurations, spherical and deformed. The latter involves excitations across the Z=64 sub-shell gap into the proton $1h_{11/2}$ orbit and, therefore, near N=82, requires considerable energy. However, in the Federman-Pittel picture [10], as further elaborated by Heyde et al [11], the monopole p-n residual interaction, which is strong between spin-orbit partners such as $\pi 1h_{11/2} - \nu 1h_{9/2}$, acts to lower the $\pi 1h_{11/2}$ energy. This lowers the energy of the deformed configuration as it likewise eradicates the gap at Z=64. At some point in adding neutrons (N=90 in particular) the deformed configuration crosses the spherical to become the equilibrium configuration. At this point the deformation suddenly jumps from $\beta \sim 0$ to nearly saturation values. Such a rapid switching on of deformation, and the concept of a level crossing, is intimately related

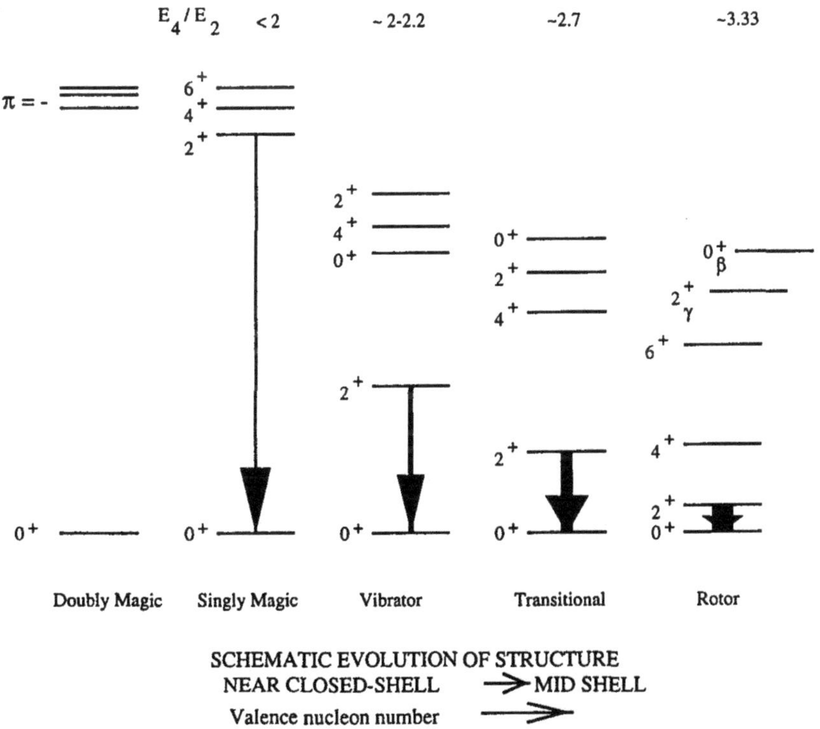

Figure 7. Schematic illustration of structural evolution in nuclei and some of the paradigms that characterize it.

to the idea of a first order phase transition. Naturally, in finite nuclei, the abruptness of any possible phase transition will be muted. Nevertheless, there has been recent evidence [12,13] that nuclei near Gd, such as Sm, show behavior very similar to a real phase transition. One can even identify a kind of order parameter, a control variable and a critical point [13,14]. Moreover, the nucleus ^{152}Sm shows evidence of phase coexistence [12]. Further discussion of this new development is beyond the scope of this paper.

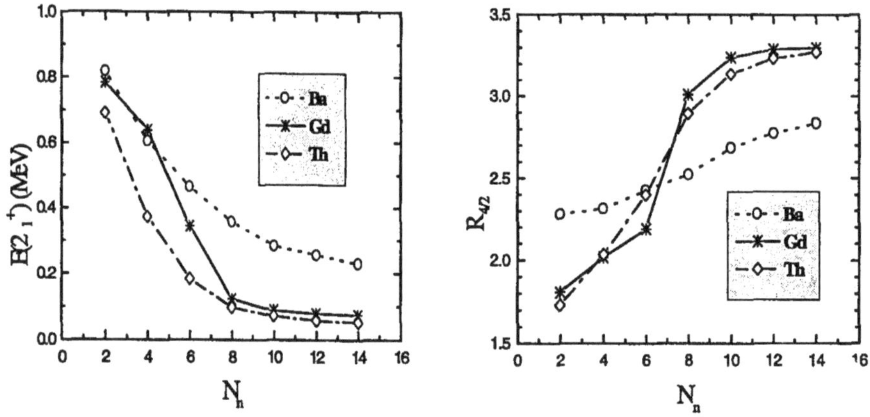

Figure 8. Evolution of structure in Gd, Th, and Ba in terms of two simple observables plotted against valence neutron number. Based on Ref. 9.

Now let us consider Th in which the onset of deformation is thought to be gradual due to a gradual build-up of coherence without any sub-shell gaps or any separation of the Hilbert space into two partially distinct spaces (as in Gd). As seen in Fig. 8, $E(2_1^+)$ drops gradually. However, the behavior of $R_{4/2}$ is surprising. Its rise is nearly as abrupt as in Gd.

This can be easily understood by recognizing that $E(2_1^+)$ and $R_{4/2}$ sample somewhat different physics: $E(2_1^+)$ is sensitive to the coherence in the wave function and, once deformation sets in, is proportional to ${}^1/_{\mathcal{I}}$, where $\mathcal{I}$ is the moment of inertia. Hence $E(2_1^+)$ continues to decrease as β grows. In contrast $R_{4/2} \to 3.33$ as soon as deformation sets in, *independent* of its magnitude. In the rotational energy expression for $R_{4/2}$ the moment of inertia cancels out leaving $R_{4/2}$ insensitive to the actual magnitude of β.

Hence a gradually changing 2_1^+ energy and a sharply changing $R_{4/2}$ can signal the gradual evolution of structure to more and more deformed shapes. However, if both change abruptly, one suspects a crossing of configurations.

The Ba (N<82) nuclei present a third morphological class. Here, both $E(2_1^+)$ and $R_{4/2}$ change gradually and neither reaches the extreme values of good rotors as in the Gd or Th regions. The reason is that a new degree of freedom enters, namely axial asymmetry. The value $R_{4/2} = 3.33$ only applies to *symmetric* rotors. For deformed nuclei with large $< \gamma >$ values, $R_{4/2}$ is substantially < 3.33. In fact for $< \gamma > \sim 30°$, $R_{4/2} \sim 2.5$. Also with finite $< \gamma >$ values, $E(2_1^+)$ increases for the same quadrupole deformation β. Hence in Ba the gradual changes in $E(2_1^+)$ and $R_{4/2}$ are due to the gradual onset of deformation along with large $< \gamma >$ values. (In fact Ba is γ-soft: the nucleus oscillates in γ from $\sim 0°$ to $\sim 60°$.) With decreasing neutron number (increasing N_n) the deformation continues to increase and $< \gamma >$ decreases (to perhaps $< \gamma > \sim 20°$) leading to the gradual drop in 2_1^+ energy and rise in $R_{4/2}$.

These three examples of evolutionary morphology are guidelines only and may be somewhat difficult to distinguish in new regions: the differences between them are quantitative only. But hopefully this embryonic classification scheme will encourage further study of different mechanisms and pathways for spherical-deformed transition regions. Not only would this be valuable in known nuclear regions near stability, but it could prove critical in new regions of exotic nuclei that will become accessible in the next decade but where data will be much more limited.

3 Studies With Stable Beams and Targets

Those issues that are of greatest current interest in nuclear structure near stability will also, of course, continue to be of interest in exotic nuclei. However, they will be more difficult to address in the latter. Therefore, the initial thrust of the study of exotic nuclei is to get the first, most basic, data on newly accessible nuclei. As we will discuss below, with RNB intensities orders of magnitude weaker than for stable beams, we will initially have to be content in many cases with just such basic data. However near stability, we have access to higher data rates, and hence to experimental techniques, such as coincidence experiments with large detector arrays, or Doppler techniques for lifetime measurements, that for many reasons(count rate, solid angle, etc.) are unavailable or much more difficult and limited in scope with RNBs. Therefore, we are able to study a richer variety of questions and excitation modes. In this section we comment briefly on a few of these to give an illustration of the possibilities.

3.1 High Spin States

Nuclei at high angular momentum display a rich variety of phenomena that have been actively studied in recent years. The best known is probably that of superdeformation which was discovered [15] about 15 years ago in ^{152}Dy and has now been studied in a variety of nuclei ranging from A~60 upwards. Superdeformation entails ellipsoidal distortions corresponding roughly to a 2:1 ratio of major to minor axis lengths, that is, to a quadrupole deformation of β ~0.6. (The predicted, and long sought, phenomenon of hyperdeformation corresponds to an extremely elongated 3:1 ratio.)

In nuclei rotating at low angular momentum there is little difference if a nucleon orbits with or opposite to the rotational motion. This is the origin of the two-fold degeneracy of Nilsson orbits and of the energy gain in pairing in which two nucleons in the same orbit travel in time reversed orbits coupled to total two-particle angular momentum zero. However, at high rotational frequency, as in most superdeformed states, there is a distinct energy gain to be had by breaking pairs of zero-coupled nucleons and aligning both their angular momenta parallel to the rotation axis. This allows formation of a state of total angular momentum $\overline{J} = \overline{J}_{Rot} + \overline{J}_{part}$ with a significantly lower $\overline{J}_{Rot}$. That is, since E(rot)~J(J+1), there will be some angular momentum at which energy will be gained in the "rotation aligned" state compared to the ground state since $|J_{Rot}|$ can then be lower for a given $|J_{Total}|$. As angular momentum further increases, successive nucleon pairs recouple from $J_{pair} = 0$ to $J_{pair} = (2j_{nucleon} - 1)$. Superdeformed states are often therefore states with no pairing condensate and moments of inertia that approach the rigid body values rather than the lower values resulting from the presence of pairing. The study of pairing collapse in high spin states has long been an active area of study.

These superdeformed bands, populated from above, lead to cascades of γ-rays connecting states of successively lower angular momenta. Consider the rotational energy formula E(rot)~ $J(J+1)$. Gamma ray energies correspond to level energy differences, usually between states with $\Delta J = 2$. The rotational formula then gives $E_\gamma(J) \equiv E(J) - E(J-2) \sim 2J - 2$. From this we get that successive γ-ray energies themselves, such as $E_\gamma(J)$ and $E_\gamma(J-2)$ differ by a constant. Hence, a sequence of γ-ray transitions shows up as a spectrum of peaks equally spaced in energy which, instantly, provides a visual signature of a rotational band. Moreover, the spacings between successive γ-rays is proportional to the inverse moment of inertia and therefore to the ellipsoidal deformation of the nucleus. A beautiful example of such a spectrum with constant γ ray spacings is seen in Fig. 9 for γ ray energies above

$\sim$ 800 keV. The contrast of this remarkable example of rotational behavior in nuclei with the jumble of γ-ray lines below 800 keV, which connect states in different bands of normal deformation, is striking indeed.

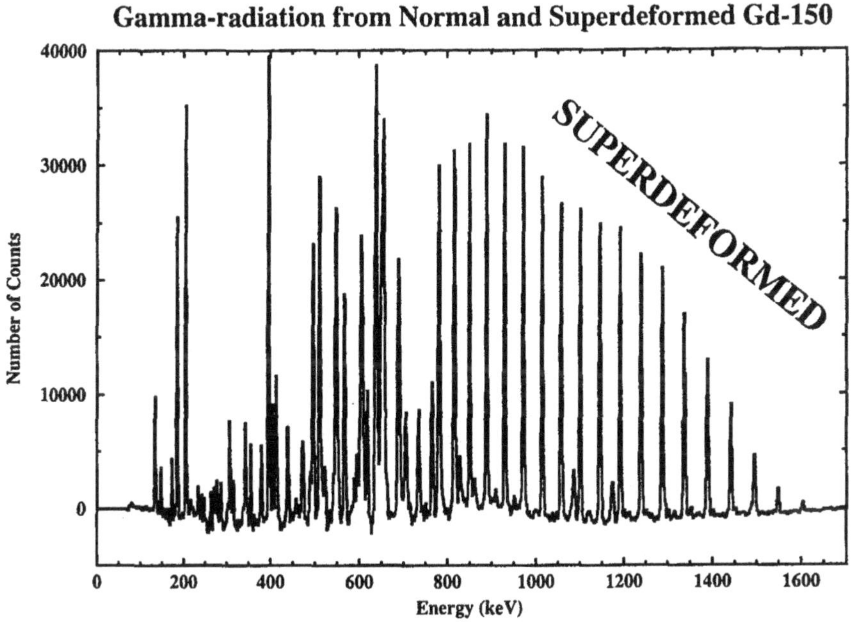

Figure 9. Gamma ray spectrum of normal and superdeformed states in ^{150}Gd. Based on Ref. 16. I am grateful to C.W. Beausang for this figure.

Aside from access to exotic nuclear shapes such as superdeformation and hyperdeformation the study of nuclei at high spin allows a fascinating window on the coupling and decoupling of particle and collective motion. We have just seen an example of this in terms of pairing collapse but it has a more general aspect as well. In the study of nuclei at low spin, the assumption is made that the single particle motion of individual nucleons and the rotational motion of the nucleus as a whole can be separated because of their different time scales (reflected also in their characteristic energy scales since energy and frequency are related). It is imagined that the nucleus is effectively stationary during the much more rapid orbital motion of a nucleon. This enormously simplifies model descriptions and leads to easily testable predictions of both rotational motion and of particle excitation modes. At higher spins, though, nuclear rotational periods and particle orbit times are not so different and the adiabatic approximation breaks down. This transition from non-adiabaticity to separability can be studied in heavy ion fusion evaporation reactions in which a compound nucleus is initially created at high spin (say 40-60$\hbar$), emits a few nucleons, and then γ-decays down to the ground state in a series of dipole and quadrupole cascades. In doing so, the nucleus undergoes something like 10^{11} rotations in a fraction of a microsecond, and passes through stages where there is little if any particle-rotation decoupling, to an intermediate regime, culminating in an almost perfectly adiabatic situation. We are studying effects here that take place on the 10^{-18} second time scale. Moreover, at the same time as adiabicity evolves so does shape, often from an initially spherical compound nucleus through (within a relatively few rotations) a superdeformed shape, through normally deformed shapes, to a quiescent (spherical or normally deformed) resting point.

One specific feature of rotational motion discovered about a decade ago and still not satisfactorily explained is that of identical bands [17] in adjacent nuclei. These are bands in which the energy spacings in two nuclei, for corresponding states, are virtually equal (to a few tenths of a % or less). The phenomenon is, in fact, also known in non-adjacent nuclei and at even low spin. That it is surprising is clear from the simple classical expression for the moment of inertia $\mathcal{I} \sim MR^2$ which, for nuclei, goes as $AR^2 \sim A^{5/3}$. For A$\sim$ 150, $\Delta\mathcal{I}$ (for $\Delta A = 1$)is 1%, an order of magnitude larger than the changes in $\mathcal{I}$ in identical bonds. Efforts to explain this in terms of counterbalancing pairing and deformation effects, pseudo-spin, and even supersymmetry have been proposed, with no explanation yet that accounts for the wide ranging occurrence of the phenomenon.

3.2 Low Energy Proton-Neutron Modes

In recent years the study of medium and high spin states has taken new directions. One particularly interesting phenomenon is that of "magnetic rotation" which is not rotation in the usual sense but is an effect of proton-neutron degrees of freedom. [18] The idea is shown schematically in Fig. 10, which also illustrates the origin of another label for this phenomena - shears bands. Imagine independent single particle proton and neutron states with angular momenta oriented perpendicular to each other, giving a resultant total angular momentum somewhat greater than either alone. Now imagine that the proton and neutron angular momenta vectors approach each other towards coplanar alignment. The resultant total angular momentum will increase and the effect is like the closing of a shears. This effect has been observed [19] in nuclei in the Pb region. Recently, the first example in a trans-Pb nucleus, ^{205}Rn, has been discovered [20] by the Yale group.

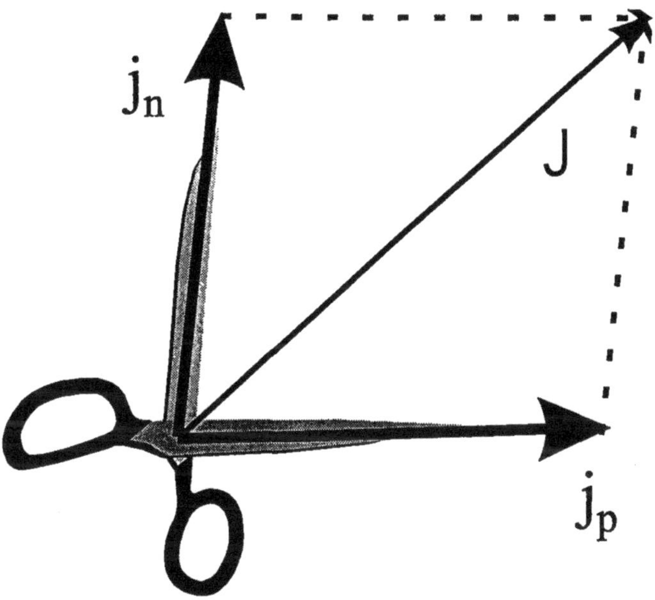

Figure 10. Illustration of the concept of "magnetic rotation". I am grateful to R. Kruecken for this figure.

Another proton-neutron excitation, in this case at low spin, is the so-called "scissors" mode, which, in deformed nuclei, consists of an oscillatory vibration of deformed proton and neutron fields with respect to each other as illustrated in Fig. 11. This mode carries a J^{π} of 1^{+} and has been discussed at length in terms of the IBA-2 model, which distinguishes proton and neutron bosons. Here the oscillatory motion is amongst the valence nucleons. As a displacement of protons against neutrons it is most easily excited by electromagnetic probes such as (γ, γ') or (e, e') reactions. Via such data, a wealth of these states has been found, typically at $\sim$ 3 MeV of excitation energy. The B(M1) strength is typically fragmented into several pieces, but the summed strength reaches collective magnitudes of roughly a nuclear magneton [21].

In spherical nuclei the scissors mode is a kind of 2-phonon mode composed of the coupling (to $J^{\pi}_{total} = 1^{+}$) of the normal, proton-neutron symmetric, 2_1^{+} state with the lowest mixed symmetry 2^{+}_{sc} state. A schematic level scheme for these excitations is indicated in Fig. 12. Since two-phonon states de-excite by the destruction of a single phonon, the 1^+_{sc} state can decay by either of two routes, to its parent one-phonon states, and with B(E2) values the same as those from the one-phonon states to the ground state. This is also illustrated in Fig. 12, where E2 transition arrows slanted in the same direction have equal B(E2) values. The 1^+_{sc} mode also decays by a direct M1 transition to the ground state.

The study of shears bands, mixed symmetry states, scissors modes, and of a new concept called chiral bands, is a very active area of current research in nuclear structure.

3.3 Multi-Phonon States

One of the most important issues currently being addressed in low spin nuclear structure is that of multi-phonon quadrupole vibrational states. Such vibrational modes at low energy are known in both spherical and deformed nuclei.

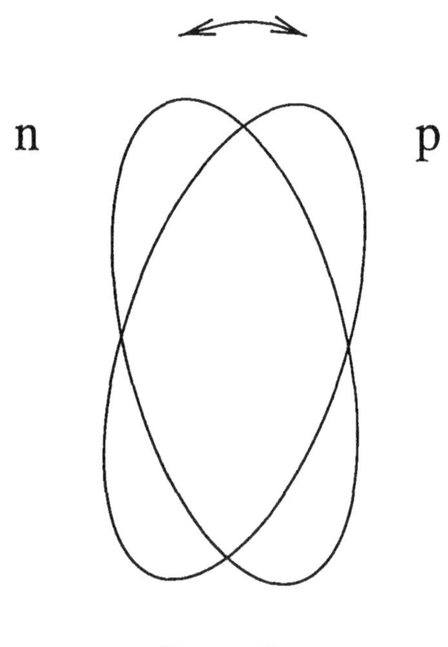

Figure 11. Illustration of the "scissors mode".

In principle, multi-phonon states can be created by superposing identical vibrational quanta. Macroscopically, since these modes are integer spin bosonic excitations, this presents no problem. However, phonon excitations must also be able to be described microscopically, in terms of particle-hole or quasiparticle fermionic excitations. Multi-phonon states therefore entail multi-particle multi-hole excitations. The interplay of the Pauli Principle and single particle motion may impact the availability of some components in the multi-phonon wave functions, thereby impacting the harmonicity and collectivity of such states, or in some cases their very existence. Indeed, until recently, the most cited microscopic models of phonon excitations in deformed nuclei predicted their almost total fragmentation.

However, starting with the work of Aprahamian [22] on the identification of multi-phonon (3-phonon) quadrupole excitations in spherical nuclei in the 1980s, the study of such collective modes has developed to become a major focus of nuclear structure studies. Numerous examples in spherical nuclei are now known, the best probably being ^{114}Cd where some states with as many as five phonons, with essentially intact collectivity, have been identified. [23]

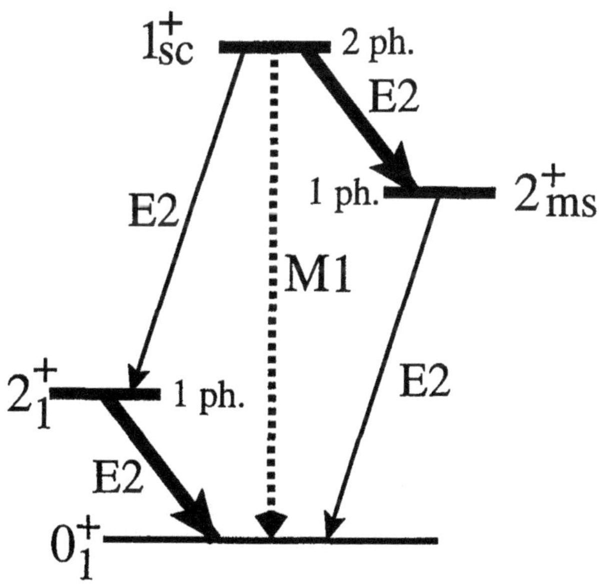

Figure 12. One and two phonon scissors mode and mixed symmetry excitations with their characteristic decay modes. I am grateful to N. Pietralla for this figure.

In deformed nuclei the situation is murkier, for several reasons. First the single phonon modes tend to occur higher in excitation energy and hence double phonon states would appear amidst a higher density of normal, non-collective levels. Also, whereas the occupation of spherical j-orbits allowed by the Pauli Principle is (2j+1), in deformed nuclei the m-substate degeneracy is broken and only two particles are allowed (in time reversed orbits) in a given deformed single particle level. Hence Pauli effects come in sooner and more strongly.

The lowest collective mode of most deformed nuclei in the γ-vibration, with angular momentum projection quantum number K=2. The two phonon mode should exist, somewhere near twice the one-phonon energy (i.e., $\sim 2\hbar\omega$), in two forms, with K=4 and K=0. Several good candidates for K=4 modes have been discovered, the first being at 2.055 MeV in ^{168}Er using the GRID technique in Grenoble to measure the absolute γ-ray transition rate (the collectivity) for the decay transition to the one-phonon γ state [24]. The situation with K=0 modes is more difficult. A few candidates are known, for example, the fourth excited 0^+ state [25] in ^{166}Er, and a partly fragmented mode in ^{168}Er [26].

Given these discoveries, nuclear models have themselves evolved and now typically predict probability densities of perhaps 30%-40% in some cases for two-phonon character in a given K=4 or K=0 state. Current experimental studies are underway in several laboratories, including WNSL at Yale, to search for additional two-phonon states or to test proposed candidates. From the Yale work, we briefly present two examples of recent results.

Figs. 13 and 14 show partial level schemes for ^{162}Dy and ^{158}Gd with new data on multi phonon $K = 0^+$ double γ vibrations. A necessary, though not sufficient, condition for a two phonon state is preferential decay to the 1-phonon state rather than to the ground state. [As we discussed above, the electromagnetic E2 operator has a selection rule $\Delta N_{ph} = \pm 1$.] Hence, one looks for branching ratios such as $R = B(E2 : 0^+ \rightarrow 2^+_\gamma)/B(E2 : 0^+ \rightarrow 2^+_g)$, or similar ones for other rotational band members, for which R$\gg$1.

The data on ^{162}Dy and ^{158}Gd illustrate two complementary ways in which the power of new γ-ray spectrometers can correct existing data and obtain much higher sensitivity. In ^{162}Dy, the $2^+(K = 0)$ level at 1453 keV had previously been assigned a decay branch to the 4^+_γ level that indicated an R value (corrected with the Alaga rules to $0^+ \rightarrow 2^+_\gamma, 2^+_g$ decays) of 10,000, an incredibly, and suspiciously, large result. An experiment [27] using the Yale YRAST Ball [28] high efficiency clover detectors in $\gamma - \gamma$ coincidence mode showed that this γ-ray in fact was misplaced in the level scheme, thereby eradicating any evidence for 2-phonon character. However the same β-decay experiment found another, weaker γ-ray, of 512 keV, that can be definitely placed between the 0^+ level at 1400 keV and the 2^+_γ level at 888 keV. The experimental difficulty of this measurement can be appreciated when one realizes that β-decay γ spectra are accompanied by enormous backgrounds of 511 keV annihilation radiation: the 512 keV line can only be seen in coincidence mode. As indicated in Fig. 13, the R value resulting from this new placement is 10±5 which plausibly points to the 1400 keV excitation as a good candidate for a $\gamma\gamma$ $K = 0$ mode.

The ^{158}Gd level scheme illustrated in Fig. 14 shows three γ rays between one of the levels of a known $K = 0$ band and the γ band. These three placements give large R values, again suggestive of 2-phonon character. Recent studies with the GRID technique at Grenoble [29] have, however, shown this scheme to be incorrect. To understand the results of the Grenoble work one has to understand that the typical energy resolution of the Ge semi-conductor detectors used almost universally for γ-ray energy measurements is about 1.5 keV for a γ ray of 300 keV. The γ ray detector at the ILL in Grenoble works on the entirely different principle of energy determination by Bragg diffraction coupled with the measurement of the γ-ray diffraction angle to enormous pre-

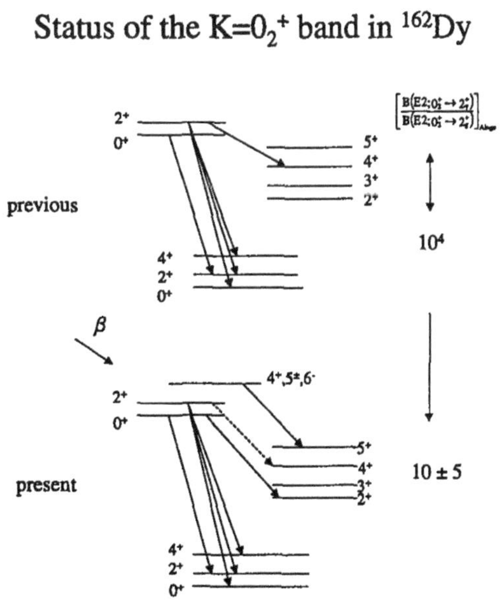

Figure 13. Partial level scheme of ^{162}Dy. Based on Ref. 27.

cision ($\sim 10^{-10}$ radians). With this device, the energy resolution at 300 keV can be just a few electron volts! The ^{158}Gd experiment [29] showed that the three transitions to the γ band shown in Fig. 14 in fact do *not* fit in energy. The deviations (and the uncertainties on these deviations) from the energy differences of the levels (established from other transitions) are 62±16, 32±25, and 32±8 eV. Hence, all positive evidence for 2-phonon character disappears.

In other nuclei, such as ^{166}Er, Garrett et. al. [25] have used the (n, n′γ) reaction to measure the level lifetimes with standard Doppler-based techniques, establishing the fourth excited 0^+ state as an excellent candidate for a $\gamma\gamma$ $K = 0$ mode. Other experiments are in progress world wide by Aprahamian, Warner, Brentano, Börner, Jolie, Yates, the Yale group, and others.

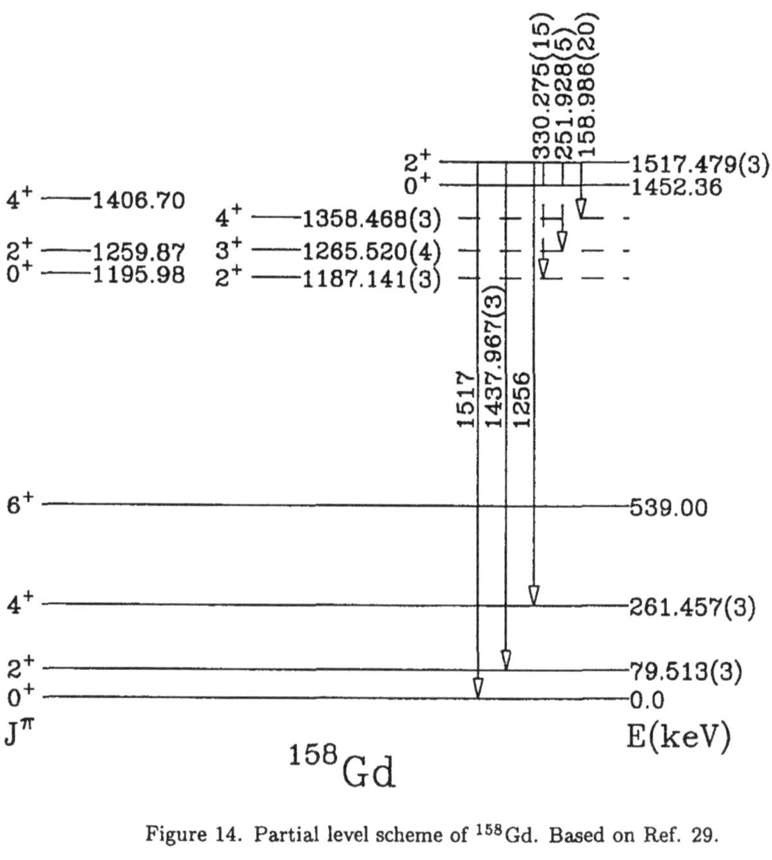

Figure 14. Partial level scheme of ^{158}Gd. Based on Ref. 29.

4 The Study of Exotic Nuclei with RNBs

History has shown that extending the frontiers of nuclear physics into the realm of neutron and proton rich unstable nuclei, and to the heaviest nuclei bound by the strong force, invariably leads to new insights into structure and structural evolution. With the rapidly developing capability to produce and use beams of exotic nuclei, our forays into this terra incognita will enjoy an unprecedented leap in quality and quantity in the first decade of the third millennium. Here, we wish to briefly outline a few themes of this research, recognizing at the same time that the most interesting and dramatic discoveries will almost certainly be those that we cannot write down or even suspect at the moment.

4.1 Motivations for the Study of Exotic Nuclei

First, it is useful to outline a few generic reasons why access to exotic nuclei almost certainly will yield exciting results, new phenomenology and new nuclear physics.

The first reason has to do with the nature of the nuclear force and the effects of the Pauli Principle in a fermionic system, as illustrated in Fig. 15.

Nucleons orbit the nuclear center of mass in quantized orbits that are characterized by specific total angular momenta, j. The Pauli Principle limits the number of identical nucleons in an orbit to (2j+1). Moreover, j values are generally rather low, seldom as large as $13/2\hbar$. Hence, the addition of more nucleons invariably entails the occupation of new orbits.

Although the nuclear force originates in the strong force, the force between nucleons in the nuclear medium is analogous to the van der Waals force between atoms – it is a remnant of a stronger interaction, barely able to bind two nucleons. The deuteron for example, has only one weakly bound state.

The weakness of the strong force between nucleons, is in fact, the reason a single particle picture of nucleonic motion has any validity at all in a nuclear environment where half the nuclear volume is occupied by its constituents: the force is too weak to scatter interior nucleons to unoccupied states. Hence collisions, in effect, seldom occur except amongst the valence nucleons and it is therefore the valence shells that dominate the structure and equilibrium shapes of nuclei.

Putting these two ideas together means that in exotic nuclei, with new values of N and Z, new valence nucleon orbits (and new combinations of proton and neutron orbits) will be occupied with, consequently, every expectation of new physics.

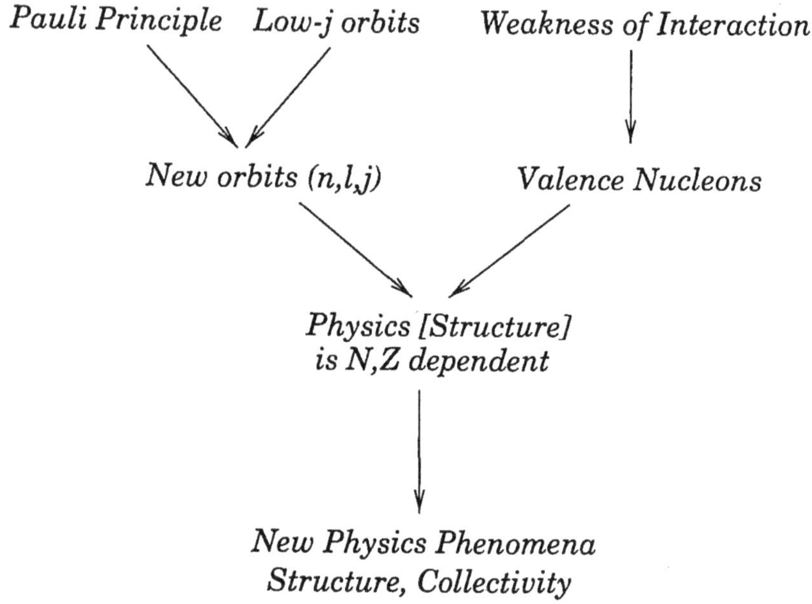

Figure 15. Generic argument for new nuclear physics in exotic nuclei.

A second reason for interest in exotic nuclei relates to the structure of near drip-line species. Here the operative concept is that of loosely bound quantal systems. Several new ingredients enter. In particular, low nucleon binding energies, (E-V), mean that nuclear wave functions, which decay as $e^{-\sqrt{2m(E-V)}}$ within the binding potential, extend to large radii.

Here we have a dramatic example of the modification of a benchmark concept of nuclear physics and the recognition that our "textbook" paradigms may only apply in the nuclei near the valley of stability that spawned them. Fig. 16 shows a simplified version of the familiar plot of binding energy per nucleon that appears in the first chapter of virtually every nuclear physics textbook. Its constancy above A$\sim$20 is used to argue that the nuclear density is approximately constant and hence that the nuclear volume scales with A. Therefore, the nuclear radius R$\sim A^{1/3}$. Now consider Fig. 17 which shows recently measured radii in light nuclei, extending in some cases drip-line to drip-line. The radii hardly scale as $A^{1/3}$ except along the N=Z line, that is, along the valley of stability in this mass region. Near the drip lines the

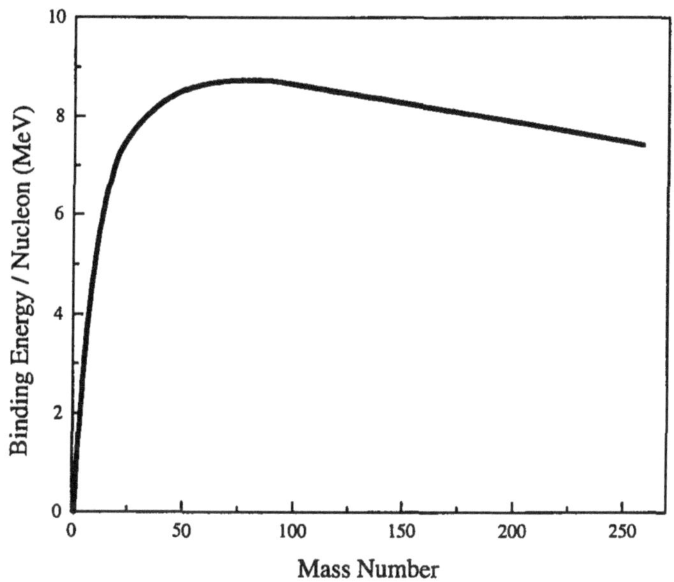

Figure 16. Simplified rendition of the traditional BE/A vs. A plot of nuclear binding energies per nucleon.

radii greatly expand. The effects of weak binding near the drip lines are most evident along the lower-right to upper-left diagonals of constant A: the radii are anything but constant.

Thus, in loosely bound nuclei the outermost nuclear wave functions are spatially extended, and the nucleon density is low. On the neutron rich side, the outer reaches of such nuclei are, therefore, low density, spatially extended, diffuse, nearly pure neutron matter – a form of matter never before encountered.

In addition, the residual interactions (e.g., pairing) amongst the weakly bound outer nucleons will involve scattering into the nearby continuum, introducing a completely new element into the nuclear Hamiltonian. The bound state shell model will, at a minimum, need to be replaced by a continuum shell model, incorporating both bound and (quasi-) free nucleons.

Finally, in neutron rich nuclei, the highly asymmetric N/Z composition means that (again, because of the Pauli Principle) the outermost neutron and proton orbits will be very different and, therefore, so will the residual valence

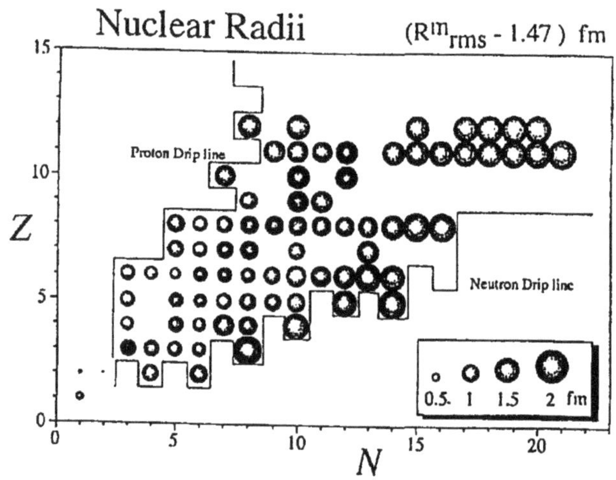

Figure 17. Nuclear radii in light nuclei. I am grateful to I. Tanihata for this figure.

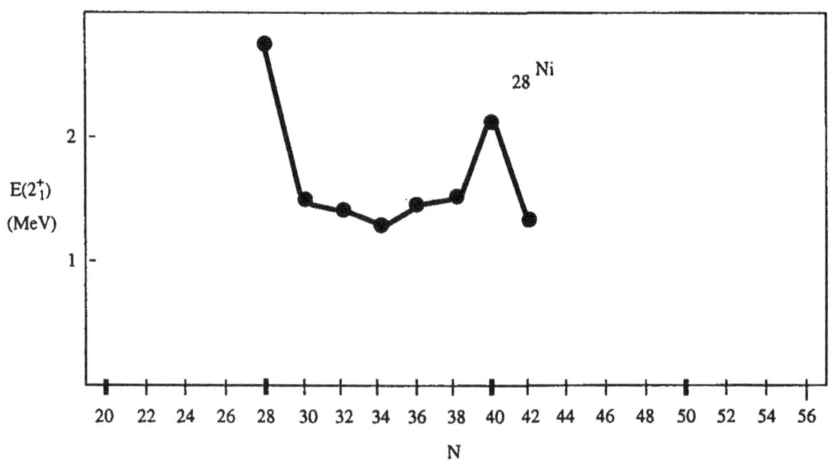

Figure 18. $E(2_1^+)$ in the Ni isotopes.

p-n interaction. It is known that, in normal nuclei, this interaction is the dominant controlling contributor to structure, and to the onset and growth of collectivity. If it is substantially different in exotic nuclei, structure and structural evolution are likely to be different.

This brings to the fore a third motivation for the study of exotic nuclei – as a part of an expanded ability to study the *evolution* of structure along greatly extended iso-chains (for technical reasons, these will generally be isotopic chains). The history of nuclear structure studies has been dominated by a "vertical" perspective in which the focus is on sequences of excited levels in individual level schemes. This is natural since actual experiments general study individual nuclear species. With the accumulation of large amounts of data on nuclei near stability over several decades, however, we are now in a position to exploit a "horizontal" perspective, looking at the evolution of structure along extended iso-chains. Access to exotic nuclei will expand by a large amount the perspective and power of such an approach.

Perhaps the most striking opportunity of this sort is the Ni isotopes. They constitute a unique element. The data for the energy of the first excited state in even-even Ni nuclei are shown in Fig. 18. With RNBs we have the possibility of studying this element from ^{48}Ni (a nucleus very near the proton drip line which has recently been discovered [30] in experiments at GANIL) to isotopes heavier than ^{78}Ni, with N>50. That means we will have data at four magic numbers (four doubly magic nuclei) – assuming magicity at the standard magic numbers (see below) – and in five major shells! Just recently, $E(2_1^+)$ was identified [30] in ^{70}Ni, the first studied Ni isotope in the N=40-50 shell. This is but the first step in forays in both directions.

A final motivation for studying exotic nuclei is not physical per se but historical: every time in the past when new nuclei have become accessible, new phenomena and new understanding have emerged. A prime example is the study near A=100 in the 1970s of fission products, which led to the discovery of the most rapid spherical-deformed transition region known and to the development of the Federman-Pittel picture [10] of the key role of the valence p-n interaction, especially in spin-orbit partner orbits. This picture underlies our current understanding of the onset of configuration mixing and the growth of collectivity and deformation in nuclei, and has spawned phenomenological approaches such as the N_pN_n scheme [31] that correlate the properties of large ensembles of nuclei into simple evolutionary trajectories. The study of the development of deformation in Sr-Zr led to the recognition of the breakdown of magicity at proton number 40. Since then, the obliteration of magicity at Z=64, as a function of neutron number, has likewise become evident [31]. In general, magicity has evolved from a robust cornerstone of our conceptual framework of nuclear structure to a fragile construct dependent on local N and Z values. The discovery of deformation and collectivity [32] in ^{32}Mg, with N=20, re-affirms this fragility.

Studies of exotic nuclei have also given us light neutron rich halo nuclei such as the first known Borromean system [33], ^{11}Li, whose outer reaches extend radially to the same dimensions as ^{208}Pb. The study of halo nuclei has proliferated with the advent of RNBs.

Finally, very recent history gives another example of pushing to the frontiers of nuclear existence and the impact of such studies. The dramatic discovery [34] of Z=118, if confirmed, extends the limits of the heaviest nuclei in a bold jump from Z=114. The very existence of these nuclei is a direct consequence of stabilization due to shell structure. Studying the masses, binding energies, and half-lives of such nuclei is already telling us much about shell structure of nucleonic systems with over 250 constituents. It is only in the last few years, for example, that our thinking about the next proton magic number past Z=82 has evolved from Z=114 towards Z=126.

In the next sections we briefly survey some of the more interesting opportunities for new research in exotic neutron rich and proton rich nuclei. See Ref. 35 for more details and for the source of some of the present material.

4.2 Neutron Rich Nuclei

Here the possibilities for new phenomena are largest, because of the longer lever arm from stability to drip line, because less is known about these nuclei (most nuclear reactions make proton rich nuclei), and because of the very asymmetric N/Z ratios on this side.

We have already referred to halo nuclei. These are systems where the outermost nucleons (usually neutrons) extend to large radii, forming a kind of neutron shell surrounding a normal core nucleus. Fig. 19 shows calculated density profiles for the best studied halo nucleus ^{11}Li. As we have noted, the outer region is essentially pure neutron matter (albeit very low density). Aside from many structure-related reasons for studying such systems, a rather fundamental one is that, if it can be appropriately probed, the halo region offers the possibility to study nuclear matter at a density intermediate between free nucleons and normal nuclear densities. The density dependence of the effective nucleon-nucleon interaction is a key concept in modern nuclear theory. Here is the opportunity to obtain a third benchmark.

In heavier nuclei, halos evolve into skins, or outer layers consisting of 10s of neutrons in a spatially extended diffuse shell. Recalling that the 1-body mean field nuclear potential is the self-consistent resultant of all nucleon

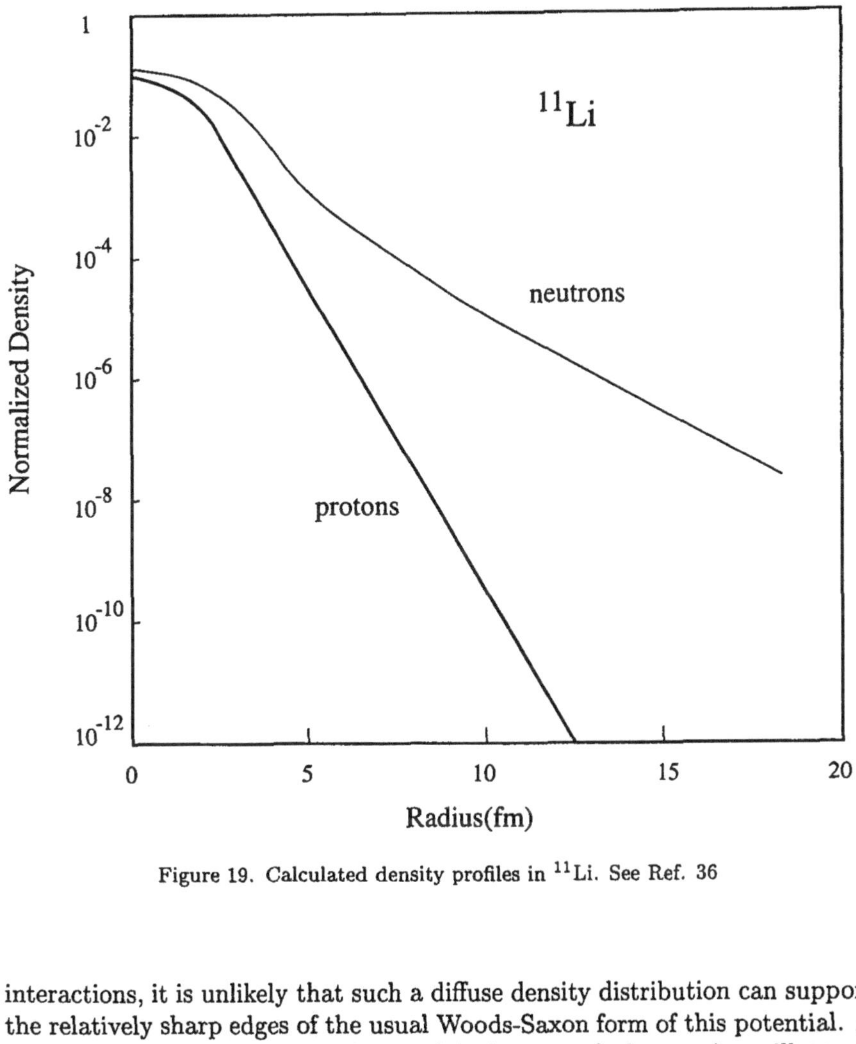

Figure 19. Calculated density profiles in ^{11}Li. See Ref. 36

interactions, it is unlikely that such a diffuse density distribution can support the relatively sharp edges of the usual Woods-Saxon form of this potential. A more rounded potential, reminiscent of the bottom of a harmonic oscillator, is more plausible [37]. Moreover, since the spin-orbit potential is surface peaked it is likely that it too may be modified (weakened). One speculative scenario for the resultant changes in single particle energies is illustrated in Fig. 20.

Such changes would radially alter nearly all aspects of structure and its evolution. Magic numbers will change or quench, and the unique parity orbit will revert towards its parent oscillator shell. As a consequence, high spin phenomena would be totally different, as well as octupole correlations. Residual interactions (e.g., quadrupole) that favor deformation will perhaps be much stronger as single particle levels differing by $\Delta\ell = \Delta j = 2$ will be close lying. On the other hand, spherical shapes may be favored since pairing may be stronger near the drip line and the p-n interaction weaker. As a result the competition between spherical-driving and deformed-driving tendencies will likely be unlike anything seen to date. It is possible that regions of deformation will be smaller in extent (in A), with smaller absolute magnitudes of the deformation, and with more oscillations between prolate and oblate shapes.

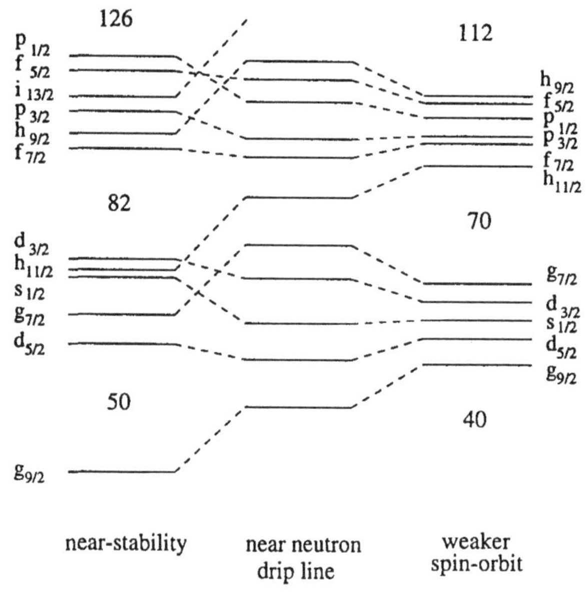

Figure 20. Possible scenarios for changes in single particle energies in neutron rich nuclei. Inspired by the ideas and calculations of W. Nazarewicz and collaborators. See Ref. 37.

The presence of a semi-decoupled skin of nearly pure neutron matter can also lead to new topologies and collective modes. One can imagine [37] mini-giant resonances in which the core oscillates against the skin, or in which a deformed core is surrounded by a spherical skin, or scissors-like oscillations of a deformed core against a deformed skin.

Finally we note, though not the purview of this paper, that the nucleosynthesis of medium mass and heavy nuclei proceeds largely by the r-process through a wealth of currently inaccessible neutron rich nuclei. The ability to sample this region, and to study waiting point nuclei, like ^{130}Cd, will benefit our understanding of the origin of these elements and help constrain theories of supernova explosions. For example, given the known r-process abundance profiles, the time-scale of these dramatic cosmic events is dictated by the summed β-decay lifetimes of the nuclei along the r-process path.

4.3 Proton Rich Nuclei

Here, a rich field of investigation remains even though many of these nuclei are already accessible. For example, on this side of stability the centrifugal barrier conspires with the Coulomb barrier to help bind protons. Thus, proton emission in nuclei at or beyond the proton drip line can be delayed, leading to half lives that depend on the angular momentum of the proton orbit and the three-dimensional shape of the nucleus. The study of three-dimensional tunneling in non-spherical systems is a fascinating one in its own right. Beyond this, the lifetimes can easily reach ms or larger, allowing us to study nuclei beyond their limits of binding by the strong force. Reversing the argument, the measurement of proton decay lifetimes can yield information (angular momentum, spectroscopic factors) on the single particle wave functions at the proton drip line.

Beyond nuclear structure proper, proton rich nuclei are of interest because the rp-process passes through these nuclei, and because tests of the Standard Model can be obtained by careful studies of superallowed $0^+ \rightarrow 0^+$ β decays in N=Z nuclei.

Perhaps the most important nuclear structure issue on this side of stability is the study of the T=0 interaction, again in N$\sim$Z nuclei. This topic, in fact, illustrates a key benefit of access to a wealth of new nuclei. It is not the number of new species per se but the ability to *choose* specific nuclei in order to focus on particular physics issues. We can sometimes, for example, exploit selected nuclei to amplify particular interactions. The T=0 interaction is present in all nuclei but, for reasons having to do with the symmetric filling of identical levels, its effects are enhanced in N=Z nuclei. The deuteron is the simplest example: the only bound state has T=0. This, in fact, was one of the earliest clues that the T=0 interaction is stronger than the T=1. In most heavy nuclei, specifically identifiable effects of the T=0 interaction are hard to isolate. However, in N=Z nuclei, the T=0 interaction has dramatic effects, in particular, on binding energies.

Masses or binding energies represent the sum effect of all nuclear interactions. Mass differences or separation energies tell us about the orbits and energies of individual outer nucleons and about shape changes with nucleon number. Double differences of binding energies can act as "filters" to isolate particular interactions. One of the most useful of these is called δV_{pn}, defined as:

$$\delta V_{pn} = \frac{1}{4}[\{B(Z+2, N+2) - B(Z+2, N)\} - \{B(Z, N+2) - B(N, Z)\}] \quad (1)$$

A little thought shows that δV_{pn} gives the average interaction of the last proton with the last neutron [38]. Fig. 21 shows empirical δV_{pn} values for all even-even nuclei. Note that the values are negative (attractive) and hence lower points on the plots correspond to stronger interactions.

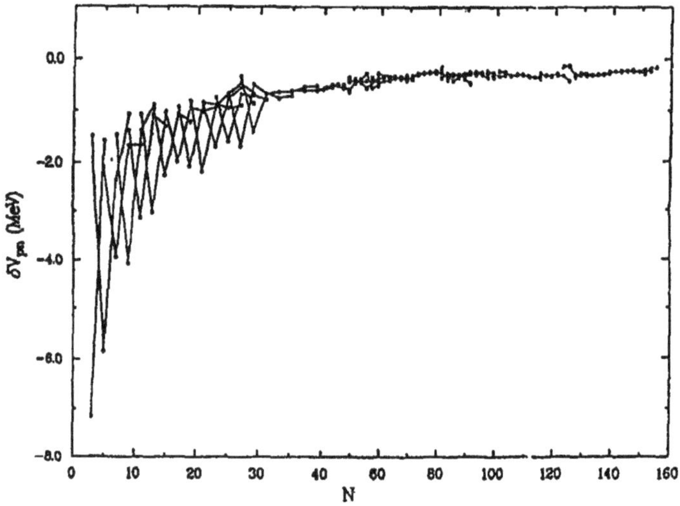

Figure 21. Empirical δV_{pn} values. From Ref. 39.

There are four interesting features to note. First, most of the points lie within a quite compact envelope that becomes gradually weaker with A. This is easy to understand. Heavier nuclei are larger and hence individual valence nucleons on average are somewhat farther apart. Also, in heavy nuclei the protons and neutrons are filling rather different orbits with less overlap. Secondly, the overall smoothness of the plot, despite the strong variations in valence orbits, shows that the average p-n interaction – averaged over the orbit occupancies of the last nucleons – is nearly constant within a region.

This justifies, in retrospect, a central assumption behind successful correlation methods such as the N_pN_n scheme. Thirdly, the micro-fluctuations that do occur – one is visible on the scale of Fig. 21 near ^{208}Pb – reflect significant structural changes such as shell closures or spherical-deformed transition regions and have been successfully explained in terms of the shell and Nilsson models [39].

The fourth feature is the most spectacular and the most of interest here – namely the huge negative spikes that occur in light nuclei. These singularities are not random but occur only for N=Z nuclei. In the s-d shell they can be quantitatively reproduced [39] with shell model calculations provided that a strong T=0 interaction (about twice the strength of the T=1) is included. The amplification of the effects of the T=0 interaction in N=Z nuclei is vividly evident in Fig. 21.

At the same time, it appears that the size of these spikes is decreasing with mass. However, there is a known variation of spike size within a given shell, and hence some of the decrease may be due to that effect. Nevertheless, it may also be that, with increasing mass, the cumulative effects of the Coulomb interaction affect the proton orbits, and even their occupancies, sufficiently that the proton-neutron symmetries in N=Z nuclei are partly broken. Only with new measurements of masses and binding energies of heavier N=Z and neighboring N=Z±2 nuclei with RNBs will it be possible to fill in Fig. 21 with N=Z points for nuclei approaching N=50.

The culmination of studies of proton rich N=Z nuclei is ^{100}Sn, a long sought goal of nuclear structure studies. Not only is this the heaviest N=Z nucleus likely to be stable, but it is a rare example of a new doubly magic nucleus and the heaviest bound doubly magic nucleus with N=Z. It is therefore of prime interest, along with its doubly magic ±1 neighbors, in order to assess magicity and to map out the shell structure around it.

At present it is not even known that ^{100}Sn is in fact doubly magic. The strong p-n interaction in N=Z nuclei can have dramatic and unexpected effects. For example, ^{80}Zr with N=Z=40, would be a candidate for double magicity but in fact is found [40] to be deformed and likely to be highly axially asymmetric as well. The possibility to march up the N=Z line though ^{84}Mo (a few data exist) to the terra incognita of ^{88}Ru, ^{92}Pd, ^{96}Cd and ultimately ^{100}Sn is thus one of the most exciting and eagerly anticipated outcomes of future studies of exotic nuclei with RNBs.

5 Conclusion

This overview has attempted to outline briefly some of the areas of high current interest in nuclear structure both for nuclei near stability and for exotic nuclei stretching to the edges of nuclear existence. Many other choices of topics could have been made. My selection has been highly personal, though hopefully representative of the interests of others as well. This field of research is vibrant, flourishing and expectant of the approaches that await with the coming of advanced facilities for the production and study of exotic nuclei.

I am grateful to a number of colleagues for especially fruitful discussions that have molded my perspective. In particular, I am grateful to F. Iachello, A. Arima, D.D. Warner, S. Pittel, P. von Brentano, W. Nazarewicz, A. Aprahamian, H.G. Börner, K. Heyde, N.V. Zamfir, A. Frank, J. Jolie, J.-Y. Zhang, and D.S. Brenner. I am particularly grateful to N.V. Zamfir, R. Kruecken, and N. Pietralla for substantial help with the figures for this report.

References

1. K. Gelbke, this volume
2. A. Arima and F. Iachello, Phys. Rev. Lett., **35**, 1069 (1975)
3. For a review of the IBM see R.F. Casten and D. D. Warner, Rev. Mod. Phys. **60**, 389 (1988)
4. F. Iachello, Phys. Rev. Lett. **44**, 772 (1980)
5. A. Metz et al., Phys. Rev. Lett, **83**, 1542 (1999)
6. F. Iachello, this volume
7. G. S. Goldhaber and J. Weneser, Phys. Rev. **98**, 212 (1955)
8. A. Bohr and B. R. Mottelson, Mat. Fys. Medd. Dan. Vid. Selsk., **24**, No.16 (1953)
9. R. F. Casten, J-Y. Zhang, and N. V. Zamfir, to be published
10. P. Federman and S. Pittel, Phys. Lett. **69B**, 385 (1977)
11. K. Heyde et al., Phys. Lett., **155B**, 303 (1985)
12. F. Iachello, N.V. Zamfir, and R.F. Casten, Phys. Rev. Lett. **81**, 1191 (1998); M. Wilhelm et al., Phys. Rev. **C57**, R1553 (1998); R. Kruecken et al., Phys. Lett. **B454**, 15 (1999)
13. R.F. Casten, D. Kusnezov, and N.V. Zamfir, Phys. Rev. Lett. **82**, 5000 (1999)
14. R.F. Casten, N.V. Zamfir, and D.S. Brenner, Phys. Rev. Lett. **71**, 227 (1993)
15. P. Twin et al., Phys. Rev. Lett. **57**, 811 (1986)
16. S. Erturk, Thesis, Univ. of Liverpool, UK, 1998

17. T. Byrski et al., Phys. Rev. Lett. **64**, 1650 (1990)
18. S. Frauendorf, Nucl. Phys. **A557**, 259c (1993)
19. G. Baldseifen et al., Nucl. Phys. **A574**, 521 (1994)
20. J.R. Novak et al., Phys. Rev. **C59**, R2989 (1999)
21. For a review of this field, see D. Zawisha, J. Phys. **G24**, 683 (1998)
22. A. Aprahamian et al., Phys. Rev. Lett. **59**, 535 (1987)
23. R.F. Casten et al., Phys. Lett. **B297**, 19 (1992)
24. H.G. Börner et al., Phys. Rev. Lett. **66**, 691 (1991)
25. P.E. Garrett et al., Phys. Lett. **B400**, 250 (1997)
26. H. Lehmann et al., Phys. Rev. **C57**, 569 (1998)
27. N.V. Zamfir et al., Phys. Rev. **C60**, 054319 (1999)
28. C.W. Beausang et al., Nucl. Instr. and Meth. (in press)
29. H.G. Börner et al., Phys. Rev. **C59**, 2432 (1999)
30. B. Blank et al., to be published; R. Grzywacz et al., Phys. Rev. Lett. **81**, 766 (1998)
31. R.F. Casten, Phys. Lett. **152B**, 145 (1985); R.F. Casten et al., Phys. Rev. Lett. **47**, 1433 (1981)
32. T. Motobayashi et al., Phys. Lett. **B346**, 9 (1995)
33. I. Tanihata et al., Phys Lett **B206**, 592 (1985)
34. V. Ninov et al., Phys. Rev. Lett. **83**, 1104 (1999)
35. "Columbus" white paper: Research Opportunities with an Advanced ISOL Facility, 1997
36. G.F. Bertsch, B.A. Brown, and H. Sagawa, Phys. Rev. **C39**, 1154 (1989)
37. J. Dobaczewski and W. Nazarewicz, Phil. Trans. R. Soc. Lond. **356**, 2007 (1998) and W. Nazarewicz, private communication
38. J.-Y. Zhang et al., Phys. Lett. **227B**, 1 (1989)
39. D.S. Brenner et al., Phys. Lett. **243B**, 1 (1990) and W.T. Chou et al., Phys. Lett. **255B**, 487 (1991)
40. C.J. Lister et al., Phys. Rev. Lett. **59**, 1270 (1987)

EQUATION OF STATE OF NUCLEAR MATTER

CLAUS-KONRAD GELBKE

National Superconducting Cyclotron Laboratory and Department of Physics and Astronomy Michigan State University, East Lansing, MI 48824-1321, USA

Experimental studies aimed at understanding the equation of state and the liquid-gas phase-transition of nuclear matter will be discussed.

1 Introduction

Little is known about the properties of bulk nuclear matter. The atomic nucleus is a mesoscopic system that can be viewed as a finite drop of a multi-component quantum Fermi liquid with strong interactions between the constituents. Bulk nuclear matter, for which finite size effects can be neglected and the thermodynamic limit is well satisfied, is likely to exist only in neutron stars and can thus not be probed by experiments. Extrapolating from experiments on finite nuclear systems to the properties of nuclear matter (and of neutron stars) presents a challenging problem.

Experimental attempts of determining the equation of state of nuclear matter and studying its predicted phase transitions can only be made by means of nucleus-nucleus collisions where finite systems of nuclear matter may be heated up and compressed temporarily. Low densities, pertinent to an expected liquid gas phase transition, may be reached during a subsequent expansion and decompression stage. The extremely high densities and temperatures, pertinent to a possible deconfinement transition to a quark-gluon plasma, can only be reached in collisions at ultra-relativistic energies.

Since nucleus-nucleus collisions produce transient systems of very short lifetime, full statistical equilibrium of all degree of freedom is unlikely to be reached in such experiments. Hence, the extraction of thermodynamic properties of bulk nuclear matter from such experiments presents a significant experimental and a formidable theoretical challenge. Since the limit of a thermodynamic freeze-out is well defined and understood, it is important to establish experimentally for which (if any!) degrees of freedom statistical equilibrium is attained and where (local) equilibrium concepts become inappropriate. In order to investigate the degree thermalization attained in nucleus-nucleus collisions, one must find observables that are sensitive to temperature, pressure, density, size, and shape of the reaction zone formed during the collisions.

In this talk, I will limit myself to energies and densities where the implications of deconfinement can be ignored. I will highlight a number of experimental approaches aimed at elucidating the thermodynamics of the liquid-gas phase transition in nuclear matter and at extracting information about its equation of state (EOS). Representative results will be highlighted and problems will be pointed out that are not yet understood. I will mainly restrict myself to results at intermediate energies (20 MeV < E/A < 1 GeV) pertinent to the liquid-gas phase transition. In this energy domain, the multifragment disintegration of hot nuclear systems is now a well-established phenomenon, as illustrated in the example [1] shown in Figure 1. Investigations of flow at higher energies will be mentioned briefly.

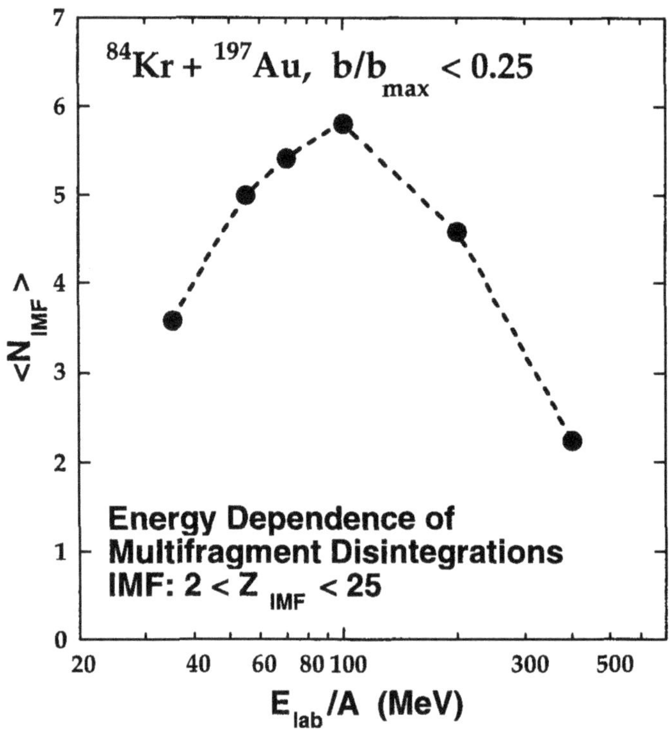

Figure 1: Energy dependence of intermediate mass fragment production in central ^{84}Kr + ^{197}Au collisions. From ref. [1].

2 Measuring Temperatures

Extractions of nuclear temperature are generally based upon comparisons of experimentally observed distributions to those predicted by statistical theory [2]. In the most naïve approach, one can determine the slopes of the kinetic energy spectra of emitted particles and extract "kinetic temperatures", T_{kin}, from fits with Maxwell distributions. These temperatures typically characterize the early phase of the reaction before the system expands and cools. The extraction of kinetic temperatures is complicated by the presence of collective flow effects that distort the shapes of the energy spectra. Alternative approaches are based upon the relative populations of states or resonances, or upon ratios of isotopic yields of emitted light fragments, yielding apparent temperatures $T_{\Delta E}$ and T_{iso}, respectively [2,3]. These temperature determinations are based upon the assumption of chemical equilibrium at freeze-out. They can be complicated by feeding from the sequential decay of excited states.

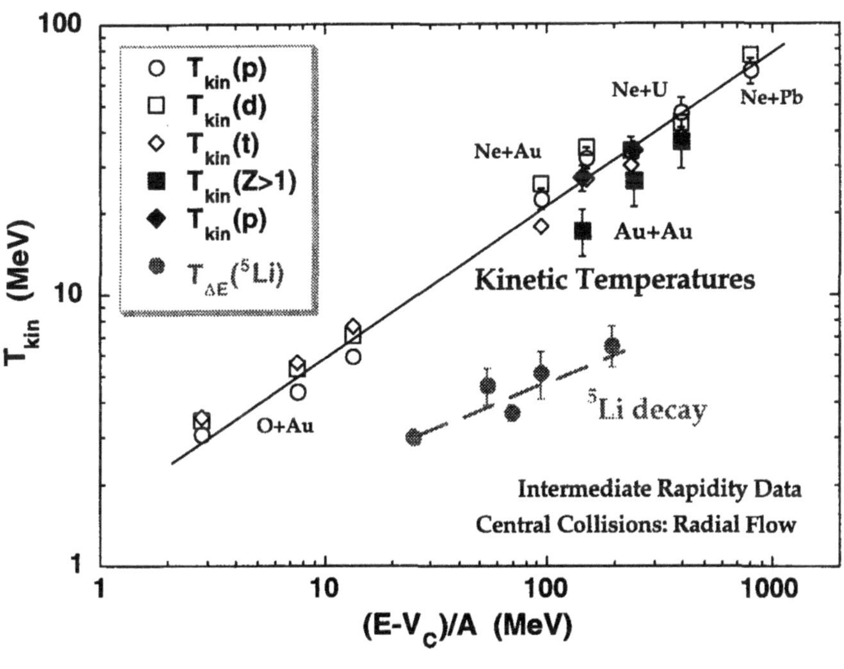

Figure 2: Energy dependence of kinetic temperatures extracted in various reactions indicated in the figure. The open points show mid-rapidity data from inclusive experiments; solid squares and diamonds show data from central Au+Au collisions that are corrected for collective radial flow, filled circles are temperatures, $T_{\Delta E}$, extracted from the relative populations of widely separated states in ^{5}Li.

Figure 2 illustrates the long-standing "temperature dilemma": For intermediate energy nucleus-nucleus collisions, measured values of T_{kin} are typically higher than measured values of $T_{\Delta E}$. Qualitatively, such a difference can be attributed to the expansion and cooling of the initially compressed nuclear system, but a quantitative analysis of the collective expansion effects could not resolve this dilemma in a consistent thermodynamic picture [4].

Much excitement has been generated by the recent extraction of a "caloric curve" [5] shown in Figure 3. The data are strongly suggestive of a phase transition, and bear some resemblance to the caloric curve of the liquid-vapor transition of boiling water. However, the analogy to boiling water is superficial at best. For the boiling of water the pressure stays constant while heat is added to the system. For the expansion phase of a nucleus-nucleus collision, on the other hand, the pressure varies while no heat is added to the system, i.e. the system evolves nearly isentropic rather than isobaric.

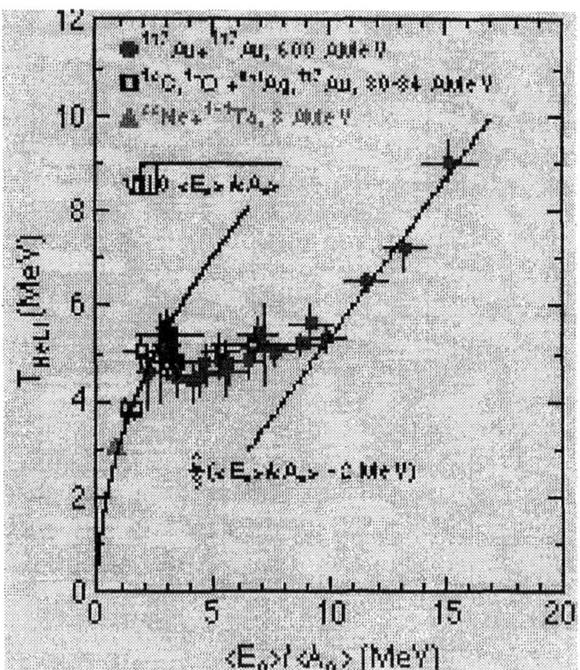

Figure 3: "Caloric curve" extracted by means of temperature determinations involving helium and lithium isotope ratios. From ref. [5].

The interpretation of the interesting data shown in Figure 3 has been challenged by many authors and is by no means straightforward. Much attention has been given to the cross-calibration of the helium-lithium isotope ratio thermometer (T_{He-Li}) employed in the region of the plateau and the temperature rise at higher excitation energy. Significant inconsistencies have emerged [6,7] between temperatures extracted from isotope ratios involving ^{3}He and ^{4}He nuclei and temperatures extracted from alternative isotope ratios or from relative populations of widely separated states, see Figure 4. These differences are presently not understood. Some, but not all discrepancies can be attributed to sequential feeding effects, others have been suggested to result from a sequential freeze-out of different degrees of freedom, from in-medium modifications, or from a strong pre-equilibrium component in the ^{3}He emission [7], which would make $T_{He\text{-}Li}$ measurements employed by ref. [5]of limited value. At this time, the validity of the caloric curve *shown in Figure 3 is questionable. The reliable determination of the caloric curve* for hot nuclear matter remains a significant challenge.

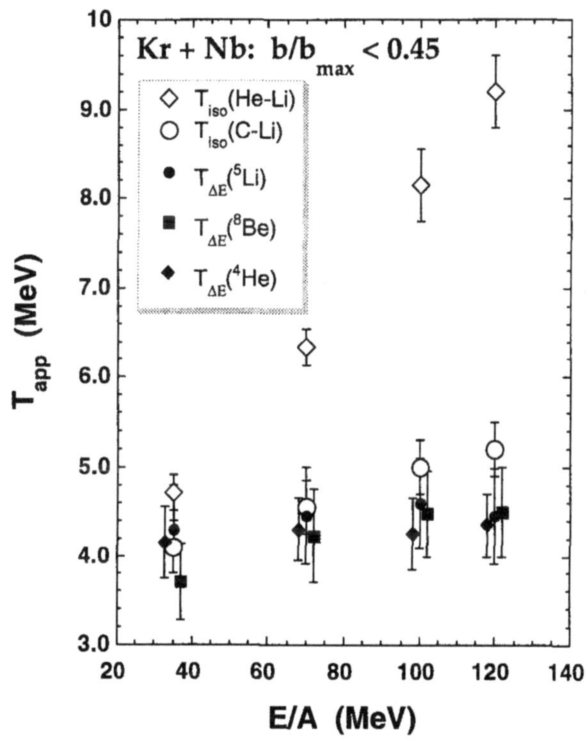

Figure 4: Temperatures extracted from the relative populations of states and from isotope ratios for central Kr+Nb collisions at various beam energies. From ref. [7].

3 Observing Radial Expansion

Evidence for an expansion of the reaction zone comes from two observations: the detailed shapes of energy spectra [8], especially those of light nuclei, and the multiplicity of intermediate mass fragments [9].

In nucleus-nucleus collisions at intermediate and high energies, the energy spectra of emitted particles are influenced by two major effects: a mass-independent, random, near-thermal energy component ($E_{th} \propto kT$) and a mass-dependent, collective component ($E_{coll} \propto mv^2_{coll}$). The superposition of these two effects produces characteristic shoulder-arm shaped energy spectra [8,10] with a strong dependence on the mass of the emitted particles. Many detailed investigations have been performed in recent years [4], and the systematics of the radial collective flow that characterizes the expansion of the hot reaction zone is now rather well established experimentally. Figure 5 illustrates the experimental trends observed in collisions between heavy nuclei over a broad range of incident energies. As noted in the preceding section, consideration of collective expansion velocity components alone cannot resolve the "temperature dilemma" illustrated in Figure 2.

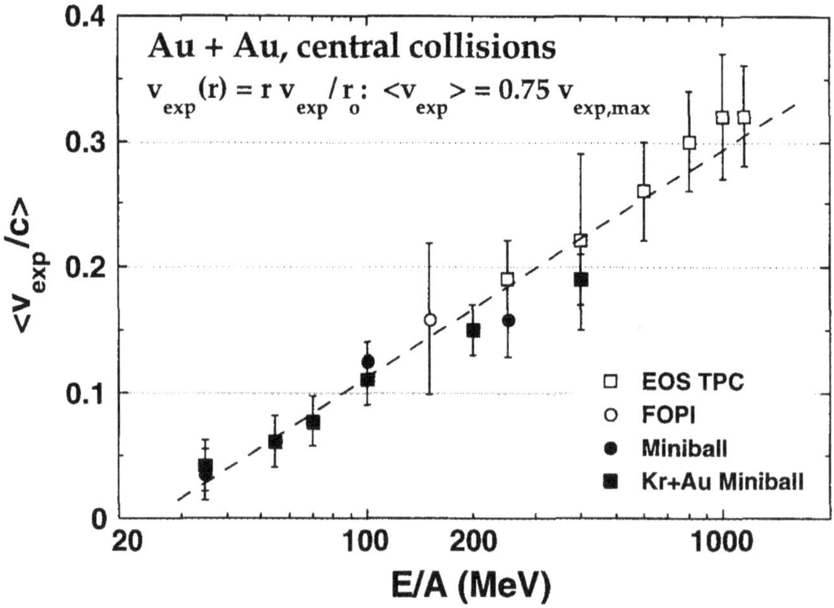

Figure 5: Systematics of collective radial expansion velocities observed in intermediate-energy Au+Au collisions.

Expansion and cooling of the nuclear reaction zone have a profound effect on the copious emission of intermediate mass fragments (IMF: $3 \leq Z_{IMF} \leq 25$), which may be related to a liquid-gas phase transition. Early work [9] has shown that standard compound nucleus evaporation models significantly underpredict the observed IMF multiplicities. Much better agreement with observations can be obtained if one assumes statistical emission from subnormal nuclear density, see for example Figure 6. The figure shows the mean IMF-multiplicity, $\langle N_{IMF} \rangle$, as a function of the charged particle multiplicity, N_C. The dotted and dashed curves indicate the region accessible by evaporation models and the solid curves indicate the region accessible by the expanding evaporating source (EES) model of ref. [11].

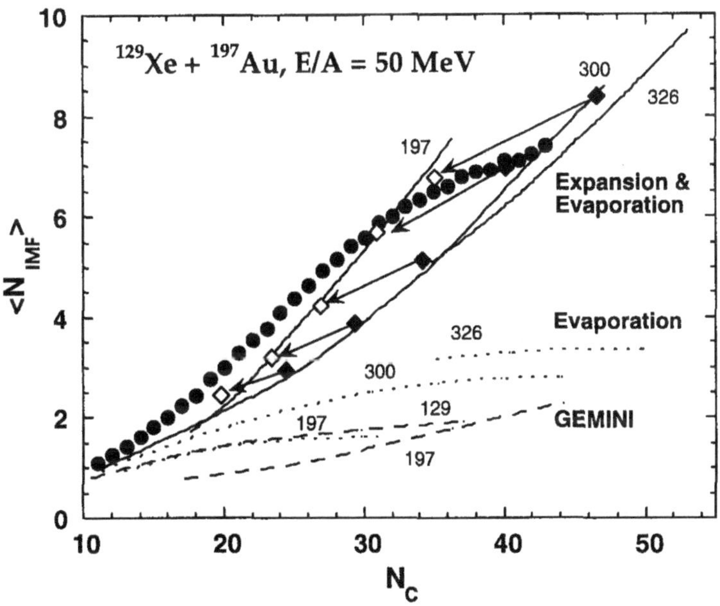

Figure 6: Solid points: relation between $\langle N_{IMF} \rangle$ and N_C for ^{129}Xe + ^{197}Au at E/A = 50 MeV. Calculations are discussed in the text. Numbers indicate the assumed mass of the decaying system. Diamond-shaped points illustrate instrumental distortions, representing unfiltered and filtered predictions, respectively, for an expanding source. From ref. [9].

Both rate equation approaches [11] and microcanonical treatments [12,13] of a low-density freeze-out scenario have met with considerable success in reproducing many features observed in multifragment emission processes. Illuminating discussions of the similarities and differences of the various models can be found in refs. [14,15]. Significant insight has been obtained from the expanding evaporating source (EES) model [11] that simulates the competition of cooling by expansion and particle emission. This model predicts a sudden onset of complex fragment emission once the system has reached subnormal density and temperatures of the order of 3-5 MeV, rather independently of the initial temperature of the reaction zone, and thus points towards a qualitative understanding of the temperature dilemma.

Experimental evidence for nuclear expansion and complex fragment emission in intermediate energy nucleus-nucleus collisions is on rather solid ground. However, a consistent theoretical treatment of nucleon and cluster emission during the full compression-decompression cycle encountered in a nucleus-nucleus collision is not yet available. The correct quantum mechanical treatment of the growth of density fluctuations and fragment formation remains a formidable challenge to theory.

4 Measuring the Space-Time Characteristics

Information about the space-time characteristics of the emitting source can be obtained from two-particle correlation functions, $1+R(\mathbf{q},\mathbf{P})$, defined as the ratio of the two-particle coincidence yield and an appropriately chosen "background" yield. Here, $\mathbf{q}$ and $\mathbf{P}$ denote the relative and total momentum vector of the particle pair. The shape of the correlation function depends on the final state interactions and quantum statistics of the coincident particles and is thus sensitive to the spatial separation of the emitted particles, which in turn depends on the source size and the emission time scale. These arguments form the qualitative basis of the technique of intensity interferometry [16-19].

As an example, Figure 7 shows longitudinal ($\mathbf{q} \parallel \mathbf{P}$) and transverse ($\mathbf{q} \perp \mathbf{P}$) two-proton correlation functions measured for central ^{36}Ar+^{45}Sc collisions at E/A = 80 MeV [20]. The observed suppression of the transverse as compared to the longitudinal correlation function can be attributed to the finite emission time scale [20], and the data are rather well reproduced by calculations with the Boltzmann-Ühling-Uhlenbeck (BUU) transport model.

Despite this impressive success of the BUU model, problems have emerged for collisions at higher energy -- where BUU model calculations should, in fact, have a more solid foundation. Several independent experiments [21-23] have revealed a strong attenuation of measured two-proton correlation functions as compared to predictions, indicating that the BUU transport model predicts proton emission from too small a reaction zone, on too short a time scale, or both. For illustration, Figure 8 shows correlation functions measured (points) and predicted (solid curve) for central ^{36}Ar+^{45}Sc collision at E/A = 160 MeV [23]. At the quantitative level, this

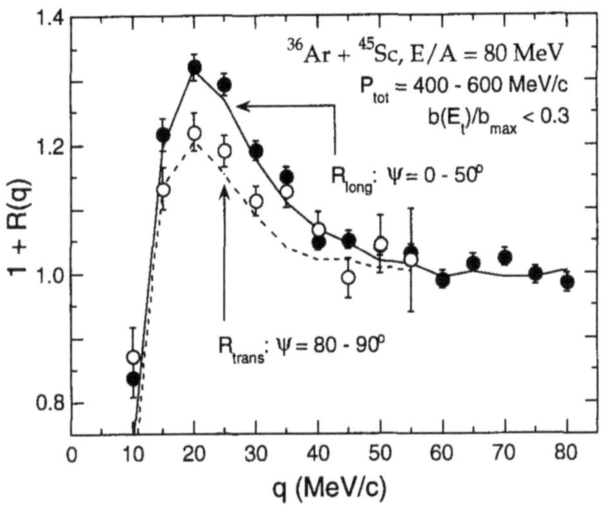

Figure 7: Longitudinal and transverse two-proton correlation functions for central $^{36}Ar+^{45}Sc$ collisions at E/A = 80 MeV. The points represent data, and the curves show BUU transport model predictions. From ref. [20].

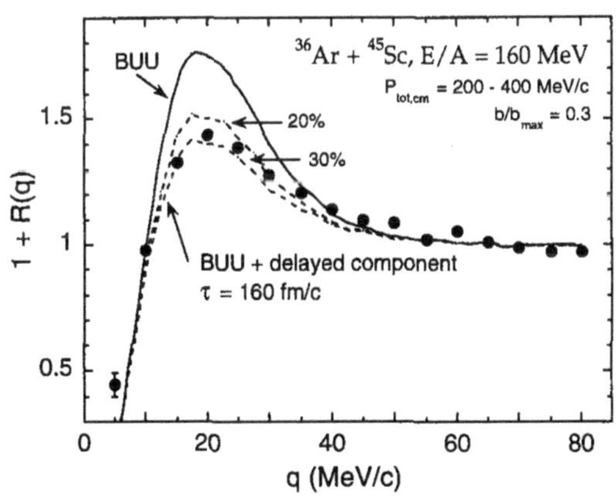

Figure 8: Two-proton correlation functions for central $^{36}Ar+^{45}Sc$ collisions at E/A = 160 MeV. The points represent data, the solid curves show BUU transport model predictions, and the dashed curves show BUU calculations in which 20% and 30% of the proton emission is artificially delayed by 160 fm/c. From ref. [23].

failure of BUU is not understood. Qualitatively, it has been suggested that some fraction of the emitted protons may not originate directly from the reaction zone, but from delayed secondary decays of highly excited intermediate mass fragments. The dashed curves in Figure 5 show that a 20-30% sequential decay contribution, delayed by some 160 fm/c, might indeed account for the discrepancy between BUU prediction and experiment.

While there is firm experimental evidence that sequential decays play an important role for nucleus-nucleus collisions in this energy domain, both magnitude and time scale of sequential decays are not yet sufficiently well understood to allow reliable and accurate predictions. Note that sequential decays also affect the emission temperatures extracted from ratios of isotopes and/or widely separated states [24,25]. Much effort is presently devoted to an improved understanding of cluster formation in highly excited states and the ensuing sequential decay.

An example of two-fragment correlation functions, measured [26] for central ^{36}Ar + ^{197}Au collisions at E/A = 50 MeV, is shown in Figure 9. To improve statistics, the correlations have been summed over several fragments and scaled to approximately correct for the varying strength of the two-fragment Coulomb interaction. The reduced relative velocity, v_{red}, is approximated [27] as $v_{red} = v_{rel}(Z_1 + Z_2)^{-1/2}$. The minimum at small relative velocity results from the repulsive Coulomb interaction between emitted fragments and can be quantitatively reproduced by many-body Coulomb trajectory calculations. The calculations indicate that fragments are emitted during a rather short time interval, $\tau < 100$ fm/c, a result that has been corroborated by many other investigations. As fragments are emitted with low initial velocity, they will not move far during this short emission-time interval. Hence, the difference between instantaneous and sequential fragment emission becomes rather blurred and difficult to resolve experimentally [28].

In principle, correlation functions can provide rather detailed information about the shape of the emitting source when narrow directional cuts are investigated for well-characterized classes of events corresponding to tight cuts on impact parameter and orientation of the reaction plane. For example, rather different correlations are predicted for spherical and toroidal emission sources [29,30]. In practice, however, such investigation have met only with moderate success due to the remaining experimental average over impact parameter and reaction-plane orientation [30]. Depending on the particle type, finite emission time intervals can produce additional averages over the time-dependent shape of an evolving source [30]. Recent theoretical progress [31] in obtaining model-independent information from two-particle correlation measurements may, however, make it worth to design high-precision experiments with much improved statistics, resolution in **q** and **P**, and stringent characterization of the impact parameter vector.

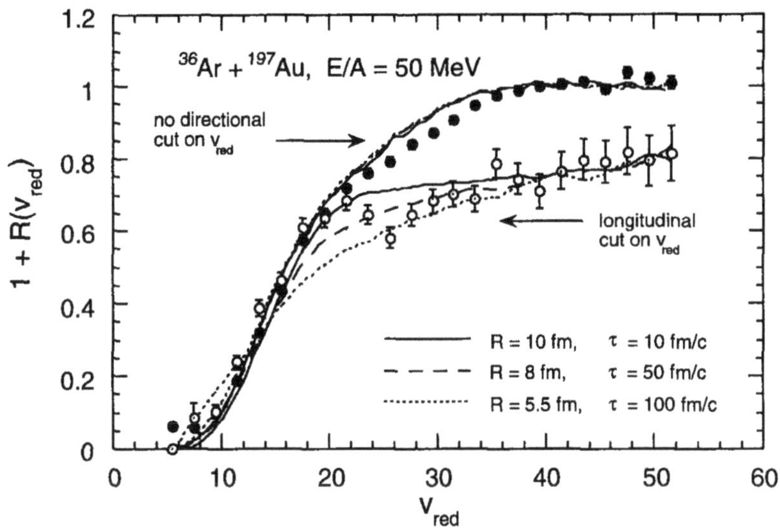

Figure 9: Two-fragment correlation functions for central Ar + Au collisions at E/A = 50 MeV without (solid points, thin curves) and with (open points, thick curves) directional cuts. From ref. [26].

5 Observing the effects of Pressure

Reaction zones formed in energetic nucleus-nucleus collisions, E/A > 100 MeV, may reach densities significantly higher than the saturation density of nuclear matter in its ground state. This is the only terrestrial situation in which such densities can be achieved and investigated experimentally to extract the equation of state (EOS) and other properties of nuclear matter.

During compression and heating, the pressure increases in the reaction zone. Eventually, an expansion to sub-normal density sets in. The ejection of nucleons and clusters during this compression-decompression cycle is influenced both by the repulsive pressure and by the attractive mean field. At low energies the attractive mean field dominates, and particles are preferentially emitted to negative deflection angles. At higher energies the pressure dominates, and deflection is largely to positive deflection angles. The expected change of sign has been confirmed experimentally in a series of experiments in which the circular polarization of gamma rays from target residues was measured in coincidence with charged particles emitted from the reaction zone [32].

Information about the mean field that governs the equation of state (EOS) of nuclear matter can be obtained from investigations of collective flow that is related to the mean nucleonic velocity field at breakup. The most commonly investigated observable is the sideward-directed flow of nuclear matter in the reaction plane [33]. This sideward-directed transverse flow F is defined as the derivative of the average transverse momentum per nucleon in the reaction plane with respect to the normalized c.m. rapidity y_n ($y_n = 1$ for the projectile; $y_n = -1$ for the target): $F = d<p_x/A>/dy_n$ at $y_n = 0$, or, alternatively, with respect to the absolute c.m. rapidity y: $F_y = d<p_x/A>/dy$ at $y = 0$.

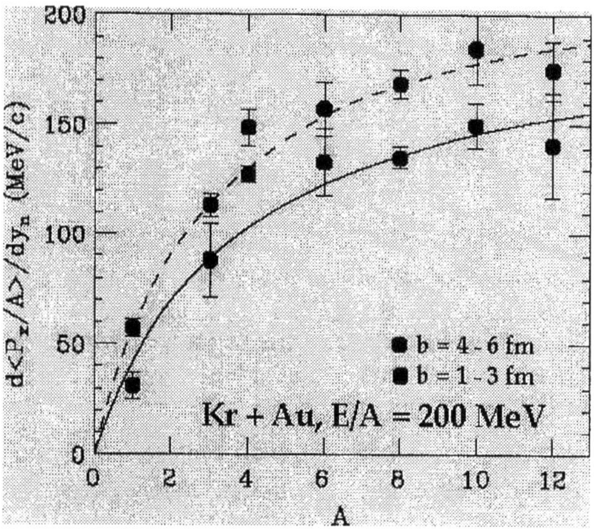

Figure 10: Mass dependence or sideward-directed flow for two cuts on impact parameter b. Here, flow was determined only for particles of fixed mass number A. From ref. [34].

Flow observables exhibit a strong dependence on impact parameter and incident energy. Due to the interplay of the collective flow velocity (which is independent of particle mass) and the randomly oriented thermal velocity (which decreases with particle mass), flow effects are typically more pronounced for heavier than for lighter emitted particles [34,35]. (A similar argument was made in the discussion of radial expansion). This effect is illustrated in Figure 10.

The strong mass dependence of collective flow observables complicates quantitative comparisons with microscopic transport calculations that do not yet incorporate complex fragment formation. In order to facilitate such comparisons, one typically constructs the "effective proton flow" by performing a charge-weighted average over the flow measured for all detected particles. The impact parameter dependence is extracted by using an empirical impact parameter filter. Figure 11 gives an example of such a comparison between experiment and theory

[34]. The data clearly favor a momentum dependent mean field, consistent with the nonlocality effects observed in nucleon-nucleus potential scattering [36]. There is some indication that in-medium cross sections are smaller than the corresponding free nucleon-nucleon cross sections, but the measurements reveal little sensitivity to the nuclear compressibility or the stiffness of the EOS. Similar conclusions were drawn from investigations [37] of the "balance energy", defined as the energy for which the average sideward directed flow vanishes due to the cancellation of the average mean-field attraction by the average repulsion from nucleon-nucleon scattering.

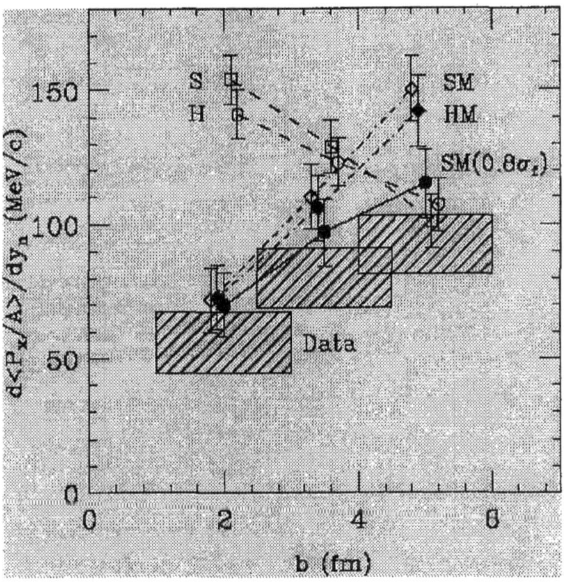

Figure 11: Impact-parameter dependence of sideward-directed flow. The crosshatched rectangles depict the effective proton flow. BUU transport calculations are shown for the following parameter sets: H – hard EOS without momentum dependence (open circles) ; S – soft EOS without momentum dependence (open squares); HM - hard EOS with momentum dependence (solid diamonds); SM – soft EOS with momentum dependence (open diamonds); SM (0.8σ_f) – soft EOS with momentum dependence and a 20% reduction in the nucleon-nucleon cross section (solid circles). Error bars indicate statistical uncertainties of the model calculations. From ref. [34].

As the quality of experimental data of increases, more detailed investigations of flow observables become possible. Of particular interest are investigations of azimuthal distributions of particles emitted at mid-rapidity [38,39] – or, more specifically, of the elliptic flow variables $v_1 = \langle\cos\phi\rangle$ and $v_2 = \langle\cos 2\phi\rangle$. Calculations performed at energies close to the balance energy indicate [39] that detailed investigations of the transverse momentum dependence of v_1 and v_2 will provide an enhanced sensitivity to the EOS and thus help reduce the existing parameter ambiguity between the EOS and the in-medium cross section. This enhanced sensitivity of elliptic flow to the EOS persists at higher beam energy [38].

To illustrate the rapidly growing systematics of flow data at higher energies, Figures 12 and 13 show the measured energy dependencies of F_y and v_2. The sideward directed flow, shown in Figure 12, increases up to E/A ≈ 0.5 GeV, and then decreases towards higher energies. A particularly intriguing effect appears in the elliptic flow data shown in Figure 13. At lower energies, these data indicate a stiff EOS, whereas a softening of the EOS seems indicated at densities of $\rho_{calc} \approx 4\rho_0$ that are calculated to occur at the higher energies. Such a softening of the EOS is expected in the presence of a phase transition to a quark gluon plasma [38,40]. While an accurate and consistent, microscopic description of these flow data is not yet available, rapid progress is being made.

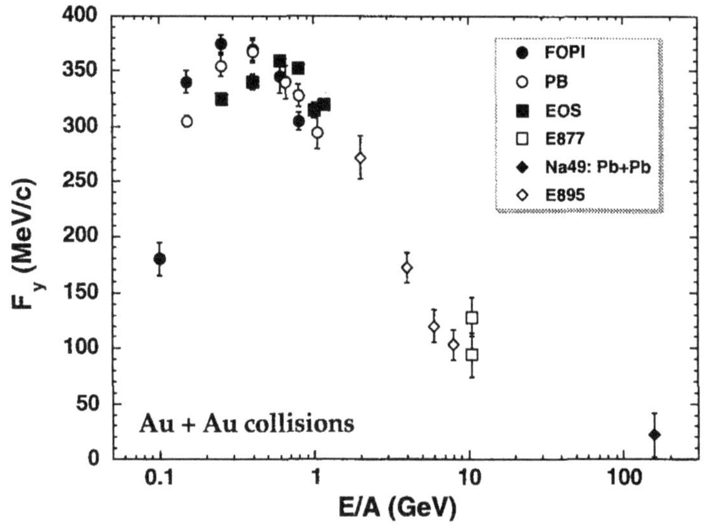

Figure 12: Measured energy dependence of sideward-directed flow.

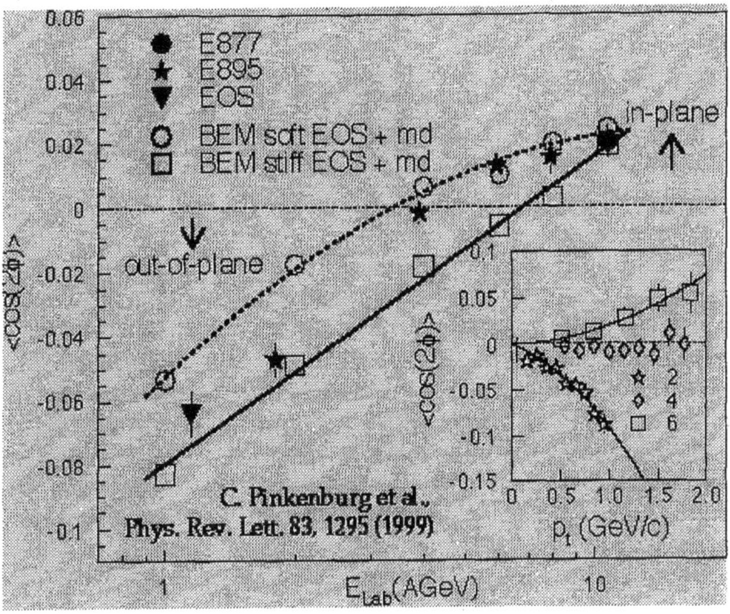

Figure 13: Observed energy dependence of the elliptic flow parameter $v_2 = \langle\cos 2\phi\rangle$. The solid points show experimental data and the open points show calculations with a soft (open circles) and a stiff (open squares) EOS. From ref. [40].

6 Summary and Outlook

Solid progress has been made in the phenomenological characterization of intermediate energy nucleus-nucleus collisions, driven by increasingly sophisticated experiments and model calculations. The beam energy dependence of multifragment disintegration has been mapped out, and the systematics of collective expansion velocities is well established. For heavy projectile-target combinations, fragment production in central collisions is maximal at incident energies of about 100-200 MeV/nucleon with a transition to vaporization occurring at higher energies.

Several methods have been employed to extract the temperatures of fragment emitting systems, but the extracted temperature values are not all consistent. As long as these discrepancies persist, the caloric curve of nuclear matter at energies pertinent to the liquid-gas transition must be viewed with great caution. Key problems that need further theoretical work are improved treatments of in-medium and final-state interactions between complex fragments and the surrounding dilute nuclear matter, and the quantitative description of the initial population as well as the sequential decay of high-lying particle-unbound states.

At energies pertinent to the liquid-gas phase transition, the major theoretical challenge remains the development of a tractable quantum transport model that allows a reliable treatment of the growth of density fluctuations and of the formation of fragments with realistic quantum properties in the exit channel. Such a dynamic theory might eventually overcome the limitations currently encountered in (over) simplified thermal freeze-out scenarios.

Substantial experimental and theoretical progress has been made in the understanding of collective flow and its sensitivity to the EOS of excited nuclear matter. At lower energies, the need for a momentum dependent mean field has been credibly established, and there are indications that the in-medium nucleon-nucleon cross-sections are smaller than the free nucleon-nucleon cross sections. At high energies, indications for an interesting softening of the EOS have been found. There is great hope that accurate measurements of the energy dependence of elliptic flow will have an increased sensitivity to the EOS. A careful delineation of the energy dependence of collective flow effects may provide meaningful constraints on the equation of state at high and low densities. (Note that it is not sufficient to determine the compressibility, K, of nuclear matter near its ground state, since different parameterizations extrapolate differently to high densities, even if K is the same.)

Additional progress, relevant for the extrapolation to the highly asymmetric nuclear matter encountered in neutron stars, can be expected by investigating reactions with nuclei of very different N/Z ratios, but fixed total mass. Such reactions can provide unique information about the symmetry-energy term of the EOS about which little is known at this time. The needed investigations can be performed with energetic beams of rare isotopes produced by projectile fragmentation at the next generation RIA (Rare Isotope Accelerator) facility currently discussed by the nuclear science community.

7 Acknowledgements

Enlightening discussions with Pawel Danielewicz, Bill Lynch, and Betty Tsang are gratefully acknowledged. This work was supported by the National Science Foundation under Grant No. PHY-95-28844.

References

1. G.F. Peaslee et al., *Phys. Rev.* **C49**, R2271 (1994).

2. D. Morrissey, W. Benenson, and W.A. Friedman, *Annu. Rev. Nucl. Part. Sci.* **44**, 27 (1994) and references given there.

3. S. Albergo et al., *Nuovo Cimento* **89**, 1 (1985).

4. W. Reisdorf and H.G. Ritter, *Annu. Rev. Nucl. Part. Sci.* **47**, 663 (1997) and references given there.

5. Pochodzalla et al., *Phys. Rev. Lett.* **75**, 1040 (1995)

6. V. Serfling et al., *Phys. Rev. Lett.* **80**, 3928 (1998).

7. H. Xi et al., *Phys. Rev.* **C58**, R2636 (1998).

8. P.J. Siemens and J.O. Rasmussen, *Phys. Rev. Lett.* **42**, 880 (1979).

9. D.R. Bowman et al., *Phys. Rev. Lett.* **67**, 1527 (1991).

10. W.C. Hsi, *Phys. Rev. Lett.* **73**, 3367 (1994).

11. W.A. Friedman, *Phys. Rev. Lett.* **69**, 2135 (1988).

12. D.H.E. Gross and Xiao-ze Zhang, *Phys. Lett.* **B161**, 47 (1985); D.H.E. Gross, *Rep. Prog. Phys.* **53**, 605 (1990); D.H.E. Gross et al., *Ann. Phys.* **1**, (1992) 467.

13. J.P. Bondorf et al., *Nucl. Phys.* **A443**, 321 (1985); *Nucl. Phys.* **A444**, 460 (1985); *Phys. Reports* **257**, 133 (1995).

14. W.A. Friedman, *Phys. Rev.* **C40**, 2055 (1989).

15. D.H.E. Gross and K. Sneppen, *Nucl. Phys.* **A567**, 317 (1994).

16. S.E. Koonin, *Phys. Lett.* **B70**, 43 (1977).

17. S. Pratt and M.B. Tsang, *Phys. Rev.* **C36**, 2390 (1987).

18. D.H. Boal et al., *Rev. Mod. Phys.* **62**, 553 (1990).

19. W. Bauer et al., *Annu. Rev. Nucl. Part. Sci.* **42**, 77 (1992).

20. M.A. Lisa et al., *Phys. Rev. Lett.* **71**, 2863 (1993).

21. G.J. Kunde et al., *Phys. Rev. Lett.* **70**, 2545 (1993).

22. S.J. Gaff et al., *Phys. Rev.* **C52**, 2782 (1995).

23. D.O. Handzy et al., *Phys. Rev. Lett.* **75**, 2916 (1995).

24. Z. Chen and C.K. Gelbke, *Phys. Rev.* **C38**, 2630 (1988).

25. M.J. Huang et al., *Phys. Rev. Lett.* **78**, 1648 (1997), and references therein.

26. T. Glasmacher et al., *Phys. Rev.* **C50**, 952 (1994).

27. Y.D. Kim et al., *Phys. Rev.* **C45**, 387 (1992).

28. R. Popescu et al., *Phys. Rev.* **C58**, 270 (1998).

29. T. Glasmacher et al., *Phys. Lett.* **B314**, 265 (1993).

30. D.O. Handzy et al., *Phys. Rev.* **C51**, 2237 (1995.

31. D.A. Brown and P. Danielewicz, *Phys. Lett.* **B398**, 252 (1997).

32. M.B. Tsang et al., *Phys. Rev. Lett.* **57**, 559 (1986); *Phys. Rev. Lett.* **60** (1988) 1479; R.C. Lemmon et al., *Phys. Lett.* **B446**, 197 (1999).

33. P. Danielewicz and G. Odyniec. *Phys. Lett.* **B157**, 146 (1985).

34. M.J. Huang et al., *Phys. Rev.* Lett. **77**, 3739 (1996).

35. F. Rami et al., *Nucl. Phys.* **A646**, 367 (1999).

36. C. Gale et al., *Phys. Rev.* **C35**, 1545 (1990).

37. R. Pak et al., *Phys. Rev.* **C54**, 2457 (1996); *Phys. Rev. Lett.* **78**, 1026 (1997).

38. P. Danielewicz et al., *Phys. Rev. Lett.* **81**, 2438 (1998).

39. Bao-An Li and A.T. Sustich, *Phys. Rev. Lett.* **82**, 5004 (1999); Yu-Ming Zheng et al., *Phys. Rev. Lett.* **83**, 2534 (1999).

40. C. Pinkenburg et al., *Phys. Rev. Lett.* **83**, 1295 (1999).

ELECTRIC DIPOLE TESTS OF TIME REVERSAL SYMMETRY

NORMAN F. RAMSEY

Lyman Physics Laboratory, Harvard University, Cambridge, MA 02138 USA

If there is either T or P symmetry, there can be no electric dipole moment for particles whose orientations are fully specified by the orientations of their spin angular momenta. As a result, there have been extensive searches for electric dipole moments as tests of these symmetries. CP non-conservation has been observed, which implies non-conservation of T if CPT is conserved, but it has only been seen in the decay of the artificially produced K_L^0. The history of the searches for neutron and atomic electric dipole moments is summarized. The earliest experiments utilized a neutron beam magnetic resonance apparatus, but the present neutron experiments study ultra cold neutrons trapped in bottles whose walls totally reflect the neutrons. Recent atomic and molecular searches are sensitive both to particle electric dipole moments and to other related T non-conserving quantities. The various neutron, atomic and molecular experiments are described. The results of these searches are compared with the predictions of theories that account for the known CP non-conservation in the decay of the K_L^0. The experiments directly set limits to T-odd, P-odd interaction terms, but through electroweak radiative corrections they also set limits to T-odd, P-even interactions.

1 Introduction

It is an honor to speak at the One Hundredth Anniversary of Gregory Breit, who contributed so much to Twentieth Century physics.

As a student at Cambridge University in 1936, I first learned of the beautiful Breit-Rabi formulae for variations of atomic energies and effective magnetic moments with changing external magnetic fields. These formulae provided the basis for I. I. Rabi's elegant measurements of nuclear spins, hyperfine structures and magnetic moments. This research attracted me to Rabi's laboratory where I had the good fortune to arrive just a few months before he invented the molecular beam magnetic resonance method so I wrote the first Ph.D. thesis using magnetic resonance. Even after the resonance method was invented, the formulae of Breit and Rabi were the basis for interpreting our resonance spectra.

As a post-doc, I was doing neutron and proton scattering experiments, I again found that many of the fundamental theoretical papers for that research were written by Gregory. Later when I was a Columbia and we were puzzled why Nafe and Nelson's measurement of the atomic hydrogen hyperfine structure differed from the predictions of the Dirac theory, Gregory Breit was the first to suggest that it could be due to an electron anomalous magnetic moment. From that time on, my subsequent research frequently led to theoretical papers by Breit.

The electric dipole moment of a particle is defined as follows. Let the coordinate z be measured from the center of mass of a particle and let ρ_{JJ} be the electric charge density inside the particle, whose angular momentum is **J** with quantum number J and whose orientation state is given by m = J relative to the z axis. The scalar magnitude d_n of the vector dipole moment **d** is then defined by

$$d = \int \rho_{JJ} z \, d\tau \qquad (1)$$

where dt is a differential volume element. If the particle is charged, this definition implies that the center of charge of the particle is displaced from the center of mass if $d \neq 0$. If, on the other hand, the particle has no net charge, the definition implies a greater positive charge in one hemisphere and a correspondingly greater negative charge in the other. The electric dipole moment is ordinarily expressed in units of e cm where e is the charge of the proton. An electric dipole moment **d** in an electric field E will have an interaction energy $W = - \mathbf{d} \cdot \mathbf{E}$.

It is easy to see(1-4) from the following argument that the electric dipole moment must vanish if there is symmetry under either the parity transformation (P) for which $\mathbf{r} \rightarrow - \mathbf{r}$ or the time reversal transformation (T) for which $t \rightarrow - t$, provided there is no near degeneracy. Since the orientation of the particle can only be specified by the orientation of its angular momentum, the dipole moment **d** and the angular momentum **J** must transform their signs the same way under P and T if **d** is to have a non-zero value and if there is to be P and T symmetry; if they did not, either of those transformations would produce a neutron of oppositely signed electric dipole moment contrary to the assumption of those symmetries and no degeneracy. But **d** changes sign under P whereas **J** does not, so **d** must vanish if there is P symmetry. Likewise **d** does not change sign under T but **J** does, so **d** must vanish if there is T symmetry. More rigorous proofs of the above can be given(1-4). If one makes the usual assumption of CPT symmetry (where C is the charge conjugation transformation, $Ze \rightarrow - Ze$), the existence of an electric-dipole moment would also imply a failure of CP symmetry. In molecules, electric dipole moments do exist and are attributed to degenerate states rather than to a failure of T symmetry; but degeneracy for the neutron would contradict the well-established fact that neutrons obey the Pauli exclusion principle. Since the existence of a particle electric dipole moment would contradict the assumptions of both parity and time reversal invariance, increasingly sensitive searches for an electric dipole moment have been carried out for a number of years to test these assumptions.

The earliest electric dipole searches, and still some of the most sensitive, have been those with neutrons, initially with neutron beams and later with ultra-cold neutrons stored in chambers with reflecting walls. With the increased interest in the theories of T non-conservation, there have been a number of recent experiments to measure electric dipole moments and related T non-conserving properties of various atoms and molecules.

2 Neutron Beam Experiments

In 1950, at a time when parity conservation was almost universally believed, Purcell and Ramsey (3) pointed out that such a belief must be based on experiment and that there was then little experimental evidence for it. They concluded that a search for a particle electric dipole moment would provide such a test since such a moment is forbidden under parity symmetry. However they also noted that an electrically charged particle even with an electric dipole moment would ordinarily have no electric dipole interaction with an electric field since $\langle E \rangle = 0$ if the particle is not accelerated or acted upon by other forces. As a result past experiments were not sensitive in detecting electric dipole moments. On the other hand tests with a neutron are sensitive since the neutron has no electric charge and is consequently not accelerated out of the observation region in a strong electric field. A neutron electric dipole moment can be sensitively measured in a neutron beam magnetic resonance experiment for the. If the neutron has a magnetic dipole moment μ_M and an electric dipole moment $\mathbf{d_n}$, the dipole moments will contribute an additional energy $-\mu_M \cdot \mathbf{B} - \mathbf{d_n} \cdot \mathbf{E}$ if the neutron is in a magnetic field $\mathbf{B}$ and an electric field $\mathbf{E}$. Therefore, if the neutron makes a transition from an orientation state with m = -1/2 to m = +1/2, the Bohr frequency ν_0 for the transition (which in this case is the neutron Larmor precession frequency) is given by

$$h\nu_0 = 2\,\mu_M \cdot \mathbf{B} + 2\,\mathbf{d_n} \cdot \mathbf{E} = 2\,\mu_M\,|B| \pm 2\,d_n\,|E| \quad (2)$$

where the upper sign is for **B** and **E** parallel. Therefore, the neutron electric dipole moment is observed by the change in the Larmor precession frequency of the neutron when the relative directions of the electric and magnetic fields are reversed.

Smith, Purcell and Ramsey (4) constructed a neutron beam magnetic resonance experiment with a strong electric field between the two separated oscillatory fields (5) and measured the change in resonance frequencies when the relative orientations of the electric and magnetic fields were reversed. When they started this experiment, the lowest limit that could be set by any existing observations was $d_n = 3 \times 10^{-18}$ e cm and in that first experiment they lowered the limit to $d_n = 5 \times 10^{-20}$ e cm but found no electric dipole moment.

In 1956 Lee and Yang (6) proposed parity non-conservation in the weak interaction as an explanation of an anomaly in the K meson decay. The proposed parity non-conservation in the weak interaction was confirmed in 1957 by Wu et al.(7) in the angular distribution of the electrons in the decay of polarized ^{60}Co and by others. Although the discovery of the failure of parity symmetry eliminated that symmetry as a fundamental argument against the possibility of a particle electric dipole moment, Landau (8) and others (9) argued that even with the failure of P symmetry there should still be conservation of CP and T (where C is the charge exchange operator) and that T symmetry alone would eliminate the possibility of a particle electric dipole moment. However, at that time Ramsey wrote a letter (10) pointing out that the T symmetry assumption must be based on experiment and that a search for a neutron electric dipole moment was a particularly sensitive test of T conservation; Jackson, et al. (11) also emphasized the need for an experimental basis for the assumptions of T and CP symmetry. In 1964, Christenson, Cronin, Fitch and Turlay (12) showed experimentally there was a failure of CP conservation in the decay of the K_L^0 and hence a failure of T if TCP were conserved. The subsequent theories that accounted for the observed CP violation in the K_L^0 predicted a non-zero value for the neutron electric dipole moment, with the initial predictions being usually larger than 10^{-23} e cm. As a result there was a series of experiments (9), mostly by the neutron beam magnetic resonance method, which successively lowered the experimental limit on the neutron electric dipole moment. The most sensitive of the neutron beam experiments (6) was that of Dress, Miller, Pendlebury, Perrin and Ramsey (13). Neutrons from the Grenoble reactor at 180 m/s were polarized and analyzed by reflection from magnetized iron mirrors between which there was a 180 cm long region with a 17 G magnetic field and a 100 kV/cm electric field. Neutron reorientation transitions were induced by the method of separated oscillatory fields (14,15). No electric dipole moment was found and the result was equivalent to an experimental electric dipole moment of $(40 \pm 150) \times 10^{-26}$ e cm.

3 Trapped Neutron Experiments

In the neutron beam experiments so far described, the neutrons had velocities of about 80 m/s and the neutron resonance widths were typically 80 Hz. It was early realized by Ramsey,(9,16) Zel'dovich (17) and Shapiro (18) that neutron electric dipole moment measurements could be made on neutrons slower than 7 m/s stored in suitable bottles by total reflection of the neutrons at the bottle surface. Such measurements are more sensitive because of the narrower neutron magnetic resonance and are free from uncertainty due to the effective magnetic field **E** x **v**/c due to the motion of the neutrons through the applied electric field. The realization of measurements with bottled neutrons was delayed for many years due to the unavailability of neutrons with such low velocity. However, experiments with stored neutrons are now being successfully carried out at both the Insitut Laue-Langevin in Grenoble, France (20,27) and the Petersburg Nuclear Physics Laboratory in Gatchina, Russia.(19,28).

In the Harvard-Sussex-Rutherford-Washington-ILL experiments at Grenoble (20,27), the ultracold neutrons from the liquid deuterium moderator at the reactor core are further slowed by total reflection from the blades of a rotating turbine (21) and then transmitted to the resonance apparatus through a guide tube. The resonance apparatus is similar to that in Figure 1. Since the indicated magnetized foil totally reflects the neutrons of one polarization while transmitting the other, it serves as a polarizer for the incident beam and an analyzer for the returning beam. The neutrons are stored in a 20 liter storage cell composed of a hollow upright quartz cylinder that is closed at each end by coated aluminum electrodes. After filling for about 10 seconds, a shutter is closed, storing the neutrons for approximately 130s. The neutron density is about 2 cm^{-3}. The oscillatory field is applied in two short initial and final coherent pulses. After the shutter is opened the neutrons again pass through the foil, which this time serves as an analyzing foil which transmits only neutrons whose spin orientation has not changed. Consequently the numbers of neutrons reaching the UCN detector provide a measure of the neutron magnetic resonance. The neutrons are counted at the steepest part of the resonance curve as the directions of the electric and magnetic fields are successively reversed. From the correlation of the detected neutron intensity with the relative directions of the electric and magnetic fields, a limit can be set to the neutron electric dipole moment from Equation (2). The magnetic and electric fields were about 1 μT and 4.5 kV/cm.

The principal uncertainty in the early results came from possible magnetic field changes that might occur when the electric field is reversed. For this reason the current from the high voltage generator was carefully monitored and kept at a low value. In addition, three rubidium magnetometers were located as close as possible to the neutron storage volume to monitor possible magnetic field changes. However, these magnetometers did not always vary together in the same way and none of them were in the same volume as the neutrons. As a result, in the latest experiments, the magnetic field was monitored with a ^{199}Hg magnetometer occupying the same volume as the neutrons, as shown in Fig. 1. The ^{199}Hg is optically pumped to an orientation state parallel to **B** and introduced into the main volume. The ^{199}Hg spin is flipped 90° at the beginning of a run so that the spin precesses around **B** at a frequency proportional to **B** and this frequency can be accurately measured by the intensity modulation of a circularly polarized light at the pumping frequency in the direction perpendicular to **B**. When the ^{199}Hg magnetometer is used to adjust the neutron measurements for magnetic field variations, most of the previously observed fluctuations disappeared. Typically, the neutron counting rate is 30,000 per cycle with each measurement cycle taking 210s.

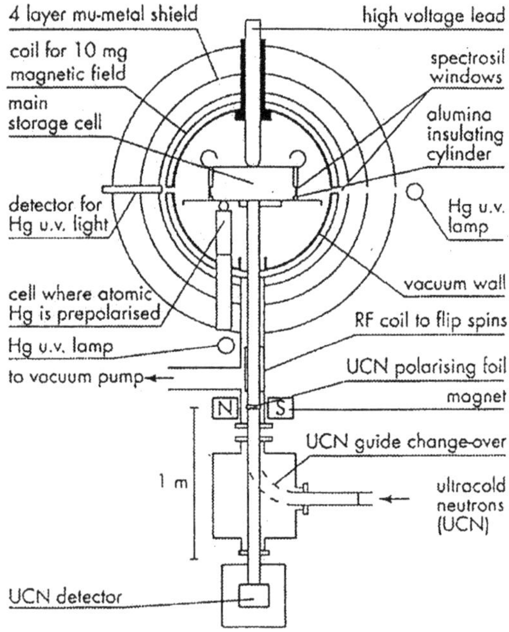

FIGURE 1. Schematic diagram of new Grenoble apparatus. The apparatus used in the most recently completed experiment is similar to that in the diagram except that no Hg magnetometer was provided.

The use of stored ultra-cold neutrons offers two particularly important advantages over the original neutron beam experiments. The resonance curve with stored neutrons is 7000 times narrower than in the best neutron beam experiments. Furthermore, in the neutron beam experiments a large fraction of the running time had to be devoted to eliminating the **E** x **v**/c effect, which effectively vanishes in the neutron bottle experiments since the average value of v is very small compared to c when the neutrons enter and exit at the same port with a storage time of 80s.

The Gatchina group (19,28) uses a similar stored neutron method with the principal difference being that their neutrons are stored in a double chamber with the high voltage electrode in the middle so the electric fields in the two chambers are in opposite directions.. This helps to cancel the effects of drifting external magnetic fields but does not compensate as well for local magnetic effects due to sparks and leakage currents from the applied electric field. So far the Gatchina group has not used separate magnetic field monitors in the same volume as the neutron.

The latest published result of the Grenoble group (27) is

$$d_n = (-1\pm3.6) \times 10^{-26} \text{ e cm} \tag{3}$$

where the estimated statistical error precedes the systematic error. This null result translates to an upper limit of $| d_n | < 7.5 \times 10^{-26}$ e cm to a 95% confidence limit. The latest published result for the neutron electric dipole moment as measured by the Gatchina group (28) is .

$$d_n = (+2.6 \pm 4.2 \pm 1.6) \times 10^{-26} \text{ e cm} \quad (4)$$

where the first error is from statistics and the second from systematics and the authors conclude $|d_n| < 11 \times 10^{-26}$ e cm. (95% C.L.) The combination of the Grenoble and Gatchina experiments together corresponds to an upper limit of

$$|d_n| < 7 \times 10^{-26} \text{ e cm to a 95\% C.L.} \quad (5)$$

As mentioned in the introduction the electric dipole measurements also provide limitations to the T-odd, P-even interactions. The experimental limit on d_n in Equation (5) provides the following limits to the T-odd, P-even parameters defined by Conti and Khriplovich (57,58).

$$\beta_{qe} < 10^{-6} \quad \text{and} \quad \beta_{qq} < 0.3 \times 10^{-6}. \quad (6)$$

Both the Grenoble and the Gatchina groups are planning to increase the sensitivity of their experiments Ringo and his associates (54) have discussed a quite different neutron experiment utilizing an interferometer with spatially separated paths.

As a longer term project, Golub, Lamoreaux, Pendlebury and others (55) are planning a neutron electric dipole moment experiment with the neutrons in a totally reflecting bottle filled with superfluid helium. A dilute solution of polarized ^{3}He will be used as a polarizer and detector since ^{3}He absorbs neutrons only when the total spin is zero (neutron and ^{3}He spins antiparallel). They hope to increase the sensitivity of the limit by a factor of 100 or more.

4 Atomic and Molecular Electric Dipole Moment Experiments

The previously stated theorem of Purcell and Ramsey (3,4) that electric dipole moments of electrically charged particles are ordinarily unobservable does not preclude measurements of atomic electric dipole moments since atoms have no net electric charge. However the Purcell-Ramsey theorem has been generalized by Schiff to what is sometimes called the Schiff Theorem, which states that an assemblage of particles that are nonrelativistic, interact only electrostatically and are such that the electric dipole moment distribution within each particle is the same as its charge distribution will have no permanent electric dipole moment even if the constituent particles do so. The important part of Schiff's paper, however, is not the theorem but the recognition that in real atoms the theorem does not apply due to relativistic effects, presence of non-electrostatic forces, spin-orbit interactions and finite size effects. As first pointed out by Sandars and others (30,42,53), in some paramagnetic atoms an electron electric dipole moment can induce an atomic electric dipole moment 600 or more times bigger than that of the atom. As a result, electric dipole moment measurements with atoms and molecules provide sensitive tests for an electron electric dipole moment and for other T-nonconserving interactions to be discussed later.

For atomic electric dipole moments a relation analogous to Equation (2) applies except that the 2's in Equation (2) should be replaced by 1/F where F is the quantum number for the total atomic spin angular momentum **F** = **I** + **J** where **I** and **J** are the nuclear and resultant electronic angular momenta. Consequently, just as with neutrons, the atomic electric dipole moments can be measured by the shift in the magnetic resonance frequency. The resonance pattern for atoms has been observed in different ways, sometimes with a single atomic beam (30-33,35), sometimes with two atomic beams moving in opposite directions (37) and sometimes with circularly polarized lasers beams used for the optical pumping and detecting of atoms stored in bottles with suitable wall coatings (34,36,38,39,40,46,).

The earliest atomic experiments were the 1960's cesium atomic beam experiments of Sandars, Lipworth and their associates (30-31) at Brandeis and Oxford. Sandars (30,31) showed that due to relativity and the admixture of a nearby p state, an electron electric dipole moment in some paramagnetic atoms can induce an atomic moment several hundreds times larger. In this fashion they set an experimental upper limit (30,31) on the electron electric dipole moment of 300×10^{-26} e cm.

There was no work on atom or molecule electric dipole moments in the 1970's, but since the mid 1980's a number of such experiments have recently been completed. With a molecule instead of an atom, Hinds and Sandars at Oxford (32) have set limits on T violating parameters in TlF with multiple resonance transitions. Wilkening, Ramsey and Larson at Harvard (33) improved the sensitivity of these results. Even greater sensitivity has been achieved in later experiments at Yale by Hinds and his associates (35) using a jet beam of TlF.

Fortson, Heckel, Lamoreaux and their collaborators (34,38,39,47) at Seattle have studied ^{199}Hg with the apparatus shown schematically in Fig.2. Each of the two vapor filled cells functions as an optically pumped oscillator based on the Larmor precession of the nuclear spins about B_z. The phase of the precession is detected by the modulation of the transmitted light intensity. The phase shift of this modulation with reversal of the applied electric field measures the electric dipole moment of the atom.

Hunter and his collaborators at Amherst (36), have set a low limit on the electric dipole moment of the electron using optical pumping of a ^{133}Cs cell.

Commins (37) and his associates at Berkeley have used the method of separated oscillatory fields and two counter propagating atomic beams as in Figure 3 to measure the electric dipole moments of the ^{205}Tl atom and of the electron. The counter-propagating beams are used to minimize the **E** x **v** effect. Polarized laser beams were used as state selector and analyzer.

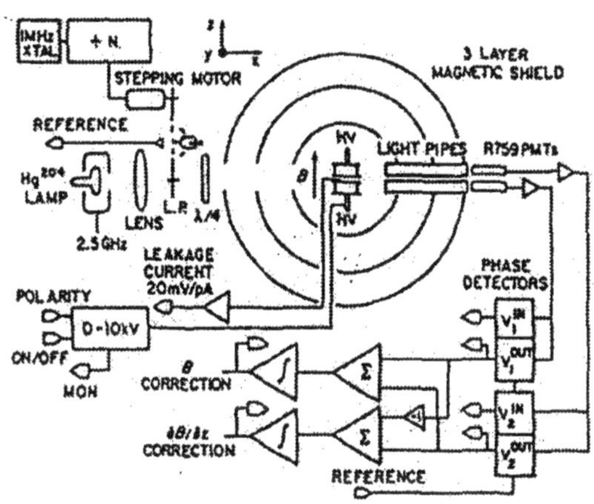

FIGURE 2. Schematic Diagram of the ^{199}Hg Apparatus (39).

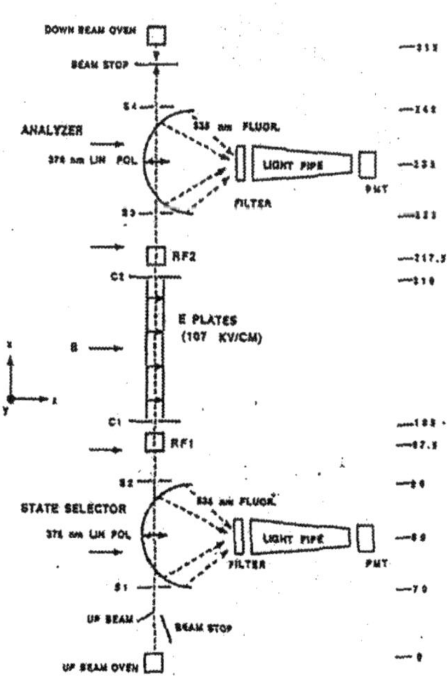

FIGURE 3. Schematic Diagram of the ^{205}Tl Apparatus (37).

Chupp (40) and his students at Harvard and Michigan used ^{129}Xe polarized by spin exchange collisions to determine the atomic electric dipole moment.

No T non-conserving interactions have been seen so far in any of these experiments, but low experimental limits have been set on electric dipole and other T non-conserving parameters as shown in Table I. For Cs and Tl there have also been improved calculations (42) of the sensitivity enhancement factors used in determining the limit to the electron electric dipole moment from that of the atom.

Although electric dipole moments of the constituent particles can produce resultant electric dipole moments of atoms, other T non-conserving interactions between elementary particles also can give rise to atomic or molecular electric dipole moments, especially in heavy paramagnetic atoms where the effects are enhanced (30,41). A resultant electric dipole moment can be due to a nuclear dipole moment distribution from hadronic interactions between quarks, to a semi-leptonic interaction from a T violating force between electrons and nucleons or to a purely leptonic interaction from an intrinsic electric dipole moment of the electron There are now extensive theoretical calculations (41-44) of the these different contributions to a net electric dipole. There are also several recent reviews of both theories and experiments on atomic and molecular electric dipole moments (45-48). The primary experimental results from the atomic experiments are electric dipole moments of the atoms, but these can be interpreted in terms of T-non conserving interactions between the constituents of the atom and its nucleus.

Khriplovich and others (41-44) have shown that the experimental limits on the various T non-conserving interactions can be parametrized in terms of the atom electric dipole moment d, the electron dipole moment d_e, the proton electric dipole moment d_p, parameters Q_S (Schiff moment), η and η_q of the hadronic interaction terms $4\pi e Q_S \sigma \bullet \nabla \rho_e$, $i\eta(\bar{n}\gamma_5 n)(\bar{n}\, n)G_F/\sqrt{2}$ and $i\eta_q(\bar{q}\gamma_5 q)(\bar{q}\, q)G_F/\sqrt{2}$ and the parameters C_T and C_S of the semi leptonic interaction terms $iC_T(\bar{e}\ \gamma_5\ \sigma_{\mu\nu}e)(\bar{n}\ \sigma^{\mu\nu}n)G_F/\sqrt{2}$ and $iC_S(\bar{e}\ \gamma_5\ e)(\bar{n}\, n)G_F/\sqrt{2}$. The neutron and atomic experimental limits on these T non-conserving parameters are given in Table 1. In each of the atomic experiments the directly measured quantity is the limit to the atomic electric dipole moment and all of the other limits are based on calculations using the atomic limit.

All of the groups doing atomic and molecular tests of T non-conservation hope to improve their sensitivities by factors of 10 to 100 in the next few years. Chupp's group for example is planning to operate both a ^{129}Xe and a ^{3}He maser simultaneously in the same cavity so the ^{3}He can serve as a magnetometer while they look for parity non-conserving effects in ^{129}Xe. Hinds has identified YbF as a

TABLE 1. Upper Limits (95% Confidence Levels) on T Non-conserving Interaction Parameters.[47] The Parameters Are Defined in this Section Except For q_{QCD}, $e_{q/e,susy}$, ε_{Higgs} and χ^{lr}, Which are Defined by Barr[53] and the T-odd,P-even Parameters Which Are Defined by Khriplovich and Conti.[57,58] This Table Is Based in Part on Tables Prepared by Fortson[39] and Barr.[53]

System Units References	n 27.28	^{199}Hg 39	TlF 35.33	^{205}Tl 37.59	^{133}Cs 36	^{129}Xe 40
d (x10^{-26} e cm)	<7	<0.087	{d_p<16,000}	<230	<2,000	<21
Hadronic Parameters:						
Q_S (x10^{-11} efm^3)		<2.2	<100	<23,000		<8,000
η (x 10^{-6})		<1,600	<20,000			
η_q (x 10^{-6})	<31	<3.4				
θ_{QCD} (x10^{-10})	<3	<13	<60	<3,000	<4,000	<300
$\varepsilon_{q,susy}$	<0.01	<0.007	<0.08			<1.3
$\varepsilon_{e,susy}$				<0.012	<0.5	
ε_{Higgs} (x 1/tan β)		<0.7		<0.7		
χ^{lr}	<0.01	<0.002				
Semileptonic Parameters:						
C_T (x 10^{-6})		<0.013	<0.5			
C_S (x 10^{-6})		<0.68	<20	<0.3	<20	
Leptonic Parameters:						
d_e (x 10^{-26} e cm)		<6.1	<40	<0.3	<9	<20,000
T-odd,P-even Parameters:						
β_e (x 10^{-6})		<0.48	<3	<0.02	<0.7	
β_{qe} (x 10^{-6})	<1					
β_{eq} (x 10^{-6})		<0.14	<1	<0.00 7	<70	
β_{qq} (x 10^{-6})	<0.3					
I m(C_T + $C_{T'}$)		<0.004	<0.02		≤0.00015	<0.005
Im(CP + $C_{P'}$)		<0.17	<1		<0.007	<0.23

particularly sensitive molecule for molecular searches for the electron electric dipole moment.

In addition, several groups with no results to report as yet have started different and potentially sensitive electric dipole searches for T non-conserving interactions. Ezhov and his associates (61) at Gatchina plan to use jet molecular beams of TlF and other molecules observing the interference at zero magnetic field of coherent states in of two degenerate levels (Hanle effect) rather than magnetic resonance. Hansch, Weis and their associates (62) in Garching have observed Ramsey fringes in nonlinear Faraday rotation spectroscopy with a thermal Rb atomic beam using separated pumping and probing laser beam regions and they now plan to use such a spectrometer to search for permanent electric dipole moments of atoms. In quite different experiments Hansch, Weis and their collaborators (63) at Bonn propose to search with magnetic resonance for permanent electric dipole moments of paramagnetic atoms implanted in solid ^4He. Hinds at Sussex University is planning experiments with TlF and TbF. Doyle at Harvard plans a YbF experiment with buffer gas trap loading and evaporative cooling. A particularly promising new experiment is one by DeMille of Yale using a metastable excited paramagnetic state of PbO with the hope of reducing the experimental limit on the electric dipole moment of the electron by a factor of more than 100.

5 Theoretical Implications

No electric dipole moments have been observed in any neutron, atom or molecule so far, but the experiments have placed low experimental limits on the moments and on possible internal T non-conserving interactions. The present limits on electric dipole moments are extremely low. A neutron electric dipole moment of 10^{-26} e cm corresponds to a displacement of a proton charge by only 10^{-26} cm or 10^{-13} of the neutron diameter; if the neutron were scaled up to the dimensions of the earth such a displacement even on the scale of the earth would be only 0.001 mm.

For comparing the sensitivity of neutron electric dipole measurements with limitations on time reversal symmetry from nuclear reaction studies, Boehm (22) uses a simple dimensional argument to obtain the corresponding value of the T violating coupling constant g_T and finds that an electric dipole moment 10^{-26} e cm corresponds to a g_T of 10^{-6} which is far below the limit set by nuclear reaction studies.

The earliest predictions for the neutron electric dipole moment were exactly zero based on assumed parity symmetry. After parity conservation was shown to fail in the weak interaction (6,7) a prediction of zero was continued on the basis of assumed time reversal symmetry. Subsequent to the observed failure (12) of CP conservation in the decay of the K_L^0 most theories predict non-zero values for the neutron electric dipole moment. The predictions for d_n range from 10^{-19} e cm to less than 10^{-33} e cm. Most of the early theories, which attributed the known CP non-conservations to electromagnetic theory or to new milliweak or millistrong forces, predicted values for the neutron electric dipole moment larger than 10^{-23} e cm. These theories have now been mostly abandoned due in part to the disagreements between the predictions and the above experimental limits.

In the most popular current theories (23-26, 48-53) for predicting electric dipole moments, the T non-conservation comes from non-trivial physical complex phases in the theories of the particle fields.

(a) Strong QCD theories originate from the Lagrangian of QCD (45,48) which can have can have a P and T violating topological density term $\bar{\theta}(g^2 G^a_{\mu\nu} G^{a\mu\nu})/(32\pi^2)$. This term predicts the neutron electric dipole moment d_n to be approximately equal to $2.3 \times 10^{-16}\ \bar{\theta}_{QCD}$. e cm where $\bar{\theta}_{QCD}$. is the strong CP violating parameter. Consequently, the neutron measurements placing a 95% confidence limit on the neutron electric dipole moment of $d_n < 9 \times 10^{-26}$ e cm imply $\bar{\theta}_{QCD} < 4 \times 10^{-10}$. For ^{199}Hg the theory predicts the electric dipole d_A to be approximately $0.0065 \times \bar{\theta}_{QCD}$. So the smaller atomic Hg electric dipole moment limit of $d_A < 0.063 \times 10^{-26}$ e cm actually provides the weaker restriction $\bar{\theta}_{QCD} < 10 \times 10^{-10}$. All other atomic experiments so far are much less sensitive than the neutron experiments in limiting $\bar{\theta}_{QCD}$. The incredibly low limit on $\bar{\theta}_{QCD}$ is often called the strong QCD problem and requires an explanation. Peccei and Quinn (64) have provided a mechanism to explain the small value of $\bar{\theta}_{QCD}$, but it implies (65) the existence of a Goldstone boson called the axion with a mass of approximately 200 keV but later modifications permit a much wider range of mass values (45,48,64-73). Although there have been many axion searches, no axion has been discovered so far.

(b) Theories (23) that attribute CP violations to the exchange of Higgs bosons. Most models with three Higgs doublets and maximal CP violation predict values greater than 10^{-25} e cm. Weinberg (23) points out that, with the present neutron electric dipole moment value, neutral Higgs exchange can not be maximally CP violating.

(c) Theories (24,25) that attribute CP non-conservation to the standard model of quantum chromodynamics with a complex phase factor δ in the Kobayashi-Maskawa matrix (24,48) and with the strong phase θ_{QCD} taken to be zero usually predict values between 10^{-31} and 10^{-33} e cm, well below the next experiments. Most theories that go beyond the standard model, however, have more parameters and room for more complex CP non-conserving phases; they are usually consistent with larger electric dipole moment values that should be within the reach of future experiments.

(d) In a class of left-right symmetric models (48,51), the predicted moments are (1 to 10) x 10^{-26} e cm. The present experimental limits on (e) Flavor mixing models predict d_n to be of the order of 1 x 10^{-26} e cm.

(e) Supersymmetry (SUSY) permits the transformation of fermions into bosons and vice versa and adds many more possible particles which in turn provides the opportunity for more CP non-conserving phases. As a result T-violating phases occur naturally in supersymmetric theories. In some SUSY theories that account for the observed CP violation in the K^0 - $\bar{K}^0$ system, the parameter $\varepsilon_{q,susy}$ should be of order 1 if supersymmetry is broken near the electroweak energy scale (53), whereas by both the neutron and ^{199}Hg experiments in Table I $\varepsilon_{q,susy}$ < 0.01. This may indicate that $\varepsilon_{q,susy}$ is not the only source of CP violation (48).

(f) Cosmological theories (26) that account for the baryon-antibaryon asymmetry and relate the neutron electric dipole moment to the entropy of the universe generated since bariosyntheses are generally consistent with a value smaller than 10^{-26} e cm.

The atomic and molecular experiments on electric dipole moments and related T non-invariant parameters have been extensively analyzed and reviewed by theorists (43-53). Although the standard model theoretical limit for the neutron electric dipole limit is beyond the reach of any proposed experiment, most of the theories that go beyond the standard model can account for values that could be observable in future experiments. Since no one yet knows the way in which CP is not conserved, it is impossible to predict whether atomic or neutron electric dipole moment experiments will be the first to show a failure of T symmetry. For some theories the neutron experiments provide the most sensitive tests while other theories are more sensitively probed by atomic experiments. For example, from Table 1 it is apparent that the neutron experiments are now the most sensitive for limiting $\bar{\theta}_{QCD}$, whereas the atomic experiments are needed for C_S and C_T. The neutron and atomic experiments are complementary and both need to be vigorously pursued.

References

1. Ramsey, N. F., *Molecular Beams*. Oxford Univ. Press (1956,1985).
2. Golub, R. and Pendlebury, J. M., *Contemp. Phys.* **13**, 519 (1972).
3. Purcell, E. M. and Ramsey, N. F., *Phys. Rev.* **78**, 807 (1950); Ramsey, N. F., *AIP* , **270**, 179 (1993).
4. Smith, J. H., Purcell, E. M. Ramsey, N. F., *Phys. Rev.* **108,** 120 (1957).
5. Ramsey, N. F., *Physics Today* **33**, 25 (1980).
6. Lee, T. D. and Yang, C. N., *Phys. Rev.* **104**, 254 (1956).
7. Wu, C. S., Ambler, E. Hayward, B. W., Hoppes, D. D. and Hudson, R. R., *Phys. Rev.* **105**, 1413 (1957).
8. Landau, L., *Nucl. Phys.* **3**, 127 (1957).
9. Ramsey, N. F., *Ann. Rev. Nucl. Part. Science* **32**, 211 (1982) and 40, 1 (1990). See also reference (70). These articles contain full lists of references.
10. Ramsey, N. F., *Phys. Rev.* **109**, 225 (1958).
11. Jackson, J. D., Treiman, S. B. and Wyld, H. W. Jr., *Phys. Rev.* **106**, 517 (1957).
12. Christenson, J. H. Cronin, J. W. Fitch, V. L. and Turlay, R., *Phys. Rev. Lett.* **13**, 138 (1964).
13. Dress, W. B., Miller, P. D., Pendlebury, J. M., Perrin, P. and Ramsey, N. F., *Phys. Rev. D* **15**, 9 (1977).
14. Ramsey, N. F., *Phys. Rev.* **78**, 695 (1950).
15. Ramsey, N. F., *Phys. Today* **33**, 25 (1980).
16. Ramsey, N. F., *Rev. Sci. Inst.* **28**, 57 (1957).
17. Zel'dovich, Y. A. B., *Zh. Eksp. Teor. Fiz.* **36,** (1959); *JETP* **9**, 1389 (1959).
18. Shapiro, F. L., *Usp. Fiz. Nauk.* **95**, 145 (1968); *Sov. Phys. Usp.* **11**, 345 (1968).
19. Altarev, I. S. Barisov, Uy. V., Brandin, A. B., Egerov, A. I., Ezhov, V. F., Ivanov, S. N., Loboshov, V. M., Nazarenko, V. A., Porsev, G. D., Ryabov, V. L., Serebov, A. P., Taldaev, R. R. and Shulgina *Nucl. Phys.* **A341**, 269 (1980)..
20. Pendlebury, J. M., Smith, K. F., Golub, R., Byrne, J., McComb, T. J. L., Sumner, T. J., Burnett, S. M., Taylor, A. R., Heckel, B., Ramsey, N. F., Green, K., Morse, J., Kilvington, A. I., Baker, C. A. and Clark, S. A., *Phys. Lett.* **136B**, 327 (1984).
21. Steyerl, A., Nagel, H., Schreiber, F. X., Steinhauser,, K. A., Gläser, W., Ageron, P., Astruc, J. M., Drexel, W., Gervais, R. and Mampe, W., *Condensed Matter* **50**, 281 (1983) and ILL Report ICSU/AB:07.
22. Boehm, F., *Comments Nucl. Part Phys.* **11**, 251 (1983).
23. Weinberg, S., *Phys. Rev. Lett.* **37**, 367 (1976) and 63, 2333 (1989).
24. Kobayashi, M. and Maskawa, K., *Prog. Theor. Phys.* **49**, 652 (1973).
25. Nanopoulos, D. V., Yildiz, A. and Cox, P., *Ann. Phys. (NY)* **127**, 126 (1980).
26. Hinds, E. A. and Sandars,P. G. H., *Phys. Rev.* **A21**, 471 and 480 (1980); Wilkening, D. A., Ramsey, N. F., and Larson, D. J., Phys. Rev. A29, 425 (1984).

27. Smith, K. F., Crampin, N., Pendlebury, J. M., Richardson, D. J., Shiers, D.,, Green, K., Kilvington, A. I., Moir, J., Prosper, H. B., Thompson, D., Ramsey, N. F., Heckel, B. R., Lamoreaux, S. K., Ageron, P., Mampe, W. and Steyrl, A., *Phys. Lett.* **B234**, 191 (1990).
28. Altarev, I. S. Barisov, Uy. V., Borovikova, N. V., Kolomensky, E. A., Lasakov, M. S., Ivanov, S. N., Loboshov, V. M., Nazarenko, V. A., Pirozhkova, A. N., Serebov, A. P., Sobolev, Yu. V., Shulgina, E. V., and Yegerov, A. I., *Phys. Lett.*, **B276**, 242 (1992).
29. Ramsey, N. F., *Acta. Phys.Acad. Sci. Hungerica*, **55**, 117 (1984).
30. Sandars, P. G, H,. Lipworth, E. and Player, M. A., *Phys. Rev. Lett.* **13,** 718 (1964); *Phys. Lett,* **14**, 194 (1965), **19**, 1396 (1967) and **22**, 290 (1966); and *J. Phys.* **B3,** 1620 (1970).
31. Weisskopf, M. C., Carrico, J. P., Gould. H., Lipworth, E. and Stein, T. S., *Phys. Rev. Lett.* **21**, 1645 (1968).and **24**, 1091 (1970).
32. Hinds, E. A. and Sandars, P. G. H., *Phys. Rev.* **A21**, 471 and 480 (1980).
33. Wilkening, D. A. Ramsey, N. F. and Larson, D. J., *Phys. Rev.* **A23**, 425 (1984).
34. Vold, T. G., Raab, F., Heckel, B., and Fortson, E. N., *Phys. Rev. Lett.* **52**, 2229 (1984), *Nucl. Instrum. Methods,* **A284**, 43 (1989).
35. Cho, D., Sangster, D., and Hinds, E. A., *Phys. Rev. Lett.* **63**, 2559 (1989); Hinds, E. A., *Rencontre de Moriond.* **26**, 389 (1991).
36. Murthy, S. A., Krause, D., Jr., Li, Z. L., and Hunter, L. R., *Phys. Rev. Lett.* **63**, 965 (1989); Hunter, L. R., *Rencontre de Moriond.* **26**, 395 (1991)
37. Abdullah, K., Carlberg, C., Commins, E. D., Gould, H., and Ross, S. B., *Phys. Rev. Lett.* **65,** 2347 (1990); *ICAP* **12**, 442 (1991); *AIP*, **270**, 34 (1993); *Physica Scripta*, T46, 92 (1993).
38. Lamoreaux, S. K., Jacobs, J. P., Heckel, B. R., Raab, F. J., and Fortson, N., *Phys. Rev. Lett.* **59**, 2275 (1987).
39. Jacobs, J. P., Klipstein, W. M., Lamoreaux, S. K., Heckel, B., and Fortson, E. N., *Phys. Rev. Lett..* **71**, 3782 (1993); *Phys. Rev..* **A52**, 3521 (1995).
40. Hoare, R. J., Oteiza, E. R. and Chupp, T. E.; *ICAP* **13**, 73 (1993) and *AIP* **270**, 84 (1993). Oteiza, E. R., Ph.D. Thesis Harvard University (1993).
41. Bouchiat, C., *Phys. Lett.* **57B**, 284 (1975); Hinds, E. A., Loving, C. E., and Sandars, P. G. H., *Phys. Rev. Lett.* **62B**, 97 (1976).
42. Martensson-Pendrill, A. M., *Phys. Rev. Lett.* **54**, 1153 (1985); Lindroth, E. and Martensson-Pendrill, A. M., *Journ Phys.,* **B23**, 3417 (1990).
43. Flambaum, V. P., Khriplovich, I. B. and Sushkov, O. P., *JETP* **60**, 873 (1984) and *Nucl, Phys.* **A449**, 750 (1986); Khriplovich, I. B., in *Atomic Physics* **11**,113 (1989) [edited by Haroche, Gay, J. C., and Grynberg, G.,World Scientific, Singapore, 1989]; Khatsymovsky, V. M., Khriplovich, I. B., and Yelkhovsky, A. S., *Ann. Phys.* **186**, 1 (1988); Khatsymovsky, V. M. and Khriplovich, I. B., Phys. Lett. **B296**, 219 (1992).

44. Johnson, W. R., Guo, D. S., Idrees, M., and Sapirstein, J., *Phys. Rev.* **A32,** 2093 (1985); **34**, 1043 (1986); Martensson-Pendrill, A. M., Ynnerman, A., and Oster, *P.,J. Phys.* **B22**, 2447 1989).
45. Barr, S. M. and Marciano, W. J., in *CP Violation*, edited by C. Jarlskrog (World Scientific, Singapore, 1989); Barr, S. M. and Zee, A., *Phys. Rev. Lett.* **65**, 21 (1990).
46. Hunter, L. R. Science, **252**, 73 (1991).
47. Fortson,N., private communication.
48. He, X. G. McKellar, B. H. G. and Pakvasa, S., *Int.J. Mod. Phys.*, **A4**, 5011 (1989) and private communication.
49. Wolfenstein, L., *Ann. Rev. Nucl. Part. Phys.* **35**, 137 (1986).
50. Ellis, J., *Nucl. Inst. & Methods in Phys. Research* **A284**, 33 (1989); and Nature, **344**, 197 (1990).
51. Mohapatra, R. N. and Pati, J. C. *Phys. Rev.* **D11**, 566 (1974).
52. Ellis, J., Gaillard, M. K., Nanopulos, D. V. and Rudaz, S., *Nature* **293**, 41 (1981).
53. Barr, S. M., *Int. J. Mod. Phys.* **A8**, 209 (1993) and private communication.
54. Friedman, M. S., Peshkin, P., Ringo, G. R. and Dombeck, T. W., *Proceedings of Third Symposium on Fundamentals of Quant. Mechs.*, **QM3**, 43 (1989).
55. Golub, R. and Lamoreaux, S. K, *J. Physique* **44**, L321 (1983) and *Phys. Reports* **237**, 2 (1994).
56. Pendlebury, J. M., *Nucl. Phys.* **A546**, 359 (1992).
57. Khriplovich, I. B., *Nuclear Physics*, **B352**, 385 (1991).
58. Conti, R. S. and Khriplovich, I. B., *Phys. Rev. Lett..*, **68**, 3262 (1992).
59. Commins, E. D., Ross, S. B., DeMille, D. and Regan, B. C., Private communication to be published in *Phys. Rev. A.*
60. Schiff, L. I., *Phys. Rev.* **132**, 2194 (1963).
61. Ashkinadzi, B. N., Ezhov, V. F., et al., *Petersburg Nuclear Physics Institute Preprint* (1992).
62. Weis, A., Schuh, B. Kanorsky, S. I. and Hansch, T. W., *Technical Digest, EQEC 93*, Florence, Italy (1993) and *Optics Comm.* **100**, 451 (1993)
63. Arndt, M., Kanorsky, S. I., Weis, A. and Hansch, T. W., *Phys. Lett.* **A174**, 298 (1993).
64. Peccei, R. D. and Quinn, H. R., *Phys. Rev. Lett.* **38,** 1440(1977)
65. Weinberg, S. , *Phys. Rev. Lett.* **40,** 223 (1978).
66. Sikivie, P. , *Phys. Rev. Lett.* **52,** 695 (1984) and 1995 Moriond Conference.
67. Jones, C. and Melissimos, A., *Cosmic Axions.* Published by World Scientific (Singapore 1990).
68. Sikivie, P., Van Bibber K., Raffelt, G. and Matsuki, S., 1995 Moriond Conference.66.Wilczek, F., *Phys. Rev. Lett.* **40,** 279 (1978).
69. Wilczek F., *Phys. Rev. Lett.* **40,** 279 (1978).

70. Ramsey, N. F. , *Atomic Physics.* **14,** 3 (1994) (AIP Conference Proceedings 323).
71. Turner, M. S., *Phys. Rep.* **197,** 67 (1990).
72. Raffelt, G. G., *Phys. Rep.* **198,** 1 (1990).
73. Turner, M. S., *Phys. Rev. Lett.* **66,** 5 (1991).

LABORATORY STUDIES OF ASTROPHYSICAL NUCLEAR REACTIONS

C. ROLFS

Institut für Physik mit Ionenstrahlen, Ruhr-Universität Bochum, Germany

1 Introduction

Investigations during the last 60 years have shown[1,2] that we are connected with distant space and time not only by our imaginations but also through a common cosmic heritage: the chemical elements that make up our bodies. These elements were created by nuclear reactions in the hot interiors of remote and long-vanished stars over many billions of years. Their nuclear fuels finally spent, these giant stars met death in cataclysmic explosions, scattering far and wide the atoms of heavy elements synthesized deep within their cores. Eventually this material, as well as material lost by smaller stars during red-giant stages, collected into clouds of gas in interstellar space; these, in turn, slowly collapsed giving birth to new generations of stars, thus leading to a cyclic evolution that is still going on. The present picture is that all elements from carbon to uranium have been produced entirely within stars during their fiery lifetimes and explosive deaths. A few of the lightest elements were formed before the stars even existed, during the birth of the universe itself. In addition, a few of the most reactive light elements appear to have been synthesized in intergalactic space by cosmic rays. Thus, theories of nucleosynthesis have identified the most important sites of element formation and also the diverse nuclear processes involved in their production. The detailed understanding of our cosmic heritage combines astrophysics and nuclear physics, and forms what is called nuclear astrophysics. In turn, nuclear reactions are at the heart of nuclear astrophysics: they influence sensitively the nucleosynthesis of the elements in the earliest stages of the universe and in all the objects formed thereafter, and control the associated energy generation, neutrino luminosity, and evolution of stars. A good knowledge of the rates of these reactions is essential to understanding this broad picture. In the present talk, I will first present a brief outline of nuclear astrophysics, for those who may not be familiar with the subject, and then discuss a few of the most recent developments in the field.

In a stellar plasma the constituent nuclei are usually in thermal equilibrium at some temperature T. Occasionally they collide with other nuclei, whereby two different nuclei can emerge from the collision, $1+2 \rightarrow 3+4$ or $A(x,y)B$, i.e., a nuclear reaction. If the nuclear reaction Q-value, $Q = (m_1 + m_2 - m_3 - m_4)c^2$ (with m_i = nuclear masses), is positive, there is a net production of energy in the reaction for each event. Such reactions are most important for the energy production in stars. Of equal importance is the intrinsic nuclear probability that a given reaction will take place. This probability, expressed as an energy-dependent cross section $\sigma(E)$, determines how many reactions occur per unit time and unit volume: hence, together with the Q-value, $\sigma(E)$ provides important information on nuclear energy production in stars.

The cross section of a nuclear reaction is of course governed by the laws of quantum mechanics where, in most cases, the Coulomb and centrifugal barriers arising from nuclear charges and angular momenta strongly inhibit the penetration of one nucleus into another. This barrier penetration leads to a steep energy dependence of the cross section $\sigma(E)$. Other energy-dependent effects, such as resonances and their mutual interference effects, also play important roles and the energy dependence of $\sigma(E)$ can be quite complex. It is the challenge to the experimentalist to make precise $\sigma(E)$ measurements over a wide range of energies, as our fragmented knowledge of nuclear physics prevents us from predicting $\sigma(E)$ on purely theoretical grounds.

2 Nuclear Astrophysical Rates

In hot stellar matter the energies of the moving nuclei can be described by a Maxwell-Boltzmann distribution, $\Phi(E) \propto E\exp(-E/kT)$, where T is the local temperature and k the Boltzmann constant. Folding the cross section with this energy (or velocity) distribution leads to the nuclear reaction rate per pair of nuclei[3]:

$$\langle \sigma v \rangle = (8/\pi\mu)^{1/2}(kT)^{-3/2} \int_0^\infty \sigma(E)\exp(-E/kT)dE, \tag{1}$$

where v is the relative velocity of the pair of nuclei, E is the center-of-mass energy, and $\mu = m_1 m_2/(m_1 + m_2)$ is the reduced mass of the entrance channel. In order to cover the different evolution phases of stars, i.e. from main-sequence stars ($T \approx 10^7 K$) to supernovae $T \approx 10^9 K$), one must know the reaction rates over a wide range of temperatures, which in turn requires the availability of $\sigma(E)$ data over a wide range of energies.

For the important class of charged-particle-induced reactions, there is a repulsive Coulomb barrier in the entrance channel of height $E_c = Z_1 Z_2 e^2/r$, where Z_1 and Z_2 are the integral nuclear charges of the interacting particles, e is the unit of electric charge, and r is the nuclear interaction radius. Due to the tunneling effect through the barrier, $\sigma(E)$ drops nearly exponentially with decreasing energy (Fig. 1):

$$\sigma(E) = S(E)E^{-1}\exp(-2\pi\eta), \tag{2}$$

where $\eta = 2\pi Z_1 Z_2 e^2/hv$ is the Sommerfeld parameter (h = Planck constant). The function $S(E)$, defined by this equation, contains all the strictly nuclear effects, and is usually referred to as the nuclear or astrophysical S-factor. If (2) is inserted in (1), one obtains[3]

$$\langle\sigma v\rangle = (8/\pi\mu)^{-1/2}(kT)^{-3/2}\int_0^\infty S(E)\exp(-E/kT - b/E^{1/2})dE, \tag{3}$$

with $b = 2(2\mu)^{1/2}\pi^2 e^2 Z_1 Z_2 h$. Since for nonresonant reactions $S(E)$ varies slowly with energy, the steep energy dependence of the integrand in (3) is governed primarily by the exponential term, which is characterized by a peak near an energy E_0 that is usually much larger than kT. The peak is frequently referred to as the Gamow peak; for a constant $S(E)$ value over the energy region of the peak, one finds $E_0 = (bkT/2)^{2/3}$. This is the effective mean energy for a given reaction at a given temperature. Approximating the peak by a Gaussian function, one finds an effective width $\Delta = 4(E_0 kT)^{1/2}/3^{1/2}$. Thus, nuclear burning takes place predominantly over the energy window $E_0 \pm \Delta/2$, the stellar thermal energy range for which information on $\sigma(E)$ must be obtained.

Due to the steep drop of $\sigma(E)$ at subcoulomb energies, it becomes increasingly difficult to measure $\sigma(E)$ as E is lowered. Although experimental techniques have improved significantly over the years[3], extending $\sigma(E)$ measurements to lower energies (with a low-energy limit at E_L, corresponding to a reaction yield in a detector of roughly 1 event per hour), it has not been possible to measure $\sigma(E)$ at stellar thermal energies, as $E_o/E_c \approx 1/100$ for hydrogen-burning reactions (i.e. *pp*-chain and CNO-cycles) in main-sequence stars such as our sun. Instead, the measured energy dependence of $\sigma(E)$ at higher energies ($E > E_L$) must be extrapolated to stellar energies, using the $S(E)$ factor defined in (2). Such an "extrapolation into the unknown" can lead to considerable uncertainty. At energies lower than E_L there might be a change of reaction mechanism, or of the centrifugal barrier, or there might be a contribution of narrow or subthreshold resonances to $\sigma(E)$ at stellar energies. The danger of such extrapolations was strikingly demonstrated in the case of $^2H(d,\gamma)^4He$, for example, where new low-energy data increased the extrapolated values by a factor of 1000, mainly due to a change in the centrifugal barrier ($d-$ to $s-$ waves). For another example, the observed solar neutrino fluxes in the existing neutrino detectors are not consistent with the current "standard" picture of the microscopic processes in the sun. A possible solution for this solar neutrino problem may be found in one or more of the areas of neutrino physics (e.g., the recent experimental evidence for neutrino oscillations at Super-Kamiokande), solar physics (models), or nuclear physics. In view of the important and fundamental conclusions regarding non-standard physics, which may be deduced from the results of present and future solar neutrino observations, it is of utmost importance to place all of the predictions on a solid basis. The nuclear physics component of this problem involves the cross sections of the hydrogen-burning reactions extrapolated to the relevant solar energies. Yet another example is the reaction $^{12}C(\alpha,\gamma)^{16}O$, which takes place in the helium burning phase of red giant stars, at $E_0 \approx 0.30MeV$. It has been found theoretically that this reaction influences sensitively not only the nucleosynthesis of the elements between C and Fe but also the evolution of stars, the dynamics of supernovae, and the kind of remnant (neutron star or black hole) of a supernova explosion. At present, useful but still inadequate $\sigma(E)$ data are available only above $E_L = 1.3MeV$. New experimental approaches are needed to reduce the uncertainties of the extrapolations.

3 The LUNA Project

Low-energy studies of thermonuclear reactions in a laboratory at the earth's surface are hampered predominantly by background effects of cosmic rays in the detectors. Conventional passive or active shielding around the detectors can only partially reduce the problem of cosmic ray background. The best solution is to install an accelerator facility in a laboratory deep underground. As a pilot project, a 50 kV accelerator facility has been installed in the Laboratori Nazionali del Gran Sasso (LNGS), where the flux of cosmic-ray muons is reduced by a factor of 10^6 compared with the flux at the surface[4]. This unique project, called LUNA (Laboratory for Underground Nuclear Astrophysics), was designed primarily for a renewed study of $^3He(^3He,2p)^4He$ at low energies, aiming to reach the solar Gamow peak at $E_0 \pm \Delta/2 = 21 \pm 5keV$. This goal has been reached with a detected reaction yield of about 1 event per month at the lowest energy, $E = 16keV$, with $\sigma \approx 20fb$ or $2 \times 10^{-38}cm^2$. Thus, the cross section of an important reaction of the pp-chain has been directly measured for the first time at solar thermal energies (Fig. 1); in principle, extrapolation is no longer needed in this reaction. The work demonstrated the research potential of LUNA and that all of the experimental requirements in such low-rate, time-consuming experiments can be filled. Installing larger facilities at LNGS in the near future will open the possibility of improving our knowledge of other key reactions by shifting their measured low-energy limit E_L significantly closer to E_0 or even to within the Gamow peak.

In the extrapolation of $\sigma(E)$ using (2), it is assumed that the Coulomb potential of the target nucleus and projectile is that resulting from bare nuclei. However, for nuclear reactions studied in the laboratory, the target nuclei and the projectiles are usually in the form of neutral atoms of molecules and ions, respectively. The electron clouds surrounding the interacting nuclides act as a screening potential: the projectile effectively sees a reduced Coulomb barrier. This in turn leads to a higher cross section, $\sigma_s(E)$, than would be the case for bare nuclei, $\sigma_b(E)$. There is an enhancement factor, $f_{\rm lab}(E) = \sigma_s(E)/\sigma_b(E) \approx \exp(\pi\eta U_e/E)$, where U_e is the electron-screening potential energy (e.g., $U_e \approx Z_1 Z_2 e^2/R_a$, with R_a an atomic radius). Note that the $f_{\rm lab}(E)$ increases exponentially with decreasing energy. For ratios $E/U_e > 1000$, shielding effects are negligible, and laboratory experiments can be regarded as essentially measuring $\sigma_b(E)$. However, for $E/U_e < 100$, shielding effects begin to become important for understanding and extrapolating low-energy data. Relatively small enhancements arising from electron screening at $E/U_e \approx 100$ can cause significant errors in the extrapolation of cross sections to lower energies, if the curve of the cross section is forced to follow the trend of the enhanced cross sections, without correction for the screening. Note that for a stellar plasma, the value $\sigma_b(E)$ must be known because the screening in the plasma will be quite different from that in the laboratory nuclear-reaction studies, i.e., $\sigma_{plasma}(E) = f_{plasma}(E)\sigma_b(E)$, and $f_{plasma}(E)$ must be explicitly included for each situation. A good understanding of electron-screening effects in the laboratory is needed to arrive at reliable $\sigma_b(E)$ data at low energies. Experimental studies of reactions involving light nuclides have shown the expected exponential enhancement of the cross section at low energies (Fig. 1). However, the observed enhancements were in all cases significantly larger than could be accounted for from available atomic-physics models[4]. This situation is disturbing because if the effects of electron screening are not understood under laboratory conditions, they are most likely to be not fully understood in a stellar plasma. A solution to the laboratory puzzle might be found in one (or all) of the following areas: the assumed energy-loss predictions from stopping-power codes at low energies, the assumed nuclear-reaction models at energies far below the Coulomb barrier, and the assumed atomic-physics models. All of these areas require additional experimental and theoretical efforts. An improved understanding of laboratory electron screening may also help eventually to improve the corresponding understanding of electron screening in stellar plasmas. It is in the nature of astrophysics that many of the processes and most of the objects one tries to understand are physically inaccessible. Thus, it is important that those aspects that can be studied in the laboratory be rather well understood. The electron-screening project addresses one such aspect.

4 Other Rapidly Developing Experimental Techniques

In explosive scenarios with temperatures of the order of $10^9 K$ (e.g., early universe, novae, and supernovae), where E_0 approaches E_c, nuclear burning times can be greatly reduced, even down to seconds[1,2,3]. If the half-life of a radioactive nuclide is longer than or of the same order as the burning time, that nuclide will be involved in the nuclear burning and thus will contribute to the composition of the final nuclear ashes. Since H and He are the most abundant species in these scenarios, the burning consists predominantly of (p, γ) and (α, γ) capture reactions involving these radioactive nuclides. If the half-life of a nuclide is longer than a day or so, it may be made into a radioactive target: e.g. $^7Be(p,\gamma)^8B$ with $T_{1/2}(^7Be) = 53d$. However, in a great majority of interesting cases, the half-lifes are too short to make targets: e.g. $^{13}N(p,\gamma)^{14}O$ with $T_{1/2}(^{13}N) = 10$ min. In this case, a possible method is to create the radioactive nuclides in an accelerator, separate them, accelerate them in a second accelerator, and allow the Radioactive Ion Beam (RIB) to interact with a H or He target (inverted kinematics). All of this must be achieved in a time shorter than the decay lifetime of the radioactive nuclei. A growing number of laboratories have already produced RIB's of astrophysical interest or are in the stage of technical development of such beams. For the study of these capture reactions, the detection of the capture γ-rays recoil nuclei (via their radioactive decay signals) leads in general to efficiencies far below 100%. In addition, the radioactivity of the RIB seriously hampers the detection methods. Only $p(^{13}N,\gamma)^{14}O$ has been successfully studied so far via γ-ray spectroscopy, because of an unusually large cross section arising from a strong, broad resonance. For all other capture reactions studied so far, only upper limits for $\sigma(E)$ have been derived at the relevant energies, $E_o \pm \Delta/2$, due to more typical (i.e., small) cross sections. Thus, future experiments require a significantly improved detection efficiency, with a detector that is less sensitive to background. Since the recoil nuclei are kinematically confined to a forward cone of a few degrees, a Recoil Mass Separator (RMS) can be used to filter the recoils from the RIB and to focus all recoils onto a suitable detector for particle identification. Of course, in most RMS's, one must make a choice of the charge state of the recoils; however, there is usually a charge state representing about 50% of the recoils produced and thus the capture reactions can be studied with an approximately 50% detection efficiency. Such an RMS is in preparation at the 4 MV Dynamitron tandem accelerator in Bochum, called ERNA (European Recoil separator for Nuclear Astrophysics), for the study of $^4He(^{12}C,\gamma)^{16}O$, and other key reactions. Although this reaction does not involve RIB's a similar RMS should be appropriate for work with RIB's.

It should be pointed out that the capture reactions $^7Be(p,\gamma)^8B$ and $^{13}N(p,\gamma)^{14}O$ have also been studied using the alternative method of Coulomb dissociation (Coulomb excitation into the continuum). However, there remain uncertainties about the interpretation and achievable precision of the method. It is to be hoped that these uncertainties can be resolved, and that this technique will eventually prove to be useful for the indirect studies of capture reactions involving both stable and radioactive nuclides. Unique applications for this technique would be the study of capture reactions involving 3 particles in the entrance channel (e.g. $^4He(\alpha n,\gamma)^9Be$), or two radioactive nuclides in the entrance channel (e.g. a neutron plus another radioactive nucleus).

Neutron-induced reactions, mainly (n,γ), play an important role in the early universe, as well as in stars where they are involved in the nucleosynthesis of the trans-iron elements via the s- and r-processes. Due to the short lifetime of the neutrons ($T_{1/2} = 10$ min), they must be produced by nuclear reactions within the star (e.g. via $^{13}C(\alpha,n)^{16}O$ or $^{22}Ne(\alpha,n)^{25}Mg$) and are quickly thermalized through elastic or inelastic scattering. Due to the absence of a Coulomb barrier in the entrance channel, the most probable reaction energy is near $E_o = kT \approx 30keV$ for $T = 3 \times 10^8 K$. Usually $\sigma(E)$ is very large and increases with decreasing E. Thus, $\sigma(E)$ measurements for the s-process can in principle be carried out directly in the relevant stellar energy region. Since the r-process involves radioactive nuclei near the neutron-drip line, the relevant $\sigma(E)$ measurements will likely require RIB-techniques (see above).

5 Summary

Impressive progress has been achieved in the knowledge of nuclear reaction rates. However, there remains much critical work to be done in the future to arrive at reliable data for many key reactions and processes. New techniques continue to be developed for this purpose, for nuclear reaction rates involving both stable and radioactive nuclei, and research in nuclear astrophysics will remain an exciting pursuit for many years to come.

Almost every topic touched on above, in this brief and necessarily selective discussion, shares some of its ancestry with the work of Gregory Breit[5]. His work on the theory of nuclear resonances, with various collaborators, including Eugene Wigner, permeates the entire structure of nuclear reaction theory. His work on Coulomb wave functions was crucial in quantitatively determining Coulomb and angular-momentum barrier penetration factors over many decades until the arrival on the scene of fast, sophisticated personal computers on almost every desk. Many are also familiar with his work on Coulomb excitation, initially by electrons, and later by protons and heavier ions. Even on the topic of electron screening effects mentioned earlier, Breit and collaborators already in the mid-1930's had to address electron screening effects in proton-proton scattering, in order to extract the true nucleon-nucleon phase shifts, because the target protons were in hydrogen molecules. Although the current screening treatments in nuclear reactions are rather different, we still need further examination of the subject. Overall, Breit's many contributions to the quantum mechanics of reaction and scattering processes are a major part of the knowledge that is today taken for granted. Finally, his career showed that he had a keen interest in and fostered the development of accelerators and the increasingly complex equipment required to get quantitative reliable answers to many questions. We can be sure that if he were still with us, he would enjoy discussing some of the more recent experimental techniques, and would continue to push for even better results.

Acknowledgments

The author thanks C.A. Barnes for comments on the manuscript.

References

1. E.M. Burbidge, G.R. Burbidge, W.A. Fowler and F.Hoyle, *Rev. Mod. Phys.* **29** (1957) 547.

2. W.A. Fowler, *Rev.Mod.Phys.* **56** (1984) 149.

3. C. Rolfs and W.S. Rodney, *Cauldrons in the Cosmos* (University of Chicago Press, 1988).

4. G. Fiorentini, R.W. Kavanagh and C. Rolfs, *Z.Phys.* **A350** (1995) 289.

5. See, for example, M. Hull, *Biographical Memoirs, National Academy of Sciences,* **74** (1998) 27 (National Academy Press, Washington, D.C.).

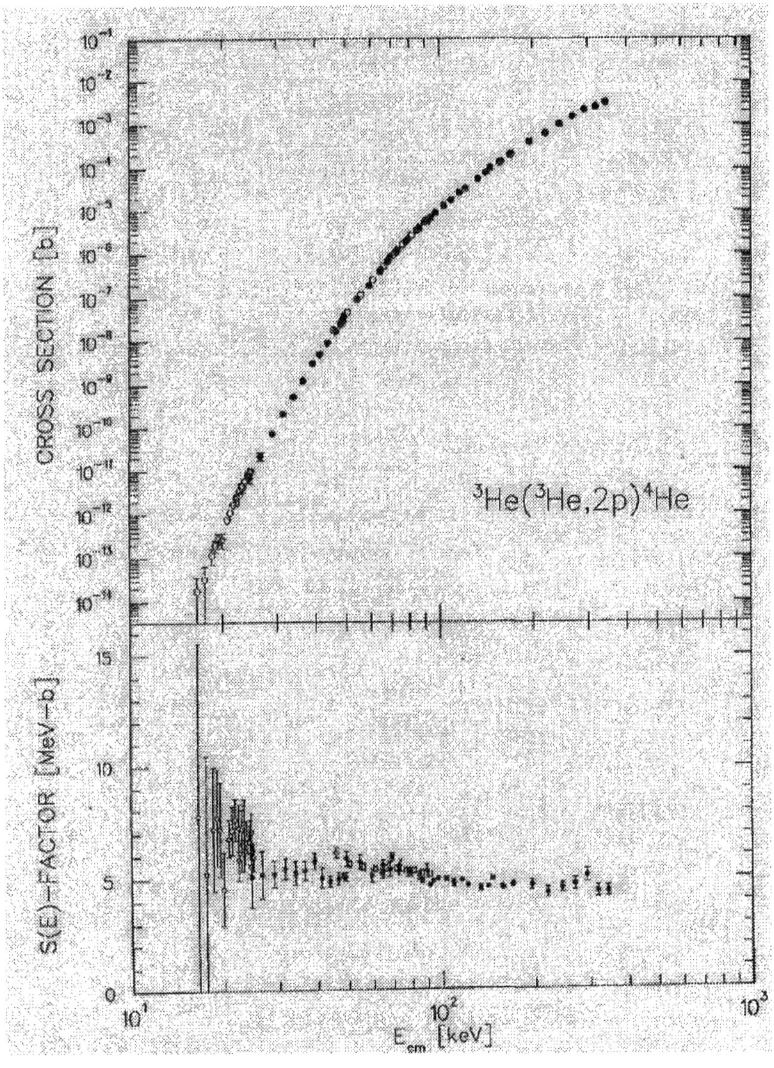

Figure 1: Energy dependence of the cross section and the astrophysical $S(E)$ factor for the $^3He(^3He,2p)^4He$ reaction. The Gamow peak is at $E_0 = 21 keV$ for solar conditions (central temperature $T \approx 15 \times 10^6 K$). The low-energy data have been obtained by LUNA. The apparent increase in the $S(E)$-factor at low energies is due to the effects of electron screening.

RELATIVISTIC HEAVY ION COLLIDER PHYSICS

John W. HARRIS
Physics Department, Yale University,
New Haven CT 06250, U.S.A.
E-mail: John.Harris@Yale.edu

The field of relativistic heavy ion physics is entering a new regime with the startup of the Relativistic Heavy Ion Collider(RHIC). A brief introduction to the field will be presented, including prospective signatures of anticipated new and exciting physics, and an overview of RHIC and its experiments which seek to uncover this new realm of physics.

1 Introduction

1.1 Relativistic Heavy Ion Physics

The motivation for studying relativistic heavy ion collisions is to understand at a fundamental level the behavior of matter at extremely high temperature and/or density. This is the study of Quantum Chromodynamics (QCD) at high energy density, which has implications for nuclear physics, astrophysics, cosmology and particle physics. The behavior of matter as a function of temperature and density (or pressure), depicted in Fig.1, is governed by its equation of state. The lower left portion of the diagram at low temperatures and near normal nuclear matter density is of interest in nuclear physics. Here normal nuclei exist and at low excitation a liquid-gas phase transition is expected to occur. This is the focus of experimental studies using low energy heavy ions. [1] At higher excitation, nucleons are excited into baryonic resonance states, accompanied by particle production and hadronic resonance formation. This region of hadronic resonance matter has been accessible in heavy ion studies at the AGS at the Brookhaven National Laboratory and at the SPS at CERN. As depicted in Fig.1, these collisions may partially traverse the transition region into the quark-gluon plasma regime. Formation of a quark-gluon plasma, a deconfined state of quarks and gluons, [2] is the major focus of relativistic heavy ion experiments. The Relativistic Heavy Ion Collider (RHIC) [3] and associated experiments will commence operation at Brookhaven in 2000 and operation with heavy ions is planned for the LHC in 2005. As seen in the phase diagram, the anticipated temperature and density trajectories at RHIC (and for LHC heavy ions) are expected to lie close to that of the early universe, while those at the AGS and SPS occur at higher baryon densities.

A quark-hadron phase transition is predicted to have occurred around ten

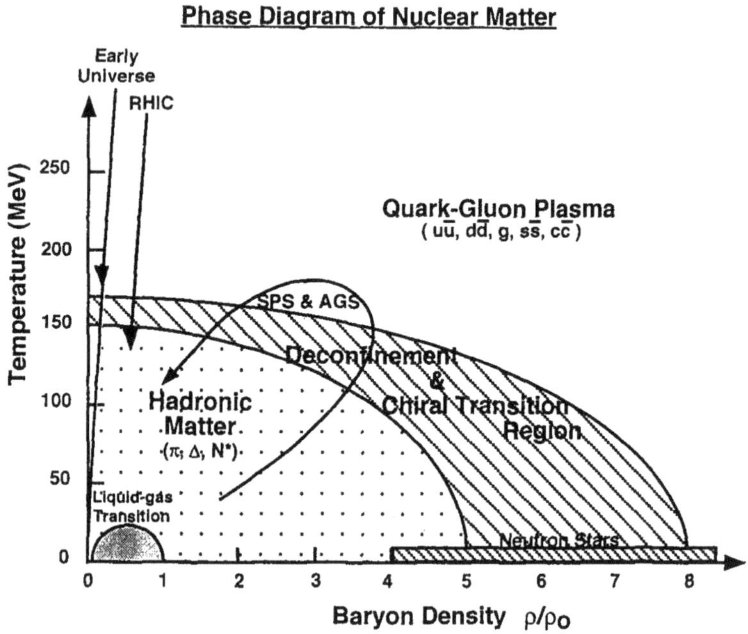

Figure 1: Schematic phase diagram of nuclear matter.

micro-seconds after the Big Bang when the universe was at a temperature of approximately 150 to 200 MeV. This transition is depicted by the crossing of the trajectory for cooling in the early Universe of the deconfinement and chiral phase transition region in the phase diagram of Fig.1. Also of interest in cosmology is nucleosynthesis and any long-term effects of possible fluctuations in the quark-hadron phase transition. The nuclear matter equation of state governs neutron star collapse and supernova expansion dynamics. The region of high baryon densities and very low temperatures is important for various aspects of stellar evolution. It has presently become the topic of much interest in the field due to what appears in theories to be the onset of color superconductivity. [4] If this is in fact the case, then there is still much to be learned about the effects of such states on steller evolution.

Collisions of relativistic heavy ions should allow an investigation of the perturbative vacuum of QCD.[5] In such studies, a better understanding of symmetry breaking mechanisms and the origin of particle masses may be found. The QCD vacuum is predicted to have a complex internal structure, made up of a quark-antiquark condensate sea. This sea possesses energy and mass, and can undergo fluctuations about its zero point. At low energy densities, quarks and gluons are confined in hadrons (baryons and mesons). Here the vacuum acts as a color dielectric, and isolated quarks are not expected. At the high temperatures anticipated in high energy collisions of heavy ions, the quark-antiquark condensate vacuum is expected to melt. The hadrons dissolve into freely propagating quarks and gluons, and the vacuum becomes a color conductor, the true perturbative QCD vacuum. This is the quark-gluon plasma (QGP) phase. At such high temperatures the QGP phase transition is accompanied by a restoration of chiral symmetry. In order to investigate the phase diagram at high energy densities (high temperature and/or baryon density), it is important to recall a few of the basic concepts of QCD, the theory of strong interactions.

1.2 Confinement and Deconfinement in QCD

Two remarkable aspects of QCD are deconfinement and asymptotic freedom. At large distances or small momentum transfer (Q^2), the strong coupling constant (α_s) is large and thus quarks are confined in colorless particles. At short distances or large momentum transfer, the coupling constant is small. [6] This is the regime of asymptotic freedom.

QCD can be calculated on a lattice by carrying out nonperturbative numerical calculations of thermodynamic variables. When two light quark flavors are incorporated in the calculation, a phase transition from a confined phase of hadronic matter to a deconfined quark-gluon plasma is predicted at a temperature of approximately 150 MeV. This transition exhibits a smooth but rapid transition (weak first order or a strong second order transition) for finite lattice sizes. Lattice QCD calculations which incorporate more than two light quarks predict a first order phase transition. [7]

1.3 Chiral Symmetry

The mass of a quark depends on the distance over which it is probed by a second quark or antiquark. Constituent quarks are dressed in a virtual cloud of quark-antiquark pairs and gluons. Since three nonrelativistic constituent quarks make-up the nucleon, the constituent quark mass should be approximately one third the mass of the nucleon, or $\sim$ 300 MeV. The shorter the distance between quarks, the smaller the mass, since the interaction weakens at short distance. Thus the current quark, which is in the regime of asymptotic freedom, is nearly bare. The best estimate for the up-quark mass is approximately 5 MeV and that for the down-quark is approximately 7 MeV. If the light quark masses reach zero, then the quarks and their composite particles obtain another symmetry, helicity.

Massless particles have helicity, which is defined as the projection of the spin of the particle onto its momentum vector. Helicity is either left- or right-handed. Since massless particles travel at the speed of light, the helicity cannot change when boosting the particle to other frames, since the boost must always be less than the speed of light. Thus, the emission and aborption of vector gluons by color charges do not change the helicity of the particle. Another way of seeing this is that the QCD Lagrangian for massless quarks factorizes into a left-handed term and a right-handed term, and is therefore chirally symmetric. Each term can be represented by the special unitary group for isospin, i.e. $SU_L(2)$ for left-handed or $SU_R(2)$ for right-handed quarks, as appropriate, where $SU_L(2) \times SU_R(2)$ is a fundamental symmetry. For example, the nucleon cannot have helicity as a conserved quantity since its mass is non-zero, and therefore we say that chiral symmetry is spontaneously broken. More technically, in this case the QCD Lagransian has an additional term which is proportional to the light quark mass times the vacuum expectation value of the quark-antiquark condensate. Since this expectation value is finite, the Lagrangian cannot be chirally symmetric.

Therefore, at low energy densities quarks and gluons are bound into colorless objects, called hadrons, and are confined. The quarks obtain large effective masses ($m_u \sim m_d \sim 300$ MeV, $m_s \sim 500$ MeV) by interactions among themselves and with the surrounding vacuum, thus chiral symmetry is broken. At sufficiently high energy densities, which we expect to create in relativistic heavy ion collisions, a phase transition should occur where quarks become deconfined, their masses decrease ($m_u \sim m_d \sim 0$, $m_s \sim 150$ MeV), and chiral symmetry is partially restored.

1.4 Space-time Evolution of RHI Collisions

Lattice QCD calculations predict that a deconfinement transition to a quark-gluon plasma should occur at sufficiently high energy densities. The quark-hadron phase transition must have existed about 10 microseconds after the Big Bang. It may exist in the cores of dense stars. Can we make it in the laboratory? This is the task of relativistic heavy ion physics, to explore QCD in the high energy density regime and to search for the quark-gluon plasma.

Various computer simulations of the space-time evolution of these collisions predict that the energy densities in collisions at RHIC and the LHC should surpass the threshold for QGP production over a sufficiently large volume of matter for as long as 5 - 10 fm/c in duration.[8,9] The strength of potential QGP signals relative to any background created by a hot hadronic phase and modifications due to final state interactions will be determined by the time evolution of the initial collision process at the parton level, the formation of a QGP, expansion, cooling, hadronization, and final state interactions. The space-time evolution of these collisions, and possible determination of whether the transient state of the QGP is attained, can only be established once measurements become available from experiments.

The nuclear stopping power determines the amount of energy in the interaction region that is available for formation of a QGP, and thus the energy density. The stopping power is defined as the degree to which the energy of relative motion of the incoming nuclei is transformed into other degrees of freedom. A measurement of the net baryon density at mid-rapidity provides a measure of the stopping power, while a measurement of the energy emitted tranverse to the incident beam will refect the energy density.

Additional information on the space-time evolution of the collisions is provided by measurements of the cross sections of various types of particles as a function of transverse momentum and rapidity. For hadrons, these spectra provide information on a superposition of scattering phenomena. This includes semi-hard and hard scattering processes which lead to hadrons at high transverse momenta after parton fragmentation, and rescattering leading to thermalization of the spectra at low transverse momenta (up to about twice the observed particles mass). There are also other processes which affect the spectra at low to intermediate momenta (up to several times the particle mass), the most notable of which is collective nuclear flow. An indication of the degree of thermal and chemical equilibration can be obtained by studying the relative abundances of the various types of particles and their spectra.

1.5 QGP and Chiral Transition Signatures

Thermodynamic Variables

One class of possible signatures of QGP formation can be classified as thermodynamic variables. This class involves determination of the energy density ϵ, pressure P, and entropy density s of the interacting system as a function of the temperature T and the baryochemical potential μ_B. Experimental observables can be associated with these variables and thus their relative behavior can be determined. If a phase transition to QGP occurs, a rapid rise in the effective number of degrees of freedom, expressed by ϵ/T^4 or s/T^3, should be observed over a small range of T. The variables T, s, and ϵ, are identified with the average transverse momentum $\langle p_T \rangle$, the hadron rapidity distribution dN/dy, and the transverse energy dE_T/dy, respectively. [10] The behavior of these observables will be established in the initial round of relativistic heavy ion collider experiments.

Electromagnetic Probes

Real photons and virtual photons, in the form of lepton pairs, provide information on the various stages of the interaction without modification due to final state interactions. These electromagnetic (EM) probes may provide a measure of the thermal radiation from a QGP, if a range of photon energy or equivalently lepton pair invariant mass, has sufficient signal from the QGP such that it can be distinguished from that of other processes. The yields for EM probes are usually small relative to background processes, which primarily emanate from EM decays of hadrons and resonances. Lepton pairs from the QGP have been predicted to be identifiable in the $1 - 1.5$ GeV invariant mass range. [11] At higher masses the Drell-Yan process dominates and is expected to exceed the thermal yield of lepton pairs. There is some indication from parton cascade calculations [12] that lepton pairs from a QGP may dominate over Drell-Yan in the mass range of $5 - 10$ GeV. This could allow one to probe in a model independent way the early evolution of the QGP. [13]

At lower invariant masses between 0.5 and 1 GeV, lepton pairs from hadronic sources are expected to dominate and provide information on possible medium modifications of hadrons at high density. [14,15] The widths and positions of the ρ, ω, and ϕ peaks in the lepton pair invariant mass spectrum are expected to be sensitive to medium-induced changes of the hadronic mass spectrum, especially to a drop of vector meson masses associated with chiral symmetry restoration. [16,17,18,19,20,21,22,23,24] A change in the K-meson mass also would affect the width of the ϕ meson. [25,26] A double ϕ peak in the lepton pair spectrum would be indicative of a long-lived mixed phase. [27]

In fact, if the thermal fireball lives significantly longer than the 2 fm/c lifetime of the ρ-meson, then the strength of the ρ-meson in the lepton pair mass spectrum is expected to increase relative to the ω. Thus, the ρ/ω ratio could serve as a "clock" for the lifetime of the thermal fireball. [28]

A hadron gas and a QGP near the critical temperature are expected to emit photon spectra that are very similar in intensity and shape. If an extremely hot QGP is formed initially, photons in the transverse momentum range $2 - 5$ GeV/c should come primarily from this QGP phase. [13,29,30] Lower momentum photons will be emitted primarily from the mixed phase and other processes.

Charmonium Suppression

The production of J/ψ particles is predicted to be suppressed in a QGP. [31] This is a result of the Debye screening of a $c\bar{c}$ pair, initially formed in the QGP by fusion of two incident gluons. [32] Less tightly bound excited states of the $(c\bar{c})$ system, such as ψ' and χ_c, are more easily dissociated and will be suppressed even more than the J/ψ. Similar arguments can be made for the heavier $\Upsilon(b\bar{b})$ system[33]. However the temperature at which the Υ ground state "melts" is predicted to be around 2.5 times the transition temperature, and that of the larger Υ' state only slightly above this temperature.

Strangeness Enhancement

A long-standing prediction for a signature of QGP formation is the enhancement of strange hadrons. The production of strange hadrons relative to nonstrange hadrons is suppressed in hadronic reactions. [34] This suppression increases with increasing strangeness content of the hadron. In a QGP the strange quark content is rapidly saturated by $s\bar{s}$ pair production in gluon-gluon reactions, resulting in an enhancement in the production of strange hadrons. [35]. Thus, multi-strange baryons and strange antibaryons are predicted to be strongly enhanced when a QGP is formed. [36,37]. Furthermore, it has been shown [37,38] that an enhanced strangeness content cannot be destroyed nor generated by interactions during expansion and freezeout.

Disoriented Chiral Condensates

The temporary restoration of chiral symmetry during a relativistic heavy ion collision could result in the formation of domains of a disoriented chiral condensate (DCC).[39] This term describes a coherent excitation of the pion field corresponding to a local misalignment of the chiral order parameter $\langle\bar{\psi}\psi\rangle$. Such domains would decay into neutral and charged pions, favoring pion ratios N_{π^0}/N_π substantially different from one-third. This could also explain why final states with a large fraction of charged pions over neutral pions, observed in Centauro events,[40] can occur with significant probability. [41,42] The observation of pion charge ratios significantly different from 1/3, or nonzero charge correlations,[43] would therefore be a direct signature of the chiral phase transition.

Domains of disoriented chiral condensate may also contribute to antibaryon production through the formation of topological defects in the chiral order parameter. [44,45] Such defects can arise at the intersection of chiral domain walls, which carry baryon number and eventually evolve into baryons and antibaryons, possible leaving a signature of the chiral phase transition in regions of phase space that are normally baryon-poor. [46]

High Pt Probes of QCD

The color structure of QCD matter can be probed by its effects on the propagation of a fast parton. [47,48] The mechanisms are similar to those responsible for the electromagnetic energy loss of a fast charged particle in matter: energy may be lost either by excitation of the penetrated medium or by radiation.

The connection between energy loss of a quark and the color-dielectric polarizability of the medium can be established in analogy to the theory of electromagnetic energy loss. [49,50,51] Although radiation is a very efficient energy loss mechanism for relativistic particles, it is strongly suppressed in a dense medium by the Landau-Pomeranchuk effect. [52] The QCD analog of this effect has recently been analyzed comprehensively. [53,54] Adding the two contributions, the stopping power of a fully established quark-gluon plasma is predicted to be higher than that of hadronic matter.

Measurements of fragmentation products of hard scattering can provide information on the matter through which the hard-scattered parton propagates. A quark or gluon jet propagating through a dense medium will not only lose energy but will also be deflected. This effect destroys the coplanarity of the two jets from a hard parton-parton scattering with the incident beam axis. [55,56] The angular deflection of the jets also results in an azimuthal asymmetry. The presence of a quark-gluon plasma is also predicted to enhance the emission of jet pairs with small azimuthal opening angles. [57] The sharp increase in the acoplanarity of di-jet events in proton-nucleus interactions observed at Fermilab [58] indicates that the interpretation of these signals is complicated by re-interaction.

2 The Relativistic Heavy Ion Collider and Experiments

Construction of the Relativistic Heavy Ion Collider has been completed, with operation for physics expected to begin in 2000. A schematic diagram of the RHIC accelerator complex at Brookhaven is displayed in Fig.2. Nuclear beams are accelerated from the tandem Van de Graaff accelerator through a transfer line into the AGS Booster synchrotron and then into the AGS, which serves as an injector for RHIC. RHIC will accelerate and collide ions from

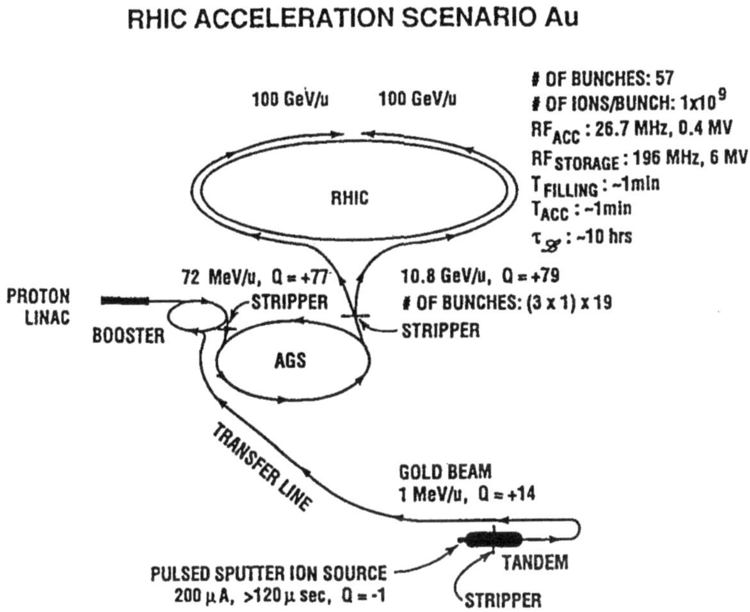

Figure 2: The Relativistic Heavy Ion Collider (RHIC) accelerator complex at Brookhaven National Laboratory. Nuclear beams are accelerated from the tandem Van de Graaff, through the transfer line into the AGS Booster and AGS prior to injection into RHIC. Details of the characteristics of proton and Au beams are also indicated after acceleration in each phase.

protons up to the heaviest nuclei over a range of energies, up to 250 GeV for protons and 100 GeV/nucleon for Au nuclei as shown in Fig.3.

Collisions of the heaviest nuclei at impact parameters near zero at RHIC are expected to produce around 1000 charged particles per unit pseudorapidity. This presents a formidable environment in which to detect the products of these reactions. The four experiments constructed at RHIC will take various approaches to search for the QGP. The STAR experiment will concentrate on measurements of hadron production over a large solid angle in order to measure single- and multi-particle spectra and to study global observables on an event-by-event basis. The PHENIX experiment will focus on measurements of lepton

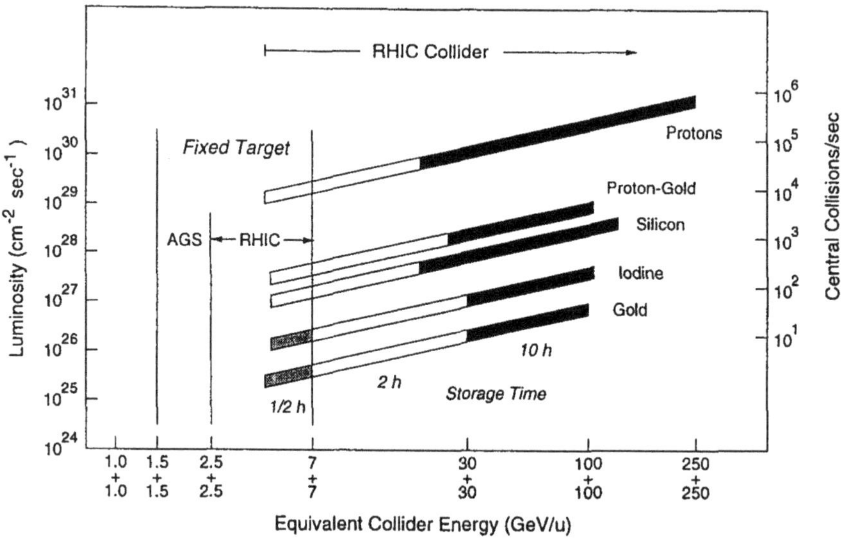

Figure 3: Operating parameters for the Relativistic Heavy Ion Collider (RHIC). Displayed are the design luminosities (left) and number of central collisions per second (right) as a function of collider energy for various combinations of nuclear beams. The anticipated average storage time before refilling the collider rings is shown for each beam combination.

and photon production and will have the capability of measuring hadrons in a limited range of pseudorapidity. In addition, there are two smaller experiments BRAHMS (a forward and midrapidity hadron spectrometer) and PHOBOS (a compact multiparticle spectrometer) which have been constructed. The collaborations, which have constructed these detector systems and which will exploit their physics capabilities, consist of approximately 900 scientists from over 80 institutions internationally. In addition to the colliding heavy ion beams, plans are underway to inject and accelerate polarized protons at RHIC in order to study the spin content of the proton. [59] STAR and PHENIX have active spin physics programs planned for RHIC.

2.1 The PHENIX Experiment

The physics goals of PHENIX (Pioneering High Energy Nuclear Interaction eXperiment) [60] are to measure as many potential signatures of the QGP as possible as a function of a well-defined common variable such as impact parameter or pseudorapidity density. PHENIX will measure lepton pairs (di-electrons and di-muons), photons and hadrons. The experiment will be sensitive to very

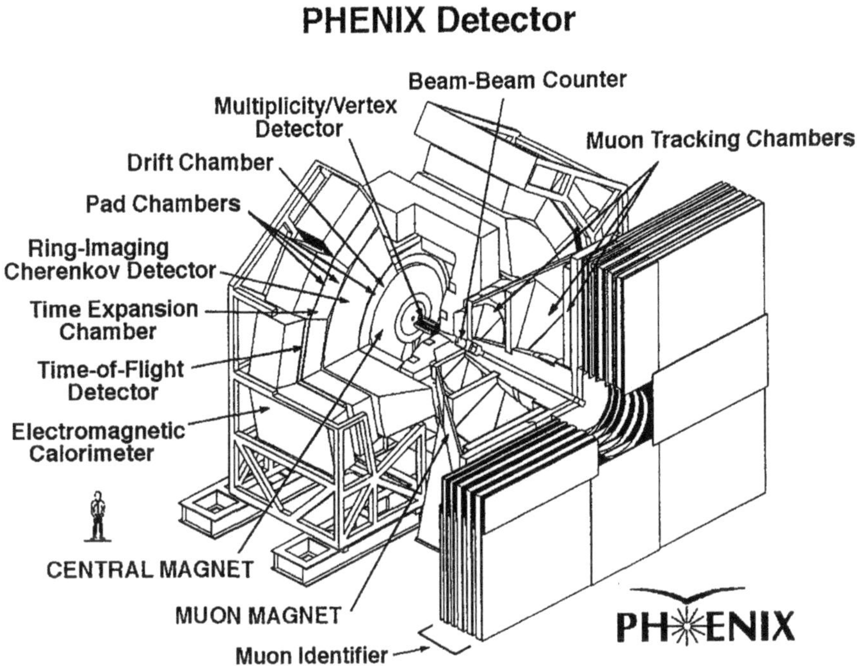

Figure 4: A diagram of the PHENIX experiment at RHIC. The beams collide along the horizontal direction in the center of the detector. A second muon arm (on upstream-side) is not shown in this diagram. The detector components are defined in the text.

small cross section processes such as the production of the J/ψ, ψ', Υ and high $p_\perp$ spectra. It will also have the capability for high rates with pp and pA collisions.

A diagram of the PHENIX detector system is displayed in Fig.4. The magnet has an axial field along the beam direction with tracking chambers and detectors for the identification of electrons, muons, photons and hadrons installed outside the field. There are two arms for dielectron measurements, each with 1 steradian acceptance at midrapidity. Each arm is equipped with a ring-imaging Cerenkov detector (RICH), a time-expansion chamber (TEC) for dE/dx measurements, time-of-flight detectors (TOF), and an electromagnetic calorimeter (EM Cal). Photons and hadrons will also be measured at midrapidity with 2 and 0.36 steradian acceptances, respectively. Separate muon spectrometers with 1 steradian acceptance each are located at forward rapidities (only one is shown in Fig.4). The magnetic field in each muon arm is transverse to the beam direction. A silicon multiplicity-vertex detector (MVD) and beam-beam coincidence counters are located near the interaction vertex. The MVD covers $|\eta| < 2.7$ about midrapidity for event selection via charged-particle multiplicity.

PHENIX will investigate the degree of deconfinement in RHIC collisions, and thus the possibility of QGP formation, by measuring the yields of J/ψ, ψ', Υ. If the J/ψ radius is larger than the Debye screening length then the QCD potential between the $c\bar{c}$ pairs, which form a J/ψ, is weakened and J/ψ formation suppressed. [31] Since the J/ψ, ψ', Υ binding and thus their radii are different, i.e. $r(\psi') > r(J/\psi) > r(\Upsilon)$, a study of the relative suppression of these is sensitive to the screening in a QGP and will help differentiate between deconfinement and possible dissociation in hot nuclear matter. PHENIX will measure J/ψ via electron-pairs near midrapidity and J/ψ, ψ', Υ via muon-pairs at forward angles. The anticipated raw muon-pair mass spectrum in the forward muon arm is displayed in Fig.5 for one-month of RHIC operation at the design luminosity for central Au + Au collisions. Peaks for the J/ψ, ψ' and Υ can be observed. With subtraction of like-sign muon-pairs, and application of kinematic cuts the peak-to-background of this spectrum is expected to improve further and the ψ' peak will become more prominent. For one year of running at the RHIC design luminosity of 2 x 10^{26} $cm^{-2}s^{-1}$, PHENIX will detect 390K J/ψ, 5.7K ψ' and 1.2K Υ for central collisions (top 10 % in centrality) of Au + Au. The mass resolution is about 100 MeV for the J/ψ and the pion rejection is somewhat better than 10^{-4} at $p > 2.5$ GeV/c. In addition to these charmonium studies, PHENIX will measure open charm ($D, \bar{D}$ decays) via unlike-sign $e\mu$ coincidences using the electron and muon spectrometers. The rate of open charm production is expected to be sensitive to details of the early stages of RHIC collisions and will provide complimentary information to the charmonium studies.

PHENIX will investigate possible chiral symmetry restoration by performing high resolution measurements of the leptonic and hadronic decays of ϕ-mesons. In a chirally-restored QGP both the ϕ-meson and kaon masses may be modified resulting in change in the mass and width of the ϕ-meson and the branching ratio between leptonic and hadronic channels. [61] The anticipated electron-pair spectrum for 8 days of central trigger Au + Au operation at RHIC is shown in Fig.6. Extremely sharp peaks for the ω, ϕ, and J/ψ are observed, with a 2-to-1 signal-to-noise ratio in the low mass resonance region. The mass resolution is approximately 4 MeV for the ϕ decaying to e+e- (and

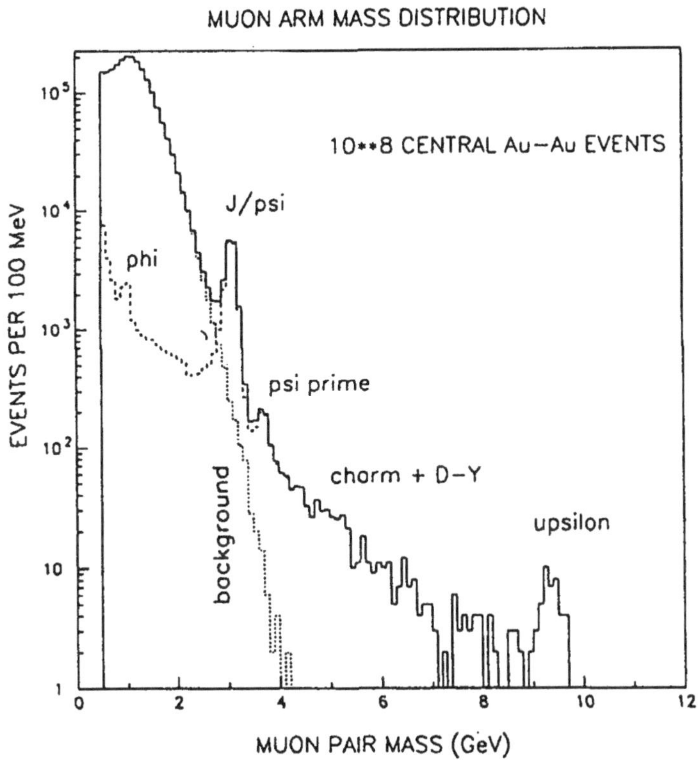

Figure 5: PHENIX muon-pair mass spectrum anticipated in one month of RHIC running for central Au + Au collisions at a luminosity of 2 x 10^{26} $cm^{-2}s^{-1}$.

$\sim$ 1 MeV for ϕ decaying to K+K- in the hadronic measurement). The pion rejection is better than 10^{-4} for $p < 4$ GeV/c.

There are various other measurements of interest that will be undertaken by PHENIX. Measurements will be made using the electromagnetic calorimeter and high resolution photon detector to determine the yield of direct photons (for $p_\perp > 0.5$ GeV/c) radiated from the nuclear interaction. An enhancement of photons with $p_\perp > 2$ to 3 GeV/c is expected because of the high gluon concentration in the early high energy-density stage of RHIC collisions. [62] PHENIX will also investigate the order of the QGP phase transition by measuring the $<p_\perp>$ of identified-charged pions, kaons and protons at mid-rapidity as a function of the energy density. By detecting both hadrons and photons, PHENIX can measure fluctuations in isospin, $\pi^\circ/(\pi^+ + \pi^-)$, which

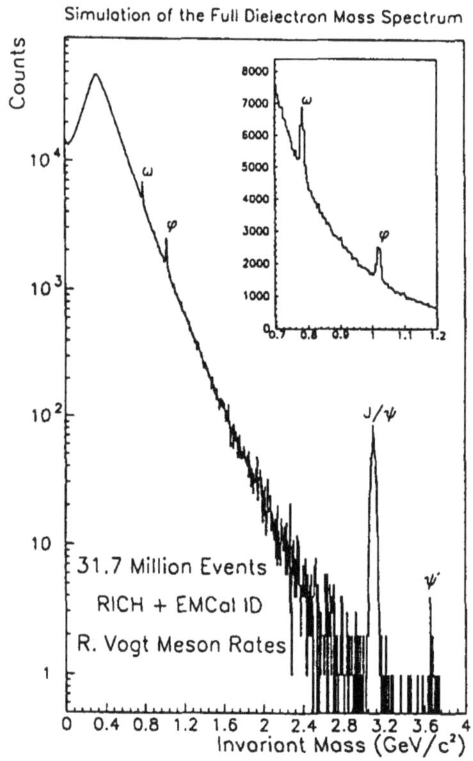

Figure 6: PHENIX electron-pair mass spectrum anticipated in 8 days of RHIC running for central Au + Au collisions at a luminosity of 2 x 10^{26} $cm^{-2}s^{-1}$.

might occur in a second order phase transition. [63] PHENIX will also study strangeness and charm production for enhancements, jet-quenching, and the space-time evolution of the system as a function of energy density.

2.2 The STAR Experiment

STAR will search for signatures of QGP formation and investigate the behavior of strongly interacting matter at high energy density in central collisions of relativistic nuclei. The STAR detector system will simultaneously measure many experimental observables to study signatures of the QGP phase transition as well as the space-time evolution of the collision process over a variety of colliding nuclear systems. [64] In addition, STAR will investigate very peripheral collisions of relativistic nuclei to study photon and pomeron interactions resulting from the intense electromagnetic fields of the colliding ions and colorless strong interactions, respectively. [65]

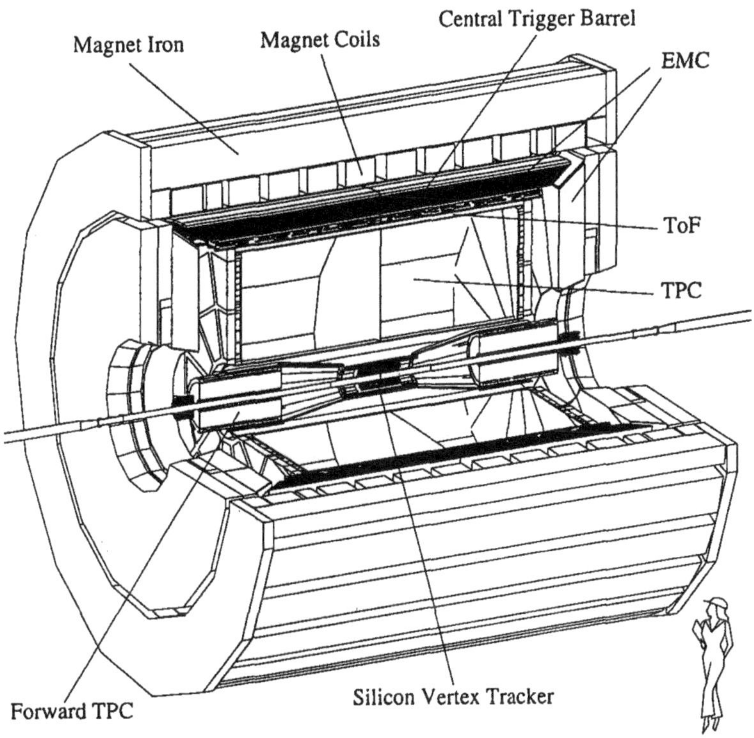

Figure 7: Layout of the STAR experiment at RHIC.

The layout of the STAR (Solenoidal Tracker At RHIC) experiment [64] is shown in Fig.7. The initial configuration of STAR consists of a large time projection chamber (TPC) covering $|\eta| < 2$, trigger detectors, a ring imaging cherenkov detector [66] covering $|\eta| < 0.3$ and $\Delta\phi = 0.1\pi$ (not shown in Fig.7), and an initial partial coverage of electromagnetic calorimetry of $0 < \eta < 1$ and $\Delta\phi = 0.14\pi$ inside a room temperature solenoidal magnet with 0.5 T magnetic field. The solenoid provides a uniform magnetic field for tracking, momentum analysis and particle identification via ionization energy loss measurements in the TPC. Measurements in the TPC and trigger detectors will be carried out at mid-rapidity with full azimuthal coverage ($\Delta\phi = 2\pi$) and azimuthal symmetry. Additional tracking detectors to be added for the run in 2001 are a silicon vertex tracker (SVT) covering $|\eta| < 1$, and a forward radial-drift TPC (FTPC) covering $2.5 < |\eta| < 4$. In addition to the tracking detectors, the solid angle of the electromagnetic calorimeter (EMC) will be increased to approximately 30% of its eventual $|\eta| < 1$ and $\Delta\phi = 2\pi$ coverage and will measure the transverse energy of events, and trigger on and measure high transverse momentum photons, particles and jets. The remainder of the EMC will be constructed and installed over the next 2 - 3 years. A time-of-flight (TOF) patch covering $0 < \eta < 1$ and $\Delta\phi = 0.04\pi$ will also be installed for 2001 operation to extend the particle identification in STAR.

For triggering STAR will utilize a central trigger barrel surrounding the outer cylinder of the TPC for charged-particle multiplicity triggering in the $|\eta| < 1$ region, a TPC endcap trigger for triggering on charged particle multiplicity over $1 < |\eta| < 2$, and zero-degree calorimeters at large pseudorapidities (not shown in Fig.7) for vetoing large impact parameter collisions with a large amount of energy remaining in the forward direction. The electronics, data acquisition and online monitoring systems for these detectors have also been installed.

Inclusive $p_\perp$ distributions of charged particles will be measured with high statistics in STAR to investigate various effects. These effects in the spectra will include collective radial flow, critical temperature at low $p_\perp$, and mini-jet attenuation [69] at high $p_\perp$. The $p_\perp$ spectra of baryons and anti-baryons at midrapidity are particularly interesting for determining the stopping power of quarks. Measurements of the net baryon number and net charge will establish the baryo-chemical potential at midrapidity. Using the silicon vertex tracking, STAR will measure the spectra of charged pions at low $p_\perp$ in order to investigate effects of the presence of a disoriented chiral condensate.

Hard scattering processes will dominate the early stage of collisions, and measurements of products of the hard scattering are necessary to understand the early evolution of the collision. The energy loss of the products of hard scattering processes in the early stages is predicted [68] to be sensitive to the medium through which they travel, and may be a direct method of observing the excitation of the medium. STAR will measure high transverse momentum ($p_\perp$) photons, charged particles and jets. These spectra should represent the energy-loss history of hard-scattered partons traversing the medium.

To determine the degree of equilibration of the system and to investigate fluctuations, STAR will measure triple differential cross sections for production of various types of particles as a function of transverse momentum ($p_\perp$), pseudo-rapidity (η) and azimuthal angle (ϕ). To obtain information on the high density phase, strange hadrons (K^+, K^-, K^o) and singly- and multiply-strange baryons ($\bar{\Lambda}, \Lambda, \Xi^-, \bar{\Xi}^+, \Omega$) will be measured in STAR over a wide rapidity interval about mid-rapidity. [67] Several examples of invariant mass spectra for strange particles measured in STAR are shown in Fig.8. An enhancement in the production of strange particles resulting from chemical equilibrium of a system of quarks and gluons was one of the first predictions for a signature of QGP formation. [35] From the measurements and ratios of strange particles and other hadrons, the baryo-chemical potential, temperature and degree of saturation in strangeness can be determined for various ensembles of events (impact parameter selections) and colliding systems in STAR. Coupled with theoretical predictions, this information can be used to determine whether a QGP has been formed. An enhancement in the strange antibaryon content[36,37] and an increased enhancement of multiply-strange baryons ($\Xi^-, \bar{\Xi}^+, \Omega$) compared to singly-strange hadrons[36] have been predicted to signify the presence of a QGP.

It has long been known that a general indicator of a phase transition is the appearance of critical dynamical fluctuations in a narrow range of conditions. Such critical fluctuations can only be seen in individual events where the statistics are large enough to overcome uncertainties due to finite particle number fluctuations. The large transverse energy and multiplicity densities at midrapidity in central collisions allow event-by-event measurement of fluctuations in particle ratios, energy density, entropy density and flow of different types of particles as a function of $p_\perp$, rapidity, and azimuthal angle. Fluctuations have been predicted to arise from the process of hadronization of a QGP. [70] As a consequence of the high multiplicities in central collision events, the slope of the transverse momentum ($p_\perp$) distribution for pions and the $\langle p_\perp \rangle$ for pions and kaons can be determined event-by-event in STAR. Thus, individual events can be characterized by a pion slope parameter T_o (effective "temperature") or $\langle p_\perp \rangle$, and a kaon $\langle p_\perp \rangle$ in order to search for events with extremely high temperature, predicted[71] to result from deflagration of a QGP. The determination of $\langle p_\perp \rangle$ for pions can be made very accurately on the single event basis in STAR in central collisions from Ca + Ca to Au + Au. For kaons, with $\sim$ 200 charged kaons per event in the acceptance for central Au + Au events, $\langle p_\perp \rangle$ can be determined accurately for single events.

Correlations between identical bosons provide information on the freeze-out geometry, the expansion dynamics and possibly the existence of a QGP. [72] STAR will measure the parameters of the pion-emitting source via pion corre-

Strangeness

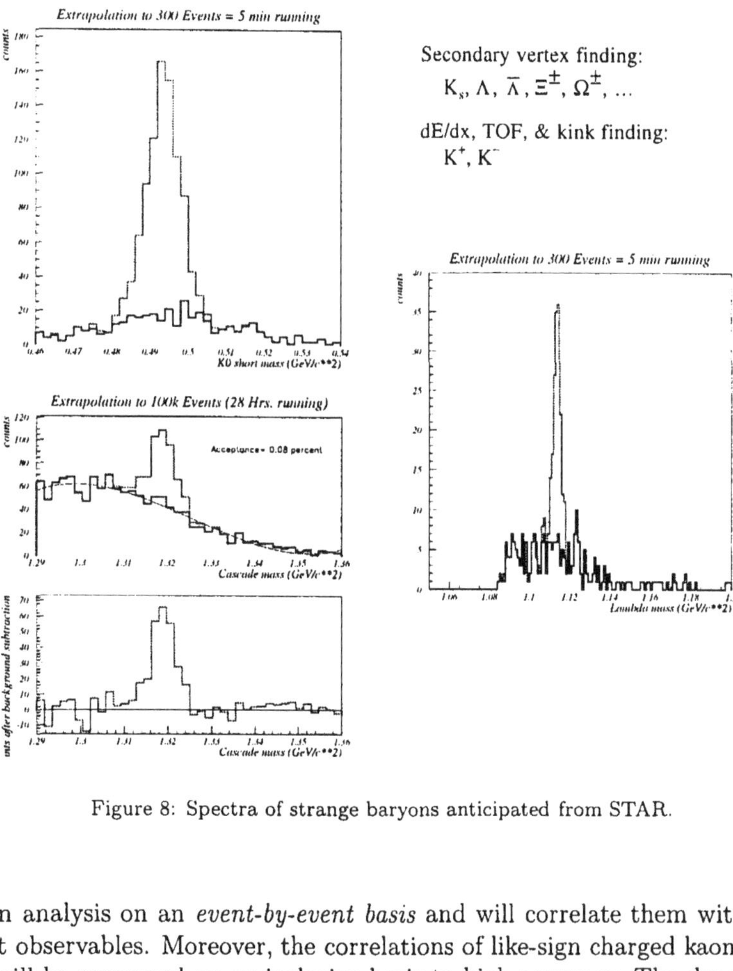

Figure 8: Spectra of strange baryons anticipated from STAR.

lation analysis on an *event-by-event basis* and will correlate them with other event observables. Moreover, the correlations of like-sign charged kaons or pions will be measured on an inclusive basis to high accuracy. The dependence of the source parameters on the transverse momentum components of the particle pairs will be measured with high statistics. The K's are expected to freeze out earlier [73] than π's in the expansion. Depending upon the baryo-chemical potential and the existence of a QGP, the K^+ and K^- freeze out at different times. STAR will also measure $K_s^o K_s^o$ correlations by reconstructing the decay topologies of $K_s^o \to \pi^+\pi^-$. [64] In this case the absence of Coulomb repulsion, as compared to like-sign charged particle correlations, will enable a more precise measurement of the large source dimensions expected at RHIC.

The production cross section of ϕ-mesons will be measured inclusively in

STAR via the decay $\phi \to K^+K^-$. Measurement of the yield of the ϕ, which is an $s\bar{s}$ pair, places a more stringent constraint on the origin of the observed flavor composition [74] than the K/π ratio and is expected to be more sensitive to the presence of a QGP. The ϕ mass, width and production rate are also expected to be extremely sensitive to changes in the quark masses [75] due to a possible chiral phase transition at high energy densities. STAR will measure $\phi \to K^+K^-$ and with lower statistics $\phi \to e^+e^-$ in order to determine the yield, mass, and width of the ϕ produced in central collisions at RHIC.

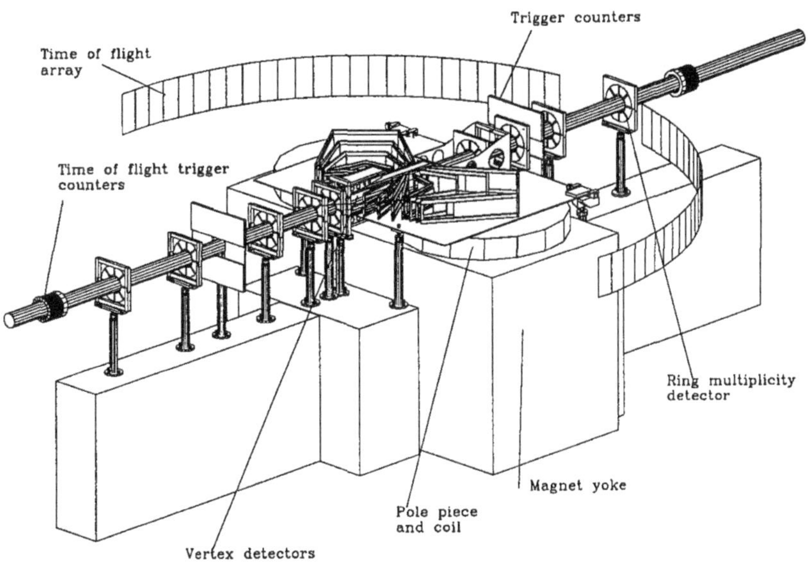

Figure 9: A diagram of the PHOBOS two-arm multiparticle spectrometer. The two arms are located on opposite sides of the beam pipe. Each arm has a 2 Tesla magnet represented by only the lower coils for ease of viewing other parts of the set-up, with silicon detector planes for tracking.

2.3 The PHOBOS Experiment

The physics goals of the PHOBOS experiment [76] are to measure single particle spectra and correlations between particles with low transverse momenta and to characterize events using a multiplicity detector. Charged particles will be measured and identified in the range $0<y<1.5$ and 15 MeV/c$<p_\perp<$600 MeV/c for pions and 45 MeV/c$<p_\perp<$1200 MeV/c for protons. The range of particles to be studied include $\gamma, \pi, K, \rho, \bar{p}, \phi, \Lambda, \bar{\Lambda}, d$, and $\bar{d}$. Particle ratios, $p_\perp$ spectra, strangeness production $(K, \phi, \Lambda, \bar{\Lambda})$ and particle correlations will be studied. An illustration of the experiment is shown in Fig.9. The top coils and magnet iron are not shown in order to see the other components. The multiplicity detector consists of silicon strip detectors and silicon pad detectors. The two magnetic arms have a field strength of 2 Tesla. Eleven layers of silicon, 5 layers of pads and 6 layers of strips, are installed in each magnetic spectrometer for tracking and momentum measurements.

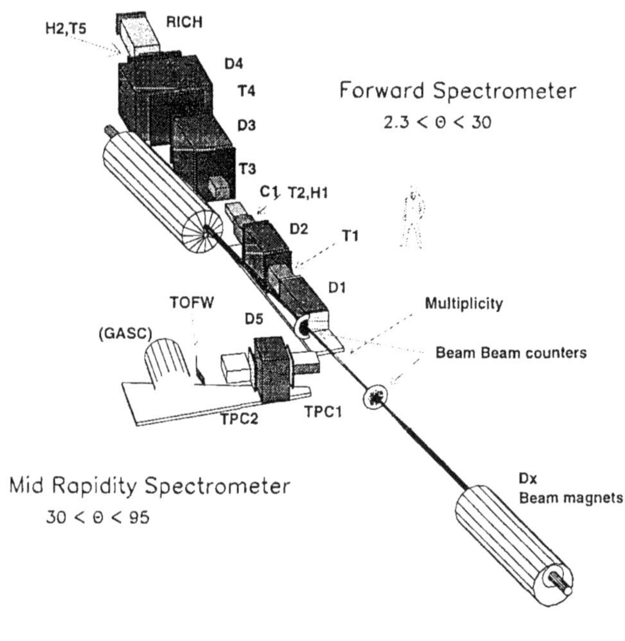

Figure 10: A diagram of the BRAHMS forward and midrapidity spectrometers in the narrow angle hall at RHIC.

2.4 The BRAHMS Experiment

The physics goals of the BRAHMS (BRoad RAnge Hadron Magnetic Spectrometers) experiment [77] are to achieve a basic understanding of relativistic heavy ion collisions at RHIC through a systematic study of particle production in AA collisions from the peripheral to the most central in impact parameter. Measurements will be performed using two high resolution magnetic spectrometers at various angles to cover both the baryon-rich fragmentation regions and the high temperature, baryon-depleted midrapidity region. A diagram of the BRAHMS forward and midrapidity spectrometers is shown in Fig.10. The spectrometers will measure and identify inclusive and semi-inclusive π, K and p, and their momenta with high statistics over a small solid angle and over a wide range of pseudorapidity ($0 < \eta < 4$) and transverse momentum. Particle identification is performed using various combinations of time-of-flight arrays, and threshold and ring-imaging Cherenkov counters. BRAHMS will measure inclusive and semi-inclusive particle spectra, and extract the net baryon densities and temperatures from spectral slopes as a function of rapidity to determine whether thermal and chemical equilibrium are reached in these collisions. It will also be able to study both high and low transverse momentum processes. Centrality will be measured using a global multiplicity detector.

Acknowledgments

The author thanks Kate Geruntho for assistance with the figures. This work was supported in part by the Director, Office of Energy Research, Division of Nuclear Physics of the Office of High Energy and Nuclear Physics of the U.S. Department of Energy under contract No. DE-FG02-91ER-40609.

1. C.K. Geblke, Proceedings of this Symposium.
2. J.C. Collins and M.J. Perry, Phys. Rev. Lett. 34, 1353 (1975); G. Chapline and L. Susskind, Phys. Rev. D 20, 2610 (1979).
3. Conceptual Design of the Relativistic Heavy Ion Collider (Report BNL 52195, 1989).
4. see K. Rajagopal and references therein, Nucl. Phys. A661 (1999) 150c.
5. T.D. Lee and G.C. Wick, Phys. Rev. D 9, 2291 (1974); T.D. Lee, Rev. Mod. Phys. 47, 267 (1975).
6. G.C. Blazey, Intersections between Particle and Nuclear Physics, 6th Conference, ed. T.W. Donnelly, AIP Press, p. 18 (1997).
7. see A. Di Giacomo and references therein, Nucl. Phys. A661 (1999) 13c.
8. K. Geiger and B. Mueller, Nucl. Phys. B 369, 600 (1992).
9. S.A. Bass et al, preprint nucl-th/9803035 (1998).
10. L. van Hove, *Phys. Lett.* B 118, 138 (1982); *Z. Phys.* C 21, 93 (1983).
11. P.V. Ruuskanen, *Nucl. Phys.* A 525, 255c (1991); *ibid.* A 544, 169c (1992).
12. K. Geiger and J.I. Kapusta, *Phys. Rev. Lett.* 70, 1920 (1993).
13. M.T. Strickland, *Phys. Lett.* B 331, 245 (1994).

14. C. Gale and J. Kapusta, *Phys. Rev.* C 35, 2107 (1987).
15. F. Karsch, K. Redlich, and L. Turko, *Z. Phys.* C 60, 519 (1995).
16. R.D. Pisarski, *Phys. Lett.* B 110, 155 (1982).
17. A.I. Bochkarev and M.E. Shaposhnikov, *Phys. Lett.* B 145, 276 (1984); *Nucl. Phys.* B 268, 220 (1986); *Z. Phys.* C 36, 267 (1987).
18. H.G. Dosch and S. Narison, *Phys. Lett.* B 203, 155 (1988).
19. R.J. Furnstahl, T. Hatsuda and S.H. Lee, *Phys. Rev.* D 42, 1744 (1990).
20. C. Gale and J. Kapusta, *Nucl. Phys.* B 357, 65 (1991).
21. Z. Aouissat, G. Chanfray, P. Schuck and G. Welke, *Z. Phys.* A 340, 347 (1991).
22. M. Asakawa, C.M. Ko, P. Lévai and W.J. Qiu, *Phys. Rev.* C 46, 1159 (1992).
23. M. Herrmann, B.L. Friman and W. Nörenberg, *Z. Phys.* A343, 119 (1992).
24. T. Hatsuda, Y. Koike and S.H. Lee, *Nucl. Phys.* B394, 221 (1993).
25. D. Lissauer and E.V. Shuryak, *Phys. Lett.* B 253, 15 (1991).
26. P.Z. Bi and J. Rafelski, *Phys. Lett.* B 262, 485 (1991).
27. M. Asakawa and C.M. Ko, *Phys. Lett.* B 322, 33 (1994); C.M. Ko and D. Seibert, *Phys. Rev.* C 49, 2198 (1994).
28. U. Heinz and K.S. Lee, *Phys. Lett.* B 259, 162 (1991).
29. D.K. Srivastava, B. Sinha, M. Gyulassy and X.N. Wang, *Phys. Lett.* B 276, 285 (1992).
30. S. Chakrabarty, J. Alam, S. Raha, B. Sinha and K. Srivastava, *Phys. Rev.* D 46, 3802 (1992).
31. T. Matsui and H. Satz, *Phys. Lett.* B. 178, 416 (1986).
32. F. Karsch, M.T. Mehr and H. Satz, *Z. Phys.* C 37, 617 (1988).
33. F. Karsch and H. Satz, *Z. Phys.* C 51, 209 (1991).
34. J.P. Bailly et al, *Phys. Lett.* B 195, 609 (1987).
35. J. Rafelski and B. Müller, *Phys. Rev. Lett.* 48, 1066 (1982). [Erratum: *ibid.* 56, 2334 (1986).]
36. J. Rafelski, *Phys. Rep.* 88, 331 (1982).
37. P. Koch, B. Müller and J. Rafelski, *Phys. Rep.* 142, 167 (1986).
38. H.W. Barz, G.L. Friman, J. Knoll and H. Schulz, *Nucl. Phys.* A 484, 661 (1988); *Nucl. Phys.* A 519, 831 (1990); *Phys. Lett.* B 254, 315 (1991).
39. J.D. Bjorken, *Acta Phys. Polon.* B 23, 637 (1992).
40. C.M.G. Lattes, Y. Fujimoto and S. Hasegawa, *Phys. Rep.* 65, 151 (1980).
41. C.S. Lam and S.Y. Lo, *Phys. Rev. Lett.* 52, 1184 (1984); *Phys. Rev.* D 33, 1336 (1986).
42. S. Pratt, *Phys. Lett.* B 301, 159 (1993); S. Pratt and V. Zelevinsky, *Phys. Rev. Lett.* 72, 816 (1994).

43. C. Greiner, C. Gong and B. Müller, *Phys. Lett.* B 316, 226 (1993).
44. T.A. DeGrand, *Phys. Rev.* D 30, 2001 (1984).
45. J. Ellis, U. Heinz and K.L. Kowalski, *Phys. Lett.* B 233, 223 (1989).
46. J.I. Kapusta and A. Srivastava, *Phys. Rev.* D 50, 5379 (1994).
47. J.D. Bjorken, Fermilab publication 82/59, Batavia (1982), unpublished.
48. B. Svetitsky, *Phys. Rev.* D 37, 2484 (1988).
49. M.H. Thoma and M. Gyulassy, *Nucl. Phys.* B 351, 491 (1991); E. Braaten and M.H. Thoma, *Phys. Rev.* D 44, R2625 (1991).
50. S. Mrówczyński, *Phys. Lett.* B 269, 383 (1991).
51. Y. Koike and T. Matsui, *Phys. Rev.* D 45, 3237 (1992).
52. A.B. Migdal, *Sov. Phys. JETP* 5, 527 (1957).
53. M. Gyulassy and X.N. Wang, *Nucl. Phys.* B 420, 583 (1994).
54. R. Baier, Y.L. Dokshitser, S. Peigné and D. Schiff, *Phys. Lett.* B 345, 277 (1995).
55. D.A. Appel, *Phys. Rev.* D 33, 717 (1986).
56. J.P. Blaizot and L. McLerran, *Phys. Rev.* D 34, 2739 (1986).
57. J. Pan and C. Gale, *Phys. Rev.* D 50, 3235 (1994).
58. M.D. Corcoran et al, *Phys. Lett.* B 259, 209 (1991).
59. Proposal on Spin Physics Using the RHIC Polarized Collider, RHIC Spin Collaboration (1992).
60. PHENIX Experiment at RHIC - Preliminary Conceptual Design Report, PHENIX Collaboration Report (1992).
61. D. Lissauer and E.V. Shuryak, Phys. Lett. B 253, 15 (1991).
62. E. Shuryak and L. Xiong, Phys. Rev. Lett. 70, 2241 (1993); K.J. Eskola and M. Gyulassy, Phys. Rev. C 47, 2239 (1993).
63. F. Wilczek, Nuc. Phys. A 566, 123c (1994).
64. Conceptual Design Report for the Solenoidal Tracker At RHIC, The STAR Collaboration, PUB-5347 (1992); J.W. Harris et al, Nucl. Phys. A 566, 277c (1994).
65. S. Klein and E. Scannapieco in Proceedings on Intersections Between Particle and Nuclear Physics: 6th Conference, ed. T.W. Donnelly, LBNL Report LBNL-40495, (AIP Press, 1997); J. Nystrand and S. Klein in Proceedings of Hadron '97, (LBNL Report LBNL-41111, BNL, 1997).
66. A Ring Imaging Cherenkov Detector for STAR, STARnote 349, STAR/ALICE RICH Collaboration (1998); ALICE Collaboration, Technical Design and Report, Detector for High Momentum PID, CERN/LHCC 98-19.
67. K. Wilson in Proceedings of Quark Matter '95 Pre-conference Workshop, ed. J. Thomas and T. Hallman, (Report No. UCRL-ID-121571, 1995).
68. M. Gyulassy and M. Pluemmer, Phys. Lett. B 243, 432 (1990)2; X.N.

Wang and M. Gyulassy, Phys. Rev. Lett. 68, 1480 (1992).
69. P.V. Landshoff, Nucl. Phys. A 498, 217 (1989).
70. M. Gyulassy, Nucl. Phys. A 400, 31c (1983); L. van Hove, Z. Phys. C 27, 135 (1985).
71. E.V. Shuryak and O.V. Zhirov, Phys. Lett. B 89, 253 (1980); Phys. Lett. B 171, 99 (1986).
72. S. Pratt, *Phys. Rev.* D **33**, 1314 (1986); G. Bertsch, M. Gong and M. Tohyama, *Phys. Rev.* C **37**, 1896 (1988); and G. Bertsch, *Nucl. Phys.* A **498**, 151c (1989).
73. K.S. Lee, M.J. Rhoades-Brown and U. Heinz, *Phys. Rev.* C **37**, 1463 (1988).
74. A. Shor, Phys. Rev. Lett. 54, 1122 (1985).
75. R.D. Pisarski and F. Wilczek, Phys. Rev. D 29, 338 (1984); T. Hatsuda and T. Kunihiro, Phys. Lett. B 185, 304 (1987).
76. RHIC Letter of Intent to Study Very Low pt Phenomena at RHIC, PHOBOS Collaboration (1991).
77. Interim Design Report for the BRAHMS Experiment at RHIC, BNL Report, (1994).

GREGORY BREIT
*JULY 14, 1899 – SEPTEMBER 11, 1981**

MCALLISTER HULL

1 Introduction

FOR NEARLY FIFTY YEARS, Gregory Breit was a leading figure in the development of physics in the twentieth century. John Wheeler in *Some Men and Moments in Nuclear Physics* wrote, "Insufficiently appreciated in the 1930's, he is today the most unappreciated physicist in America." This was written in 1979, when Gregory was in physical decline, and he probably never saw it, but if public recognition was slight (but by no means absent), he was appreciated very well (in spite of a difficult personality) by his students, collaborators, and colleagues. The range of his interests and duration of his active career made this cadre a large one.

Trained as an electrical engineer, Breit did his early work in radio, including the definition of the characteristics of early tubes and finite coils. The most important of this work, with Merle Tuve, was the use of radio to demonstrate the existence of the postulated ionosphere by receiving return signals of a pulsed radio beam sent from the earth's surface. Ranging with a pulsed signal is, of course, the principal of radar. He also inspired and worked with the production of high voltages to accelerate charged particles (protons) to use as probes of the nucleus—the first manmade probes in the United States (Cockroft and Walton at Cambridge, England, had a beam of protons and deuterons first, but at a lower voltage than the Carnegie team). With Tuve and Odd Dahl, he demonstrated the soundness of the betatron principle. Breit's work at Wisconsin included his organizing the infant theory of quantum electrodynamics (and studies of photon-photon interaction), early studies of the nucleon-nucleon interaction, and with Eugene Wigner the theory of nuclear resonances, which continues to be the basis for understanding many nuclear reactions.

During the second world war, Breit recognized very early that it would be sensible not be publish basic studies of nuclear properties, especially those of uranium and plutonium, for he among others understood the military (as well as energy) possibilities of a chain reaction in uranium. This caution

*BIBLIOGRAPHICAL MEMOIRS, NATIONAL ACADEMY OF SCIENCES, **74** (1998) PP. 27-58 (NATIONAL ACADEMY PRESS, WASHINGTON, DC)

resulted in the voluntary withholding from publication of many important papers until after the war. His own war work began with the organization of neutron studies that developed into the laboratory at Los Alamos—under Oppenheimer rather than Breit, who had gone off to the Naval Ordnance Laboratory to study degaussing of ships (for which he invented the magnetic extrapolator) as a defense against magnetic mines, and the Ballistic Research Laboratory of the Army to work on proximity fuses, exterior ballistics, and the fire control. After the war, Gregory returned briefly to Wisconsin to take up his studies of nucleon properties and nuclear reaction. He transferred to Yale in 1947, with his advanced graduate students—and one undergraduate (me). At Yale, postdoctoral associates were added to the students who came from Wisconsin, and a few new graduate students joined what came to be called the "Breit group."

In addition to his prolific personal research, Breit led and participated in the work of members of the group. He and the members;

- Studied hyperfine structure of atomic levels; nuclear magnetic moments; the isotope shift; photo-disintegration of the deuteron; polarization of Bremsstrahlung radiation; Coulomb excitation; and semi-classical treatments of quantum mechanical calculations.

- Worked on Coulomb wave functions for calculating nuclear reactions (before high speed computers made the tables unnecessary);

- Initiated some of the first computer-aided calculations of the phenomenological nucleon-nucleon intereaction, where it was shown that the strong force is charge independent, has a sharp repulsive core, and exhibits spin dependence;

- Showed that nucleon-nucleon scattering cross-sections could be expanded in phase shifts (and that the phenomenological force could be used to calculate those phase shifts); and

- Invented (theoretically) and studied heavy ion physics extensively.

2 Personal History

Gregory Breit was born July 14, 1899, in Nikolayev, Russia, some 100 kilometers northeast of Odessa. His parents, Alfred and Alexandra Smirnova Breit Schneider, operated a textbook business until Alexandra died in 1911, and the business was sold. Alfred emigrated to the United States in 1912, leaving Gregory and his sister Lubov in charge of a governess while Gregory attended the School of Emperor Alexander in Nikolayev. In 1915 Alfred instructed his children to come immediately to the United States, and with their governess they traveled by train to Archangel and then by ship to New York. They landed on July 30, 1915. Their father, now Alfred Breit, assisted them with entry formalities and took them to Baltimore, where he was living.

John Wheeler relates a story told to him by Lubov that she and Gregory were vacationing on the sea when the call to leave Russia came, and they 'came as they were.' For Gregory this meant dressed in a sailor suit with short pants; he was still wearing it when he enrolled in Johns Hopkins (at age sixteen!). Wheeler attributes some of Gregory's subsequent reticence in personal relationships to the ragging he took at the hands of his classmates for his dress—unlikely, perhaps, as the definitive reason, but a contributing influence? My wife Mary, who was one of the few women Gregory was comfortable with (she was the only person allowed to cut his hair when he became ill), attributes his personality to "middle European genes, especially prominent in Ukrainians" (she is first generation Czech).

There was an older brother, Leo, who had escaped the tsar's army through Turkey and then practiced medicine in Maryland. A deserter sought by the tsar's agents, he gave as little information about himself as possible, which along with the dropping of "Schneider" from the family name, caused Gregory some difficulties years later when he needed clearance to do war work. Gregory, on the other hand, responded to Russian recruiter in 1918 and attempted to join the Russian Army. He was rejected because he couldn't pass the physical. He had a lifelong hatred of Communist Russia: much stronger than the intense distrust that most of us had.

Gregory was supported by scholarships as he continued his formal education and was awarded three degrees by Johns Hopkins: the A.B. in 1918, the A.M. in 1920, and the Ph.D. in 1921 (when he was twenty-two!). His publications began in 1920 (some with E.O. Hurlburt), and his dissertation (Joseph S. Ames was his advisor) on the distributed capacity of inductive coils was published in 1921. For those, like me initially, who wonder at this apparently trivial topic for a Ph.D. thesis, I suggest you look up the paper (*Phys. Rev.* 17:649). It is a masterly piece of applied mathematics (a skill Breit demonstrated for the rest of his professional life). He was a National Research Council fellow at the University of Leiden in 1921-22 and at Harvard University in 1922-23.

Gregory married Marjory Elizabeth McDill on December 30, 1927, in Washington, D.C., and acquired a stepson, Ralph Wycoff, from Marjory's previous marriage. He had no biological children.

Breit knew and worked with the physicist in his field with a degree of comprehensiveness impossible today. At the symposium in his honor at Yale in 1968, over 200 of his colleagues and former students attended from around the world. Among the notable speakers were John Archibald Wheeler, Henry Margenau, Isidor I. Rabi, Victor F. Weisskopf, Hans A. Bethe, Eugene P. Wigner, D. Allan Bromley, Vernon W. Hughes (the latter two organized the symposium), Gerald E. Brown, Raymond G. Herb, and Merle A. Tuve. I also spoke, and had not been in so much fast company since Los Alamos!

Gregory Briet was elected to the National Academy of Sciences in 1939 and to the American Academy of Arts and Sciences in 1951. He was a fellow of the American Physical Society, Physical Society of London, Institute of Radio Engineers, and the American Association for the the Advancement of Science. He was a member of the American Mathematical Society, American Geophysical Union, Washington Academy of Science, the Army Ordnance Association, Sigma Xi, and Phi Beta Kappa.

Gregory died in Oregon, where he had gone with Marjory in 1973, finally to retire. When Yale's mandatory retirement policy sent Breit from New Haven in 1968, I had gotten him an appointment as distinguished professor at Oregon State University, where I was chair of physics. However, one of his former postdoctoral associates, Moti Lal Rustgi, had, through the department and President Meyerson, beat me to it, and he went to State University of New York at Buffalo (as it happened, the same friend nominated me as chair at SUNY Buffalo shortly afterward, so I became Gregory's department chairman!). Marjory wished to spend her last years near her son and daughter-in-law, Ralph and Faith Wykoff, and her grand children, so they moved into a retirement home near Salem. Gregory, as vigorous a man as I had ever known, began to decline in Buffalo, and was not professionally active in Oregon. His death in 1981 called for a two-column obituary in The New York Times.

Noting that I am writing nearly eighty years after Gregory's first publication, the reader will understand that it has been difficult to get recollections from early colleagues. Anne Herb recalls that the late Ray Herb admired Gregory above most of his colleagues at Wisconsin, and that in tapes made before his death Ray had recalled with pleasure Gregory's interest in the work he was doing with accelerators. Norman Heydenburg has "fond memories" of Breit at New York University and Wisconsin where Norm was a postdoctoral fellow. Again, the subject (in part) was accelerators and nuclear physics.

Charles Kittel followed Gregory to Washington during the war, and his work on degaussing influenced his decision to spend his distinguished career in condensed matter physics. He tells as 'folklore' that Gregory and I. I. Rabi came from Russia on the same boat, and thus knew each other long before their collaboration. Glenn T. Seaborg remembers 'many important contacts' with Gregory in 1941 as the Uranium Committee was formed (more on this later). John Wheeler, who has already been quoted here, joined Briet as a National Research Council fellow at New York University when Norm Heydenburg was there. Wheeler especially remembers Breit's "kindness" in crediting his work with joint authorship on papers (I do not consider it kindness. Breit was more conscientious than most in giving due credit, and equally demanding that it be given by others) and in helping secure a fellowship renewal that made it possible for Wheeler to work with Neils Bohr. This also was typical of Breit. He was always interested in his younger associates, and regularly helped them get positions when they left him—as Norm Heydenburg also remembers. Wheeler, in this context, credits Gregory with "saving" Eugene Wigner for Princeton by getting him an appointment at Wisconsin when Princeton was moving too slowly (Willis Lamb tells me that Wigner always said that Princeton fired him!).

The collaboration between Breit and Wigner was significant for the development of nuclear physics (see below), and Wigner returned to Princeton when it "mustered the resolution and funds" to make him an appropriate offer. Willis Lamb recalls that Breit was a referee of an early paper of his and took the trouble to write directly to Willis with useful comments and encouragement—a rare instance of a referee abandoning the customary anonymity, and again demonstrating Breit's concern for the development of young physicists. For a later generation of associates, Arthur Broyles refers to Breit as "one of the most sincere and conscientious men I have known," and recalls that Gregory interrupted his work to help Arthur decide between opportunities for postdoctoral work. I do not recall a single student or postdoctoral fellow with the group who ever left without a suitable position.

Gregory had a dark side that was the source of legends even from the New York University days. He was formally polite in the European way (the result of his upbringing in the first decades of the century!), and usually apologized—quite sincerely, I believe—after an outburst of temper. Wheeler recalls no personal incidents during his association with Breit, although he heard of them from others. I experienced only one, very early in our working relationship (I was still an undergraduate). He misinterpreted a comment of mine and when he blew up, I said something like, "I think I'll return when you are feeling better. Please call me." I walked out and back to my office. He duly called, apologized for the outburst (even more contritely when I showed him the source of his misinterpretation), and never raised his voice with me again, although we had our differences many times. We became very close friends over the succeeding years. Others of my colleagues were not so lucky. Gerry Brown, who remembers Breit as a second father, was regularly a target, and

I was present when Gregory took the hide off a graduate student who had wished him 'a good talk' at a meeting: of course his talk would be good! There is no point in detailing more examples: they occurred regularly, and were simply a fact of life for his students (and on occasions his colleagues. Allan Bromley recalls mediating a heated 'discussion' between Gregory and Henry Margenau on the occasion of Gregory's seminar. Each thought the other's talk was the 'worst ever heard.' Having just reread them, I fear both were correct!). Wheeler thinks that this irascibility was at least a partial cause of Breit's lack of recognition commensurate with his accomplishments.

Gregory's devotion to his student's intellectual development and personal welfare was equally legendary. He was available at any time for consultation, and if a student was shy, he would be invited in for a chat. Weekly group lunches were remembered by Jack McIntosh as times of terror in anticipation of Gregory's asking someone a difficult question, but we learned a great deal, including how to think on our feet! Frequent "parties" at his home, set up by Marjory (who also entertained the wives of group members—separately) were opportunities to talk physics in general. Gregory was incredibly well informed. He received hundreds of reprints a month, read most of them, and shared them with the group members according to current interest. Bromley thinks he retained everything he read, including the location (journal reference, page number, location on the page!) of the source. The sessions at Gregory's home also were opportunities for us to meet the great physicists of the time. Jack McIntosh especially remembers Werner Heisenberg, for example (as I do; someone had told him I had been at Los Alamos during the war, so the head of the 'counter project', as we called him then, asked a number of questions, most of which I couldn't answer for security reasons!).

Gregory was equally solicitous of our health. Any ailment was occasion for concern and advice. Gray Herling, one of Breit's later students, recalls being advised that exercising "a few times a week for an hour is better than several hours every few weeks": sound advice according to current practice in the fitness world!

Gregory's recreations were exercise and reading (other than physics, I mean). He had a canoe on Lake Mendota in Madison, as Anne Herb recalls. Gerry Brown has even better reasons to remember that canoe. Gregory once took him and another student out for a paddle. Gregory seemed to wish to use the leverage of the stern position to turn the canoe away from the course, and the other student soon recalled an appointment, and was put ashore. Gerry then determined to keep the course against Breit's pull, and did so. Nothing was said at the end of the afternoon, but twenty-five years later Gregory recalled the trip to Gerry, and remarked "I saw then you had some stuff in you." The canoe didn't make it to New Haven, but, as McIntosh recalls, the group "was hiked" regularly, with "Mac or Bob (Gluckstern) leading with Breit, and the rest trying to keep up." On one of these occasions, Breit sat down, removed his shoes, and changed his socks, advising the rest of us to do likewise. As usual, he had our physical wellbeing at heart! While Jack failed to take the advice, he recalls the faces at the windows of nearby houses staring out at this extraordinary sight.

Breit's recreational reading included historical fiction, as I know because when I was hospitalized in Buffalo once, he sent over several Hornblower novels. This matched my own tastes, and I have since read all of the series. But we never talked about such things. Physics was always on his mind, and like most driven and creative persons, his specialty provided most of his focus, and the line between work and pleasure was never a clear one.

His students, collaborators, and colleagues over nearly sixty years universally characterize him as a brilliant, informed, dedicated physics, equally devoted to the subject and to developing and encouraging younger persons for the field. If he was sometimes difficult, the intellectual rewards for working with him were worth putting up with his moods. I, of course, am biased. A bond of affection grew between us that lasted all his life and is still fondly remembered by me.

3 Professional History

Breit's professional positions began with an appointment as assistant professor at the University of Minnesota (1923-24). He then began an extended association with the Carnegie Institution of Washington as a mathematical physicist (1924-29, with residency at the Technische Hochschule, Zurich, in 1928) and as a research associate (1929-44). He was professor of physics at New York University (1929-34) and the University of Wisconsin, Madison (1934-47), with a visiting membership in the Institute for Advanced Study, Princeton (1935-36). During the war, he worked in the Naval Ordnance Laboratory (1940-41) and at the Metallurgical Laboratory, University of Chicago (1942). He was a member of the Applied Physics Laboratory, Johns Hopkins University (1942-43) and was head physicist at the Ballistic Laboratory, Aberdeen Proving Grounds (1943-45). His return to the Madison campus was short-lived, as he accepted a professorship at Yale in 1947. At Yale he was given the first Donner professorship in 1958, and he retired from that position in 1968 (at Yale's mandatory retirement age of sixty-eight!). He went to the State University of New York at Buffalo, from which he retired to private life in Oregon in 1973. His professional honors, in addition to membership and fellowships in a number of societies (already noted), included the award of an honorary doctorate of science by Wisconsin (1954), the Benjamin Franklin Medal in 1964, and the National Medal of Science in 1967. At various times in his career he was associate editor of *Physical Review, Proceedings of the National Academy of Sciences* and *Il Nuovo Cimento*.

A discussion of Gregory's research is made difficult by at least two circumstances. He worked in many areas and he returned to most of them over many years. He published, alone and with colleagues, some 320 papers. My list probably is not complete. I've added two references since I started this memoir! Since I may include only twenty-five items in an accompanying bibliography, you can sense the difficulty I have in doing justice to the work of this premier physicist. However, two review volumes deal with Briet's work. Volume 41/1 of the *Handbuch der Physik* (cited as HP and a number representing the order of the papers as they occur in the volume) was written by Gregory and some of his colleagues and was published in 1959, and a symposium in Breit's honor was held in New Haven on the occasion of his retirement from Yale. The proceedings of the symposium were published in 1970. The symposium volume *Facets of Physics* (cited as FP and a chapter number) has the best bibliography of Breit's work that I know. Between these two, one can find most of the items he published, as well as a hint of the breadth of his interests and the influence his work had on others.

The first research Breit published that had an impact beyond the community of physicists came out of his education as an electrical engineer. When he took a position in the Department of Terrestrial Magnetism of the Carnegie Foundation, he invited Merle Tuve to join him in attempting to demonstrate the existence of an ionized layer in the atmosphere, which had been postulated by Kennelly in the United States and Heaviside in the United Kingdom. It was typical of Breit's approach to physics that he worked on the theory of radio wave reflection from charged layers in general as he planned to measure them in the atmosphere. The idea was to measure the delay between the "ground" wave from a radio transmitter and the "sky" wave reflected from the ionized layer. Since the height of the expected layer was not known, the size of the delay could not be estimated. Breit thought that a parabolic sending antenna would be ideal for the attempt, but funding difficulties soon scorched that idea (cf. Tuve's contribution to the symposium, FP8). Station KDKA, Pittsburgh, the first licensed commercial radio station in the United States, was used in early trials with a special key Breit constructed. When the Naval Research Laboratory transmitter became available, the experimenters turned to it. It was only 13 miles away and had the schedule of a research installation! Evidence of a delay was indeed obtained, but interference made measurement difficult. Breit and Tuve thought of pulsing the transmission with a period so that the delay could be determined before the next train of waves arrived. Today we recognize that this is the central idea of radar. I was once told that Breit and Tuve noticed occasional spurious signals when they were testing their apparatus. These were finally attributed to planes flying from Washington's airport (of course, their 80m wavelength was useless in locating planes, but it could sense them), and the signals were ignored henceforth! In any event, they got their delay and were able both to prove the existence of the ionized layer and measure its height above the earth's surface. Breit and Tuve published their work in 1926 and with Odd Dahl in 1928. When Breit was given the Fellow Award of the Institute of Radio Engineers (now IEEE) in 1945, the citation read:

> For pioneering in the experimental probing of the ionosphere and giving to the world the first publication of the experimental proof of the existence of the ionosphere; and for having initiated at an early date the pulse method of probing by reflection which is the basis of modern radar.

Kittel considers this Breit's most significant work. Only in terms of public impact can one agree.

There were no more publications on the Kennelly-Heaviside layer. As early as 1923, Breit's interest had turned to the emerging field of quantum mechanics. He continued to publish interesting items on topics in radio until 1930, but increasingly his papers were devoted to problems in quantum mechanics. As the papers of Schroedinger, Heisenburg, and Dirac appeared, Breit wrote interpretive comments where he thought they could help and looked for problems with a classical analogue to study. The breadth of his interests began to appear, and it would never again be possible to discuss his work over a year or two as a package, except perhaps for his wartime activity.

The gap in Breit's publication record between 1940 and 1946 does not mean he was inactive! He served on committees on publication that persuaded American physicists to withhold any papers on properties of uranium, or the new elements beyond, that neutron bombardment could make. Breit early recognized the possibilities for both energy generation and explosive reactions with these isotopes. Papers published in 1946 with submission dates in the early 1940s are results of this policy. He chaired the fast-neutron project at Chicago until his departure for Washington, when Oppenheimer took over and transferred the program to Los Alamos. Breit had contributed some five reports on iostope separation, neutron diffusion, and chain reactions while at Chicago. At Washington, he worked first on degaussing (to protect merchant ships from magnetic mines), as we have noted, and invented the magnetic extrapolator, allowing two men to accomplish in a day a task that had taken a month before. He wrote three reports on degaussing. At the Army Ordnance Laboratory, he provided fifteen reports on the proximity fuse (a major contribution to anti aircraft success), and thirteen on exterior ballistics and fire control. Both Navy and Army ordnance departments cited Gregory for exceptional and outstanding performance during the war. In 1967 President Johnson awarded him the National Medal of Science for his work in nuclear physics and his wartime work in ordnance.

Gregory's work on the question of ignition of the atmosphere and oceans by runaway reactions initiated by thermonuclear explosions may be considered an extension of his interest in supporting the military in its role of protecting his adopted country. Some work had been done (by Bethe) before the Trinity test of the first nuclear bomb, but the "super" was expected to be a thousand times as powerful—and to involve reactions closer to the constituents of the environment. Edward Teller, according to *The New York Times* obituary, considered Breit "the most conscientious, meticulous and painstaking of physicists," and hence the ideal person to undertake this task. Under Breit's direction, we geared up at Yale for the necessary calculations and created a "secret" room to house our work. A few of us were cleared, and Gluckstern and I were made responsible for security and supervision of the work of other participants. There was a vent in the room that Breit considered a possible security leak, so I designed and built a variable pitch audio oscillator to run in the vent when we had conferences (it sounded worse than the wobbling shriek of modern emergency vehicles—for the same reason: one can discriminate against a steady sound, but not a variable one). Our work was essentially applied astrophysics, and Gregory put it all together. Subject to the favorable outcome of some new experiments (which were successful), there would be no runaway reactions from super explosions. As it happens, he was right!

In general, however, we must trace Gregory's interests in a subject over many years. For example, he interested himself in accelerators for forty years. His participation in the invention of the betatron principle had already been noted. He was the first American physicist to realize that to induce nuclear reactions with artificial sources (accelerators) would be superior to the use of naturally radioactive sources pioneered by Rutherford. Breit, with his DTM colleagues, began to look at the high voltage applications to particle acceleration in 1928. In 1929 he published his initial ideas on nuclear reactions produced by "artificial" sources of radiation. By 1936, the first proton-proton scattering experiments were published by the DTM group (Breit had moved by then, but he continued to encourage the work) using a voltage multiplier device Breit had begun. At Wisconsin, he inspired the work of Ray Herb and colleagues with Van de Graaff machines (FP7). In 1952, when the founding paper on heavy ion physics was published, the funding proposal that preceded it discussed the modifications in cyclotron operation that could accommodate the ions (the first experiments in the field were done at Oak Ridge using a cyclotron built on a wartime isotope separation magnet), provided a concept design for a heavy ion linear accelerator, HILAC, (we used matrices to follow the particles through the machine). Two HILACs were built, with design input from Bob Gluckstern, one of the authors of the heavy ion paper. Yale got one of them, and the other went to the Seaborg group at Berkeley for the study of trans-uranic elements. Trans-uranic elements are quite fascinating objects in nuclear physics, and we discussed the possible properties of super heavy ones that would be accessible to head-on heavy ion collisions with uranium, but the topic was introduced for completeness, and to attract the attention of the funding agency reviewers. It was not a central interest of Breit or the rest of us. Bromley (FP4) mentions it in his symposium article, and Breit notes the work of Seaborg's group and the Nobel institute at the end of his *Handbuch* paper (HP1). However we thought of it, the use of heavy ions in trans-uranic element studies has become an international effort. In addition to the Nobel institute, the Darmstadt group is in the field, and has made number 112! Breit's last direct concern with accelerators came in 1964. We proposed what we called a "meson factory," a linear accelerator designed to produce an intense beam of pi mesons for the study of nuclei. Legend has it that at the meeting with funding agencies, Louis Rosen stood up at the end of the Yale presentation and said, "I want a machine like Yale has proposed." Once again Gluckstern turned our conceptual design into a practical one, and the accelerator was built—at Los Alamos, not New Haven! Yale got priority in the use of the machine for a time, and Vernon Hughes made good use of it. By the time the accelerator was in operation, our theory group had begun to break up, and we never worked on the analysis of experiments. Our throw-away suggestion (I do not wish to imply that we were not serious, but only that the area was not central to our interests) this time was cancer therapy with pi mesons. The localization of energy deposition from particle ionization in matter is a function of particle mass, so that the heavier mesons will do less damage to healthy tissue as they pass through than lighter electrons or massless X rays. A Yale oncologist established the Cancer Research and Treatment Center at the University of New Mexico to pursue this idea, and it was clinically successful. It fell, however, to budget restrictions, but the excellent cancer center continues at UNM. Breit's final publication on accelerators was in 1954.

In 1928 Breit discussed the interpretation of the Dirac equation, and during the next few years developed the theory of two electron interactions, the separation of angles in the two electron problem, and calculated the fine structure of helium using the large-large component approximation to his 16 component equation. After the war, he returned to the two-fermion problem with Gerry Brown, and studied the effect of the finite size of the proton on the fine and hyperfine structure of hydrogen spectra. Over the next several years, this approach was used to obtain first order relativistic corrections to proton-proton scattering analyses. Hughes's review of the Breit interaction (FP5), as it is properly called, reports on measurements, modern extensions to higher orders of approximation, and on the study of the interaction in the fine structure of the positronium atom. The study of the Breit Interaction and its generalizations is "fundamental to atomic physics and modern quantum electrodynamics," says Hughes (FP5). As usual, Briet's pioneering work set the standard for the field.

Gregory began his study of dispersion relations in 1925 with two papers. These were followed by another in 1930 and major review articles in 1932-33. The latter papers essentially summarized the status of quantum electrodynamics then. The choice of dispersion relations was typically perceptive of Breit. Their analytic properties make it possible to obtain many useful results without having to specify details of the theory; optical level crossing, for example, was derived in these papers. It is a major aid in disentangling experimental observations. Analyticity properties were central to the use of dispersion relations in particle physics thirty years later. Breit himself applied them to nuclear processes in 1962.

In 1932 Breit started his work on the isotope effect. The shift of atomic spectra as a function of added neutrons tells us something of the behavior of the extra particle in the nucleus. As Brown notes in his symposium article (FP6), Breit's early work had settled on two effects. If the neutron is buried in the center of the nucleus, its presence pushes protons out because of the near incompressibility of nuclear matter. If the neutron is attached to the surface, it pulls protons out because of the nucleon-nucleon attractive force. Spectroscopists call these effects mass and field, respectively. Breit's work continued through the 1930s, and resumed in 1950, 1952, and 1953 with attempts to use our increasing knowledge of nuclear forces to understand the changes in the nucleus as neutrons are added. Brown's discussion of the problem in 1968 (FP6) summarized work that began when he raised the topic with Briet years earlier. It centered on an attempt to understand changes in the nucleus starting with a self-consistent description with a static potential between nucleons hence a spiritual continuation of Briet's studies. Brown showed that there is no low-level "breathing mode" excitation in the nuclei studied (calcium and lead)—and hence none in intermediate mass nuclei. He also showed that density dependence of the effective interaction is important for the isotope shift, and obtained believable information on the compressibility of finite nuclei—worthy extensions of Briet's earlier work.

Much of the work that occupied Breit during his remarkable career involved the scattering of reactions of charged particles—usually both positive (i.e., an interaction in which the repulsive Coulomb force operates). For non-relativistic energies, the Schroedinger equation applies, and is separable. The problem, therefore, is to obtain solutions of the radial part. These solutions are called Coulomb wave functions (when only the Coulomb force acts). Breit's early interest in proton-proton scattering made it necessary to deal with these functions, and in 1936 he published the first of a long series of papers on the subject with F.L. Yost and Wheeler.

The mathematical difficulty arises because the values of the angular momentum appearing in the equation are integers (in terms of the reduced Planck's constant). The confluent hypergeometric functions defined by Whittaker and Watson *Modern Analysis*, which are formal solutions to the Coulomb equation, have divergent expansions when the appropriate indices are integer. They are also not normalized properly for the physics applications. Before modern computers were available, tables of functions were needed, and they were given, even in the earliest papers. We organized the preparation of tables at Yale. For us a computer was a person using a desk calculator to obtain the values of the regular and irregular solutions of the Coulomb wave equation as defined in the original paper. When digital computers became available (large main frames at first), I found programming for them a direct translation of the steps we had set up for our human computers. We programmed in machine language then, with the only aid an assembly program, and I became proficient in octal arithmetic.

Some ten papers by Breit and colleagues written over the years from 1936 to 1959 were summarized, together with notational references to all known papers of other writers on the subject, in a *Handbuch* article (HP2). This extended effort was again typical of his approach. Whatever the physics problem needed, even some pure or applied mathematics, it was supplied. He was an excellent mathematician, and I became an adequate one working with him (it was typical of him that he never taught me any mathematics. It was only a tool, and we talked about physics).

In 1936 Breit also published his first paper on proton-proton scattering with E.U. Condon and R.D. Present (obviously there is a connection between the papers on Coulomb functions and the theory of p-p scattering). It is reliably reported that he was so anxious to get results that he would volunteer to help with the experimental runs if the experimentalist would just get on with it! Merle Tuve and Ray Herb tell of these early days at DTM and Wisconsin in papers for the symposium (FP7 and FP8). In the theoretical work, the shift in the phase of the outgoing wave with respect to the incoming one required some consideration since the Coulomb part of the interaction has an infinite range. However, since the target protons are in a hydrogen atom, the proton charge is screened at sufficient distance. Let the screening distance go to infinity and one has a means of defining the phase shift—with respect to a Coulomb wave rather than a plane one.

In 1941 Breit showed that the most fundamental description of the collision process is the scattering matrix and the phase shifts that define it. This fact was used by Hans Bethe and colleagues in the study of nuclear matter with realistic nucleon-nucleon interactions in 1967 (FP2), as I shall note later. From the first, however, the delineation of the nucleon-nucleon force was a part of the study. Initially the distance dependence of the force was arbitrary. A square well, given its ease of treatment, was favored. Both phase shifts and the early models of the attractive nuclear forces suggested charge independence of the interaction. The singlet S phase shift for n-p and p-p scattering were nearly the same and the nuclear force models to give them were alike. This result supported the similarity of binding energies of mirror nuclei, already noted. The force was short ranged from the start.

When it became possible after the war to use more sophisticated mathematical forms for the potential, the Yukawa force was introduced (i.e., a descending exponential divided by the distance, with the exponent dependent on the mass of the meson presumed to cause the force). The analyses showed that the current value of the mass was too little to explain nuclear forces; the force was too short ranged. Although the phenomenological mass in these early studies turned out to be a bit large when the pi meson was discovered, our later work accommodated the experimental mass quite handily. As we introduced larger angular momenta, it became necessary to take into account the fact that the neutron-proton interaction contains a tensor part, and this affects both the scattering matrix form, and the force potential description. The shell model contributed a spin-or-bit force, and as the data came from higher and higher energies with more intense beams, experimentalist began to measure double and triple scattering. We, of course, kept up with our formulations. In particular, we looked at one-and two-pion exchanges as determinants of terms in the potential and introduced a hard core of short range to take account of higher order pion interactions we could not safely model (or, perhaps, of quark interactions that we didn't even think about). By 300 or so MeV, we needed to look at relativistic corrections, where Breit's two-fermion equations provided a basis.

The use of high-speed (for that day) computers was essential to our work; the AEC center at NYU and the IBM laboratory at Poughkeepsie gave us valuable time to run the analyses. Breit never learned to program, but he was instrumental in introducing the use of computers into the work. He understood my wiring of plug boards to run an IBM accounting machine in the Yale business office as a differential analyzer (Metropolis had pioneered this at Los Alamos during the war, and taught me), and even designed an analogue differential analyzer using Navy surplus balltable integrators that worked, but he was happy to leave the making of stored programs, *à la* von Neumann, to his younger colleagues. Gradient searches in the multi-dimensional phase shift space produced excellent fits to as many as 2,000 pieces of data, n-p and p-p, by the early 1960s. Our formulation of the scattering matrix was verified. Yale finally got a computer center, and our graduate students became its first consultants. This made it possible for us to make a coup. I was scheduled to talk at one of the annual meetings of the American Physical Society, and an experimental group gave values of measured triple scattering parameters in an abstract for the same session. When I talked, I reported calculated values at their energy and angles that agreed within standard error with the measurements! All the work over more than thirty years by Breit and his colleagues was summarized in a presentation for the Breit symposium (FP1). As the nearly sixty papers reported with Breit's participation atests, and the care with which the work was done suggests, this effort was a major contribution to nuclear and particle physics in this century. It is, perhaps, remarkable that over the thirty years of this study, major, characteristics of the interaction didn't change. The potential remained short ranged, essentially charge independent, and the phase shifts still provide the best representation of the physics. Bethe (FP2) used the phase shifts to derive phenomenological potentials to use in studying the properties of nuclear matter, with satisfactory results—an appropriate justification of Breit's expectation in undertaking the study in the first place.

Breit is, perhaps, remembered in the physics community for his work in nuclear reactions as much as for any other of his areas of interests. In 1936, he and Eugene Wigner published papers on resonance theory, and this work has been a major component of studies since. The obvious difficulty with treating the nucleus is that it is a many-body problem with the intereaction between nucleons in a state of developing understanding. Thus, in the thirty years following the Breit-Wigner beginning, they and others have sought models that could capture properties of the nucleus (turning Wigner's black box into grey ones!) and other facets of the total picture that would be useful in organizing experimental data and revealing for the understanding of nuclear structure. Breit's papers, some fourteen between 1936 and 1963, usually dealt with specific problems rather than general formulations, but his treatments were intrinsically consistent with them, and Wigner attributes much of his own R-matrix theory to "Briet's exploratory work" (FP3). Wigner made use of the short range of the nuclear force by treating the interior of the nucleus as a "black box" and dealing analytically with the exterior—but for particles coming in at low angular momentum (i.e., more or less directly at the target nucleus). He defines a derivative matrix, the R matrix, which has obvious analytic properties. It can utilize the unitary and symmetric nature of the collision matrix readily to give sum rules and has poles that give a resonant structure. The energies and reduced widths are parameters rather than physical entities; he wished to capture the general behavior rather than detailed levels. He summarized this beautifully in his symposium paper in 1968 (FP3).

Wigner's 1968 summary thus brought forward a few years the treatment that Breit had made a major part of his own summary of the theory of resonance reactions in his monumental *Handbuch* article (HP1) of 1959; his taste for treating as much of a problem as possible with mathematical elegance attracted him to his old collaborator's approach. He also treated the compound nucleus model, the shell model, stripping and pick-up, direct interactions, selection rules, threshold behavior, the optical model, reviewed then recent self-consistent field calculations with correlations between particle positions utilized, and heavy ion physics. He also dealt with radioactive states, showing as he had noted in 1940 that these states can also be treated by resonance reaction theory. With John S. (Jack) MacIntosh (HP3), he discussed the polarization of nucleons scattered by nuclei of arbitrary spin, which, among other interests, informs the angular distribution of nuclear reactions—particularly pick-up and stripping (Breit had published a paper on angular distribution of reaction products in 1947). Throughout, the central idea of the resonance reaction connects the various models and treatments.

By the 1950s the preparation of particle beams had progressed to the point that one could conceive of stripping all the atomic electrons from nuclei heavier than helium. In addition, octuply charged oxygen 16 has the same charge-to-mass ratio as the deuterium nucleus, so machines designed for accelerating deuterium could be used to produce the same velocity for oxygen. ¿From such considerations came the initiation of heavy ion physics by Breit (the first publication was in 1952 with Gluckstern and me, but we had presented similar considerations in the the "scientific justification" for a linear accelerator proposal a couple of years earlier, and I have no hesitation in claiming that the seminal ideas were Breit's). The principal types of experiment may be classified as (a) transfer reactions, where only the surface of the target nucleus is involved, and a nucleon is transferred from one nucleus to the other (obviously the older terminology of the "pick-up" and "stripping" is subsumed by "transfer") and (b) Coulomb excitation, where closer approaches between target and projectile occur, but the possibility of treating the kinematics of the collisions classically is retained. A third type, head-on collisions, is the basis for trans-uranic element studies, and the suggestion was seminal for that field, but Briet never really pursued that idea after the initial proposal.

Coulomb excitation (an old idea that becomes powerful as a nuclear probe when the projectile is a heavy ion) became a major experimental field, and Breit with colleagues (especially Bob Gluckstern) discussed the theory in great detail over the fifteen years following the first paper—over twenty publications by my nonscientific count, the last occurring in 1967, when he looked at transfer in Coulomb excitation. As Gary Herling points out, this approach "fore-saw the coupled-channel, distorted-wave method," and he notes the continuing activity in this most productive field. The rotational levels of the target nucleus are excited, and if one wishes to study a spin reorientation, one has a probe that is otherwise unavailable. Breit summarizes the field of heavy ion theory in HP1 and with Gluckstern treats Coulomb excitation extensively (including reorientation effects) in HP4. Allan Bromley's survey of the experimental data up to 1967 (FP4), with comments on the theoretical analyses, is especially illuminating. While, as I have noted, one of the linear accelerators built as a result of our 1950 proposal was at Yale, the most productive heavy-ion experiments there were done with the Emperor tandem Van de Graaff that Bromley got for the department and used to good effect.

I have by means exhausted the areas, even by title, that Gregory initiated or advanced during his remarkable career. Perhaps the breadth and importance of his work has been suggested, however. In terms of the kind of recognition that John Wheeler felt Breit did not get, the fact that he addressed so many topics may have been a disadvantage. This review of his work may serve to redress the oversight that remark implies. Those who did not follow all of his work now have an introduction to it.

A NUMBER OF GREGORY'S colleagues and former students responded to my request for recollections. When I quoted or paraphrased their comments, I named them without a footnote. In all such cases the reference is "private communication." Ralph and Faith Wyckoff, Breit's stepson and daughter-in-law, have provided some personal items of Gregory's life that I could have gotten nowhere else. John Wheeler was especially generous with his response, and D. Allan Bromely's recollection extended into Breit's later years with affection. Mrs. H.H. Barschall responded for Heinz, who unfortunately died before he could answer my inquiry; I am grateful for her help. Anne Herb wrote of Ray's regard for Breit's interest in his early work at Wisconsin. Norman Heydenburg gave a look at Breit's early years that no one else is left to give. Charles Kittel recalled the war years when Breit bundled up the graduate students and took them to Washington, and Glenn Seaborg recalls working with Breit on the neutron project. Willis Lamb talked at length over the telephone about his interaction with Breit, both early and later in his own career. My own contemporaries, Jack McIntosh, Gerry Brown, Bob Gluckstern, and Arthur Broyles revived many memories, and Gary Herling, a later student, gave a good account of Breit's work after the Breit group began to dissolve.

The Breit symposium was organized and the proceedings edited by Allan Bromley and Vernon Hughes; I am grateful for both activities. The work of surveying Breit's contributions would have been much more difficult without the contents of the proceedings and the bibliography they included. I have referred to the papers in the proceedings by "FPx", where the x is the chapter number in *Facets of Physics* (Academic Press, Inc., New York, N.Y. 1970). The papers in volume 41/1 of *Handbuch der Physik* (Springer-Verlag, Berlin, 1959) were invaluable, and are referred to as "HPx," the x being the order of the papers as they occur in the volume.

Finally, I should note that the material supplied by the National Academy of Sciences was very helpful. I have never seen a complete vita written by Breit, but the fragments the Academy sent filled in some gaps, especially about reports during the war years. The obituary from *The New York Times* allowed a quote from Edward Teller that I otherwise would not have known.

4 Biographical Memoirs Selected Bibliography

Breit published some 320 items in his career. I have not referred directly to any of them in the memoir text; the selection below is intended to note the ones that initiated a study or made a significant advance. I have, in consideration for Academy guidelines, included only the first of the many papers on Coulomb wave functions. They're all listed in the *Handbuch* article I did with Breit (HP2). The bibliography in the symposium proceedings is the best there is, but there were one or two papers after 1968, and I found two from 1936 that were missed.

1926

With M.A. Tuve, Radio evidence of the existence of the Kennelly Heaviside layer. *J. Wash. Acad. Sci.* 16:98.

1928

With M.A. Tuve, The production and application of high voltage in the laboratory. *Nature* 121:535. With M.A. Tuve and O. Dahl. Effective heights of the Kennelly Heaviside layer. *Proc. Inst. Radio Eng.* 16:1236.

1929

The effect of retardation on the interaction of two electrons. *Phys. Rev.* 34:553 On the possibility of nuclear disintegration by artificial sources. *Phys. Rev.* 34:817.

1930

Fine structure of He as a test of the spin interaction of two electrons. *Phys. Rev.* 36:383.

1932

Quantum theory of dispersion. Parts I-V. *Rev.Mod. Phys.* 4:504. Dirac's equation and the spin-spin interactions of two electrons. *Phys. Rev.* 39:616.

1933

Isotope shift of Ti. *Phys. Rev.* 44:418. Quantum theory of dispersion. Parts VI and VII. *Rev. Mod. Phys.* 5:91.

1936

With F.L. Yost and J.A. Wheeler. Coulomb functions in repulsive fields. *Phys. Rev.* 49:174.

With E.U. Condon and R.D. Present. Theory of scattering of protons by protons. *Phys. Rev.* 50:825.

With E. Wigner. Capture of slow neutrons. *Phys. Rev.* 49:519.

1939

With L.E. Hoisington, S.S. Share, and H.M. Thaxton. The approximate equality of the proton-proton and proton-neutron interactions for the meson potential. *Phys. Rev.* 55:1103.

1940

The interpretation of resonances in nuclear reactions. *Phys. Rev.* 58:506.

1948

With A.A. Broyles and M.H. Hull, Jr. Sensitivity of proton-proton scattering to potentials at different distances, *Phys. Rev.* 73:869. With G.E. Brown. Effect of nuclear motion on the fine structure of hydrogen. *Phys. Rev.* 74:1278.

1950

Evidence concerning equality of n-n and p-p forces. *Phys. Rev.* 80:1110.

1952

With M.H. Hull, Jr., and R.L. Gluckstern. Possibilities of heavy ion bombardment in nuclear studies. *Phys. Rev.* 87:74.

1953

With M.H. Hull, Jr., Advances in knowledge of nuclear forces. *Am. J. Phys.* 21:184 (selected as one of the major contributions to the *Journal* over 30 years).

1956

a With M.E. Ebel. Nucleon tunneling in N^{14} + N^{14} reactions. *Phys. Rev.* 103:679. With R.L. Gluckstern and J.E. Russell. Reorientation effect in Coulomb excitation. *Phys. Rev.* 103:727.

1957

With V.W. Hughes. Information obtainable on polarization of μ^+ and asymmetry of e^+ in muonium experiments. *Phys. Rev.* 106:1203.

1962

With M.H. Hull, Jr., K.E. Lassila, H.M. Ruppel, and F.A. McDonald. Phase parameter representation of neutron-proton scattering from 13.7 to 350 MeV. II. *Phys. Rev.* 128:830. With M.H. Hull, Jr., K.E. Lassila, K.D. Pyatt, Jr., and H.M. Ruppel. Phase parameter representation of proton-proton scattering from 9.7 to 345 MeV II. *Phys. Rev.* 128:826.

1967

Virtual Coulomb excitation in nucleon transfer. *Proc. Natl. Acad. Sci.* U.S.A. 57:849.

WORK ON THE *SUPER* AND THE STUDY OF ATMOSPHERIC IGNITION

A BRIEF REMINISCENCE BY MCALLISTER HULL

Department of Physics and Astronomy University of New Mexico, Albuquerque, NM, 87131
USA
E-mail: machull@unm.edu

By 1951, the group of young faculty (Bob Gluckstern and me), post doctoral fellows and graduate students, was comfortably established in Sloane Laboratory, and as "The Breit Group", in the department of physics at Yale. We didn't consciously separate ourselves from others in the department, and no one seemed to put us apart, but the association with Breit gave us a special "aura"—sometimes as objects of pity. Any project that involved the group as a whole, or a significant part of it, was discussed with Bob and me, and sometime in 1951 Breit called us in to consider a request that we look at the possibility of atmospheric ignition by a "Super" – as we still called it then, after Edward Teller. At the end of my stay at Los Alamos, in the summer of 1946, we youngsters had many bull sessions about that "Super", and the possibility of controlled fusion energy. We were confident that the former would be working in perhaps five years, and the latter within ten. After all, Han Bethe had shown us how the sun made its energy, and all that was needed to do was to replace a couple of his slow beta decays with something faster. As I write, some fifty-four years later, we still haven't got fusion energy for power. Our crystal balls were cloudier than Vicky Weisskopf's were to be! But I knew some of the lingo, and had worked out the way radiation penetrates the atmosphere (for understanding the Bikini tests), so in 1951, initially drew the task of looking at opacity, a key question in the bomb itself, and in the possibility of atmospheric ignition. Most of our limited work on the bomb was checking calculations that had been done by others as the Teller-Ulam design was being put together. At least some work on atmospheric ignition had already been done as well, but Teller wanted the "most careful physicist he knew" to do the definitive calculations, so Breit got the call. I do not recall all the members of the group who worked on the project and for that I apologize – no private notes were allowed because the work was classified. But in addition to Bob Gluckstern, I recall, for reasons that will appear, that John McHale was one of the project workers. At one point I think we were all working on nitrogen reactions.

The required secrecy called for clearances. My own still in force, and of course Breit, as first head of what became the Manhattan Project, was still cleared. But we needed a secure place to work. A partition was thrown across one end of a large room in Sloane, and thus the "Secret Room" was created. It was obviously not secret from anyone in physics, but what went on inside gave it the name. Security was supplied by ADT, and only Breit, Gluckstern or I could open the door (after a suitable telephone call). The armoire housing our file cabinets also required the use of a code recognized by the ADT office to open. There was a vent near the ceiling at one end of the "Secret Room", so that the room was not secure from possible overheard conversations. Gregory, who was a stickler for security—after all he was the person that got voluntary postponement of publication of nuclear physics appears in 1940--, gave me the task of fixing the possible vent leak. I was not the carpenter that I have become since, so the solution of boarding up the hole never occurred to me—or perhaps we needed the vent for air circulation. In any case, I decided on a "high tech" solution: an audio oscillator to be placed in the vent, and to be turned on whenever we were discussing the work. I was aware that the Weber-Fechner law implied that a constant tone could eventually be heard through, so I thought we needed an oscillator that had a variable pitch (modern alarm signals work this way, but not in 1951). As a theoretician, I know how things work only in a most direct way, so I designed (and built) an oscillator with a partial disc rotating through a conducting cavity so that the resonant frequency changed periodically. To get sound in the middle of the audible range, I made the variable capacitor about the size of Lawrence's first cyclotron. The device worked, and gave off the most horrible sound that has ever been my misfortune to hear! We used it as infrequently as possible. The fact that my friends outside the group didn't cut me off from their society shows their generosity of spirit.

The file cabinets needed combinations, and Bob suggested natural constants. I told him of having startled a security officer at Los Alamos once by confidently offering to open a cabinet for which he needed access in an absent colleague's office. I tried pi, then e, then gamma, etc. One finally worked, and next day we got orders to change our combinations. So I changed mine to the seventh through twelfth digits of pi. Bob thought that was a good idea, and, as a former math club contestant in New York, knew several natural constants to a dozen decimal places and rattled off the digits. I, of course, would have to look them up in a table until I learned them. My first try the next day brought no result. We finally determined that Bob had slipped a digit. I could still look up the combination if I forgot, but had to put in a 6 instead of a 5.

The combination problem was not the only security difficulty we had. Gregory, I and Bob, in that order, were to be contacted by ADT if there was a security breach. One morning about three AM, I got a call. Come to the lab immediately! When I arrived, there were three ADT men outside the door of the "Secret Room" with drawn revolvers casually aimed at John McHale, who was sitting somewhat uncomfortably in a chair in the midst of the guns. John had been unable to sleep, and went to the lab to do some work. He forgot that the door to the room had to be opened by an authorized person in and authorized procedure—he had never been around when Bob or I opened it. I cleared things up, thanked the ADT men for their alert attention, and went back home to bed. Fortunately, that was the only incident of its kind as long as the "Secret Room" existed.

We worked for several months, with all results communicated to Gregory without copies. He wrote whatever final reports there were. I knew that we had shown that atmospheric ignition by a "Super" explosion was impossible if all the cross sections were correct. Gregory made the re-measurement of critical cross section a condition of the conclusion, and I believe they were quickly carried out. I never saw the accompanying declassified notes for a talk at Oak Ridge in 1952 until recently, but I recognized the work from discussions with Gregory as our results were coming in to be put together by him. As Teller had wanted, Gregory was the most conscientious physicist around, and our conclusions were borne out first in the Mike test that followed.

Below we include a copy of the now declassified report on atmospheric ignition by Gregory Breit and his group at Yale.

Notes for Talk

on

Atmospheric Ignition

at the ORNL Conference

in the Fall of 1952

A verbal report on the status of the atmosphere ignition problem was given at the classified conference at the Oak Ridge National Laboratory in the fall of 1952. Notes were prepared in connection with the talk. These are reproduced below. Their preparation was preceded by a study of LA-602 with checks on the main results claimed in LA-602 and additional calculations regarding effects of resonances and changed assumptions about choice of approximations in the treatment of cooling by radiation.

Plan of Talk at ORNL

(1) Subject first treated by T & k; speaker has been asked by ET and the LASL to look into matter again; it was felt by ET that the considerations of the war time period might have been too hurried and the general advance of the art justified additional thought. Plans for the runaway super have increased the importance of the question.

W. Effects & Inst.

(B) Soon after first examination of problem the speaker became convinced that assuming validity of ordinary theory of nuclear σ the calculations of T & k in LA-602 are good enough in essential respects. On the other hand it was felt that the theory of nuclear reactions might perhaps not quite good enough. Reasons:

(a) The theory is based on studies of reactions of an altogether different character from $N^{14} + N^{14} \rightarrow O^{16} + C^{12}$, etc. which appear to be important in present problem.

(b) The main cause making the σ's $<$ assumed values is the magnitude of Coulomb Barrier. For heavy particles barrier penetration is not pronounced and the assumed value of nuclear radius is important. The nuclear radius as derived from different evidence varies somewhat. It plays the role of an empirical parameter. Nuclear theory is not advanced enough to be certain that the nuclear surface is as sharp as ordinary estimates would lead one to believe. Nor is

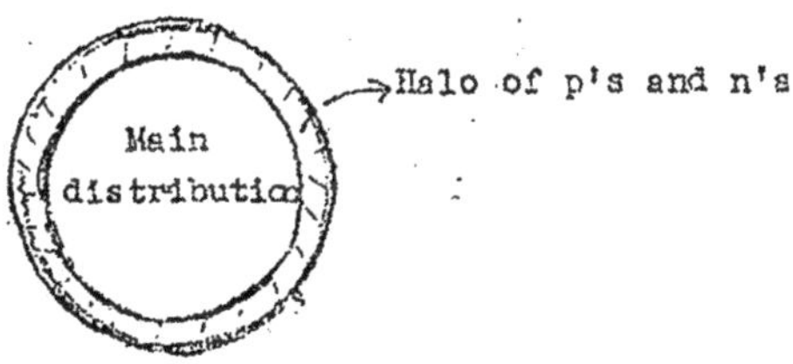

Fig. 1

it certain that there are no appreciable distortions of one nucleus by the field of the other to make colliding nuclei effectively larger. The theory of nuclear forces is still in a rudimentary stage and while it is generally agreed that very

large effects of this kind are improbable it would be unfortunate if one were mistaken and if one were the cause of disastrous consequences through overconfidence in present day nuclear theory. Much of the latter is based on incomplete evidence.

The attitude regarding what constitutes certainty in question under discussion is necessarily different from that taken in deciding what constitutes a good Ph.D. dissertation problem or even what constitutes a suitable subject for inclusion in an ordinary bomb test.

(c) Above considerations have made it desirable to urge that a program of experimental studies be undertaken for the purpose of clarifying question of correctness of estimates of $\sigma(N^{14}+N^{14})$. The Division of Research of the AEC (Pitzer-Platt) has taken a live interest in the matter. The ORNL has undertaken to carry through a program of measurements. Observations are in progress. Results are not sufficiently quantitative as yet for inclusion in present report. It is hoped, however, that before long the measurements will begin to give evidence concerning the correctness of usual ways for $\sigma(N^{14}+N^{14})$.

(3) Main features of N burning.

$$N^{14}+N^{14} \rightarrow (S_{14}^{28})^* \rightarrow Si_{14}^{28} + \gamma + 26.7 \text{ Mev}$$

$$\rightarrow Si_{14}^{27} + n + 10.6 \text{ Mev}$$

$$\rightarrow Al_{13}^{27} + p + 16.0 \text{ Mev}$$

$$\rightarrow Mg_{12}^{24} + \alpha + 17.7 \text{ Mev}$$

$$\rightarrow O_8^{16} + C_6^{12} + 10.6 \text{ Mev}$$

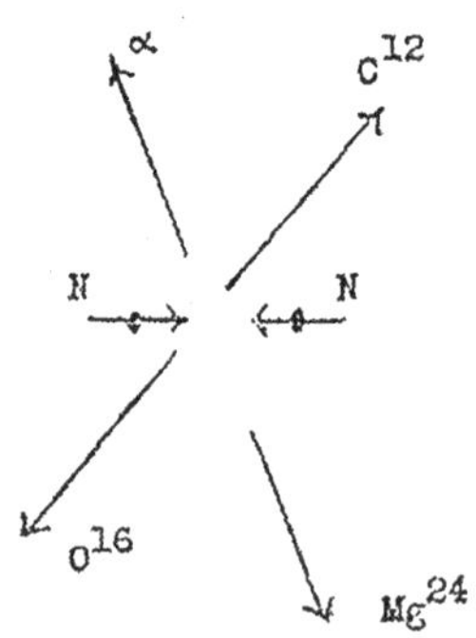

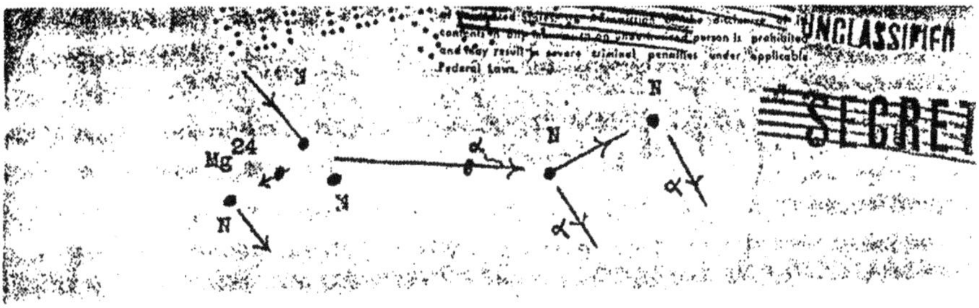
UNCLASSIFIED
SECRET
Mg24
N
N
N
N
N
α
α
α

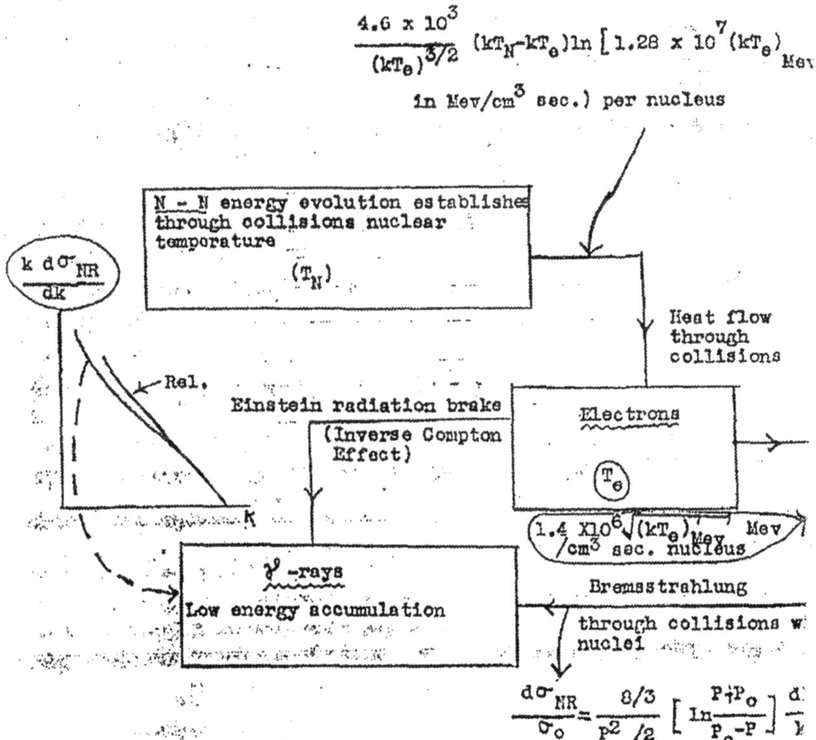
$\frac{4.6 \times 10^3}{(kT_e)^{3/2}} (kT_N - kT_e) \ln [1.28 \times 10^7 (kT_e)_{Mev}$
in Mev/cm3 sec.) per nucleus
N - N energy evolution establishes through collisions nuclear temperature
(T_N)
$\frac{k\, d\sigma_{NR}}{dk}$
Rel.
k
Heat flow through collisions
Einstein radiation brake
(Inverse Compton Effect)
Electrons
T_e
$1.4 \times 10^6 \sqrt{(kT_e)_{Mev}}$ Mev /cm3 sec. nucleus
γ-rays
Low energy accumulation
Bremsstrahlung
through collisions w
nuclei
$\frac{d\sigma_{NR}}{\sigma_0} = \frac{8/3}{P_0^2/2} \left[\ln \frac{P+P_0}{P_0 - P} \right]$

Bring out: (1) The long m.f.p. of disintegration products (e.g. $Mg^{24} + He^{4}$) requires a volume of air with $R \sim 60$ met. to perhaps $R \sim 7$ met. ($O^{16} + C^{12}$). The larger R allows accumulation of low energy γ's which lower T_e. Since the N – e energy exchange $\cong$ const. $\frac{T_N - T_e}{T_e^{3/2}} \cong$ const. $\frac{T_N}{T_e^{3/2}}$ nuclei are cooled faster and T_N is lowered.

(2) The energy evolved in the N + N reactions is too small to maintain equilibrium burning.

(3) To obtain appreciable energy evolution it is necessary to have a good number of collisions with $E > B$; accordingly $kT_N = 10$ or at least 5 Mev is necessary.

(4) To produce these T_N one needs:

(a) A gadget operating at a comparable temperature. [Even the runaway super does not approach this requirement].

(b) The amounts of material needed to supply the energy are enormous if one uses fission bombs ($\sim 10^6$ kg). This energy is available at much too low a temperature. A fission gadget supplies a large number of low energy photons which are unfavorable for reaching high T_e.

(c) A thermonuclear <u>runaway reaction</u> approaches ignition conditions more closely. Here also there is considerable safety on account of much lower T. The total available energy is reasonably within range [The energy in this case is more strongly concentrated in the K.E. of nuclei].

(d) Even though the energy could be supplied the T_N condition alone is sufficient to rule out any practical danger subject to condition of correct premises.

(e) In the fall tests there is no danger that one can see. The equilibrium reaction is being planned for a much lower temperature (~10 - 15 kev). Although one does not know all the factors with certainty it is quite unreasonable to expect them to be off to a degree which would matter.

(f) In view of (e) (above) it appears to be more important to make plans for careful and significant measurements within the next year than to obtain immediate and therefore necessarily qualitative results. It would be helpful to get the beam out of the cyclotron so as to be able to work under known conditions with reasonable certainty concerning the nature of the fragments.

Experiments of Interest for the practical problem

(a) Relative cross sections of $Si^{28} + \gamma$, $Si^{27} + n$, $Al^{27} + p$, $Mg^{24} + \alpha$, $O^{16} + C^{12}$, $N^{13} + N^{15}$.

Values of relative σ's may enable one to make more intelligent guesses regarding the absolute values.

(b) Absolute measurements of some σ's.

(c) Effect of barrier. — Important for estimating energy evolution due to collisions below the Coulomb barrier threshold.

(d) For question of energy deposition desirable to measure $\alpha - N^{14}$, $\alpha - O^{16}$ scattering could be done in Oak Ridge machine (He^{++}); also $N^{14} - N^{14}$, $N^{14} - N^{16}$, $p - N^{14}$.

January, 1953

G. Breit

BIBLIOGRAPHY OF GREGORY BREIT

1920

The calculation of detecting and amplifying properties of an electron tube from its static characteristics. Phys. Rev. 16:387.

With E.O. Hulburt. The detecting efficiency of an electron tube amplifier. Phys. Rev. 15:551.

With E. O. Hulburt. The detecting efficiency of the single electron tube. Phys. Rev. 16:405.

With E. O. Hulburt. The measurement of received radio currents with the electrometer. Phys. Rev. 15:405.

1921

The distributed capacity of inductive coils. Phys. Rev. 17:649.

Some effects of the distributed capacity of coils to ground. Phys. Rev. 18:135, Bureau of Standards Scientific Paper No. 427:521-527.

The high frequency resistance of inductance coils. Phys. Rev. 18:135. Bureau of Standards Scientific Paper no. 430.

1922

The effective capacity of multi-layer coils with square and circular section. Phil. Mag. 43:963.

The effective capacity of a pancake coil. Phil. Mag. 44:729.

A method of measuring coil capacities and standardizing wave meters. Radio Rev. 3:57.

Amplitude of electrical oscillations generated by electron tubes. Wireless World and Radio Rev. 10:517.

The propagation of fan shaped group of waves in a dispersing medium. Phil. Mag. 44:1149.

The field radiated from two horizontal coils. Bureau of Standards Scientific Paper No. 431.

Der Dipol-Rotationseffekt von Born-Lertas. Zeit. F. Physik 11:129.

With P. Ehrenfest. A remarkable case of quantization. Zeit. f. Physik 9:207. Proc. Amsterdam Acad. 25:2.

1923

Transients of magnetic field in supraconductors. Proc. Amsterdam Acad. 26:529.

Are quanta unidirectional. Phys. Rev. 22:313.

The Heisenberg theory of the anomalous Zeeman effect. Nature 112:396.

The interference of light and the quantum theory. Proc. Nat. Acad. 9:238.

Note on the width of spectral lines due to collisions and quantum theory. Proc. Nat. Acad. Sci. 9:244.

The calculation of the effective permeability of a powder. Proc. Amsterdam Acad. 25:293.

With K. Onnes, Magnetic researches XXVI. Measurements of magnetic permeabilities of chronium chloride and gadolinium sulphate at the boiling point of liquid hydrogen in alternating fields of frequency 369,000 per second. Proc. Amsterdam Acad. 26:840.

1924

A method of producing short wave length oscillations in electron tubes. J. Optical Soc. Amer. 9:709.

A singular case of electron tube oscillations. J. Franklin Inst. 197:355. Rotation in a rearfied gas from the point of view of kinetic theory. Phys. Rev. 23:608.

Polarization of resonance radiation. Phil. Mag. 47:832.

With K. Onnes. Preliminary measurements concerning the dielectric constant of liquid hydrogen and liquid oxygen and its dependence on temperature as regards the latter substance. Proc. Amsterdam Acad. 27:617.

1925

Note on the theory of optical dispersion. J. Washington Acad. Sci. 15:34. On a method of calculating the resistance of coils at very high frequencies. J. Optical Soc. Amer. 10:65.

Polarization of resonance radiation and the quantum theory of dispersion. Phys. Rev. 25:242.

The motion of free electrons in black body radiation and the interaction between radiation and matter. Phil. Mag. 49:537.

Polarization of resonance radiation in weak magnetic fields. J. Optical Soc. Amer. 10:439.

The dependence of radio fading on modulation. Phys. Rev. 25: 589. The inductance and resistance of a coil encircling the Earth. J. Washington Acad. Sci. 15:217.

Some remarks on two-coupled multiply periodic systems, the statistics of quantum theory and the theory of dispersion. J. Washington Acad. Sci. 15:269.

On the supposed conducting layer in the atmosphere and the effect of the Earth's magnetic field in radiotelegraphy. Bull, Nat. Res. Council 53:62.

Depolarizing influence of alternating magnetic fields on resonance radiation. J. Optical Soc. Amer. 11:465.

Glaser's experiments and the orientation of molecules in a magnetic field.

J. Washington Acad. Sci. 15:429. Regarding polar lights. Carnegie Inst. Wash. Year Book 24:210.

With A.E. Ruark. A proposed test of the space quantization of atoms in a magnetic field. Phil. Mag. 49:504.

With E.O. Hulburt. On the momentum imparted to electrons by radiation. Phys. Rev. 25:193.

With M.A. Tuve. A radio method of estimating the height of the conducting layer. Nature 116:357.

With A. Ellett. The depolarizing influence of a rapidly changing magnetic field on the resonance radiation. Phys. Rev. 25:888.

1926

Polarization of radiation scattered by an electronic system in a magnetic field. J. Optical Soc. Amer. 12:195.

A correspondence principle in the Compton effect. Phys. Rev. 27:362.

Diffraction by a grating. J. Washington Acad. Sci. 16:201.

The electromagnetic mass and momentum of a spinning electron. Proc. Nat. Acad. Sci. 12:451.

An application of Pauli's method of coordination to atoms having four magnetic parts. Phys. Rev. 28:334.

Properties of waves refracted by the conducting layer. Carnegie Inst. Wash. Year Book 25:218.

With M. A. Tuve, Radio evidence of the existence of the Kennelly-Heaviside layer. J. Washington Acad. Sci. 16:98.

With M.A. Tuve. A test of the existence of the conducting layer. Phys. Rev. 28:554.

1927

The length of light quanta. Nature 119:280.

A suggestion of an explanation of the long life of metastable atoms. Phys. Rev. 29:361.

Unidirectional quanta in wave mechanics. J. Optical Soc. Amer. 14:373.

A suggestion of a connection between radio fading and small fluctuations in the Earth's magnetic field. Bull. Nat. Res. Council, 61:150. Same title with mathematical appendix, Proc. Inst. Radio Eng. 15:709.

With A.E. Ruark and F.G. Brickwedde. Frequency changes of light and quantum theory. Phil. Mag. 3:13;6.

1928

With M.A. Tuve. The production and application of high voltage in the laboratory. Nature 121:535. Abstract in Carnegie Inst. Wash. Year Book 27:241.

With M.A. Tuve and O. Dahl. Effective heights of the Kennelly-Heaviside layer. Proc. Inst. Radio Eng. 16:1236. Abstract in Carnegie Inst. Year Book 27:241.

Propagation of Schroedinger waves in a uniform field of force. Phys. Rev. 32:273. Abstract in Carnegie Inst. Wash. Year Book 27:241.

The magnetic moment of the electron. Nature 122:649. Principle of uncertainty of Weyl's system. Phys. Rev. 32:570.

An interpretation of Dirac's theory of the electron. Proc. Nat. Acad. Sci. 14:553. Abstract in Carnegie Inst. Wash. Year Book 27:240.

1929

The effect of retardation on the interaction of two electrons. Phys. Rev. 34:553. Abstract in Carnegie Inst. of Wash. Year Book 28:251. Phys. Rev. 34:375.

Group velocity and long retardations of radio echoes. Proc. Inst. Radio Engin. 17:1508. Abstract in Carnegie Inst. Year Book 28:251.

On the possibility of nuclear disintegration by artificial sources. Phys. Rev. 34:817-818.

Correction to The propagation of Schroedinger waves in a uniform field of force. Phys. Rev. 34:1491.

The significance of observations of the phase of radio-echoes. Proc. Inst. Radio Engin. 17:1815. Abstract in Carnegie Inst. Wash. Year Book 28:251.

1930

Separation of angles in the two electron problems. Phys. Rev. 35:569. Abstract in Carnegie Inst. Wash. Year Book 29:293.

The fine structure of He as a test of the spin-interactions of two electrons. Phys. Rev. 36:383. Abstract in Carnegie Inst. Wash. Year Book 29:294.

Possible effects of nuclear spin on X-ray terms. Phys. Rev. 35:1447

Fine structure of He as a test of the spin interactions of two electrons. Phys. Rev. 36:383.

With M.A. Tuve and O. Dahl. A laboratory method of producing high potentials. Phys. Rev. 35:51. Abstract in Carnegie Inst. Wash. Year Book 29:295.

With E.O. Salant. Note on the frequency shifts in dispersing media. Phys. Rev. 36:871.

With F.W. Doermann. The magnetic moment of the Li_7 nucleus. Phys. Rev. 36:1262.

With F.W. Doermann. The hyperfine structure of S and P terms of two electrons atoms with special reference to Li+. Phys. Rev. 35:1732.

With M.A. Tuve and L.R. Hafstad. The application of high potentials to vacuum tubes. Phys. Rev. 35:66. Abstract in Carnegie Inst. Wash. Year Book 29:317.

1931

Anomalies in hyperfine structures. Phys. Rev. 37:1182.

Derivation of hyperfine structure formulas for one electron spectra. Phys. Rev. 37:51.

Mean value theories in quantum mechanics. Phys. Rev. 37:90. On the hyperfine structure of heavy elements. Phys. Rev. 38:463.

Book review of Gruppentheorie und Quantenmechanik by Hermand Weyl, second edition. Phys. Rev. 37:783.

On the interpretation of Dirac's matrices. Proc. Nat. Acad. Sci. 17:70. With I.I. Rabi. Measurement of nuclear spin. Phys. Rev. 38:2028.

1932

Quantum theory of dispersion. Parts I-V. Revs. Modern Phys. 4:504.

A remark on Gamow's treatment of radioactive distengration. Phys. Rev. 40:127.

Dirac's equation and the spin-spin interactions of two electrons. Phys. Rev. 39:616. Abstract in Carnegie Inst. Wash. Year Book 31:261.

The isotope displacement in hyperfine structure. Phys. Rev. 42:348.

On the interpretation of Dirac's matrices. Phys. Rev. 39:616.

With S.A. Korff. Optical dispersion. Revs. Modern Phys. 4:504.

With J.E. Rosenthal. The isotope shift in hyperfine structure. Phys. Rev. 41:459.

1933

Isoptope shift of Ti. Phys. Rev. 44:418.

Quantum theory of dispersion. Parts VI and VII. Rev. Modern Phys. 5:91.

With M.H. Johnson. Magentic interaction of a valence electron with inner shells. Phys. Rev. 44:77. Abstracts Phys. Rev. 43:1047.

With J.A. Wheeler Li fine structure and wave functions near the nucleus. Phys. Rev. 44:948.

With L.A. Wills. Hyperfine structure in intermediate coupling. Phys. Rev. 44:470.

1934

Nuclear stability and isotope shift. Phys. Rev. 46:319.

With I.S. Lowen. Polarization of fluorescence radiation. Phys. Rev. 45:120.

With J.A. Wheeler. Collision of two quanta. Phys. Rev. 45:766.

With I.S. Lowen. Radiation damping and the polarization of fluorescence radiation. Phys. Rev. 46:590.

With I.I. Rabi. On the interpretation of present values of nuclear moments. Phys. Rev. 46:230.

With J.A Wheeler. Collision of two light quanta. Phys. Rev. 46:1087.

With F.L. Yost. Capture of charged particles by nuclei due to emission of gamma-radiation. Phys. Rev. 46:1110.

1935

With L.A. Willis, Nuclear magnetic moment of Na^{23}. Phys. Rev. 47:704.

With F.L. Yost. Radiative capture of protons by carbon. Phys. Rev. 48:203. Abstract Phys. Rev. 47:508.

With F.L. Yost and J.A. Wheeler. Coulomb wave-functions. J. Terr. Mag. and Atoms. Elec. 40:443.

With E. Wigner. Note on Majorana's exchange energy. Phys. Rev. 48:918.

1936

Book review of Atomic Spectra by E.U. Condon and A.H. Shortley. Science 83:57.

With M. Ostrofsky and D.P. Johnson. The excitation function of lithium under proton bombardment. Phys. Rev. 49:22.

With F.L. Yost and J.A. Wheeler. Coulomb wave-functions in repulsive fields. Phys. Rev. 49:174.

With M. Ostrosky and W.E. Blaick. Effects of exchange forces on the excitation function of Li_7 under proton bombardment. Phys. Rev. 49:904.

With E.U. Condon. Energy distribution of neutrons slowed by elastic defects. Phys. Rev. 49:229.

With E.U. Condon and R.D Present. Theory of scattering protons by protons. Phys. Rev. 50:825. Paper delivered at the Tercentury Conference on Arts and Science at Harvard University, September, 1936.

With E. Wigner. Beta ray spectrum of Li^8. Phys. Rev. 50:1191.

1937

Approximately relativistic equations for nuclear particles. Phys. Rev. 51:248.

Approximately relativistic equations for nuclear particles. Addendum. Proof of approximate invariance. Phys. Rev. 51:778.

Third Washington Conference on Theoretical Physics. Revs. Sci. Instr. 8:141.

Book review of An Introduction to Nuclear Physics by N. Feather. Revs. Sci. Instr. 8:95.

Book review of The Structure of Atomic Nuclei and Nuclear Transformations by G.Gamow. J. Amer. Chem. Soc. 59:1770.

With J.R. Stehn and E.U. Condon. Correction to the photoelectric effect of the dueteron. Phys. Rev. 51:56.

With E. Wigner. Disintegration of Li^8. Phys. Rev. 51:593.

With J.R. Stehn. On the comparison of proton-proton and proton-neutron interactions. Phys. Rev. 52:396.

With S. Share. Relativistic effects of the Deuteron. Phys. Rev. 52:546.

1938

Approximately relativistic equations. Phys. Rev. 53:153.

Some recent progress in the understanding of atomic nuclei. Revs. Sci. Instr. 9:63.

With J.R. Stehn. Fine structure of the nuclear ground level of Li^7. Phys. Rev. 53:459.

With E. Wigner. The saturation requirements for nuclear forces. Phys. Rev. 53:998.

With J.K. Knipp. Note on K-electron capture in Be^7. Phys. 54:652.

With L.E. Hoisington. Calculation of Coulomb wave function for high energies. Phys. Rev. 54:627.

1939

With H.M. Thaxton and L. Eisenbud. Analysis of experiments on the scattering of protons by protons. Phys. Rev. 55:1018.

With L.E. Hoisington, S.S. Share, and H.M. Thaxton. The approximate equality of the proton-proton and proton-neutron interactions for the meson potential. Phys. Rev. 55:1103.

With L.E. Hoisington and S.S. Share. Effects of shape of potential energy wells detectable by experiments on proton-proton scattering. Phys. Rev. 56:884.

With C. Kittel. Note on the scattering of neutrons by protons. Phys. Rev. 56:744.

1940

The interpretation of resonances in nuclear reactions. Phys. Rev. 58:506.

Scattering matrix of radioactive states. Phys. Rev. 58:1068.

With C. Kittel and H.M. Thaxton. Note on P-wave anomalies in proton-proton scattering. Phys. Rev. 57:255.

With E. Teller. Metastability of hydrogen and helium levels. Astrophysical J. 91:215.

1941

Proton-proton scattering. Paper read at University of Pennsylvania Bicentenial Conference, September 9, 1940. Published by the University of Pennsylvania Press entitled Nuclear Physics pp. 5-19. (1941).

1946

Schematic treatment of nuclear resonances. Phys. Rev. 69:472.

1947

Scattering of slow neutrons by bond proton. I. Methods of calculation. Phys. Rev. 71:215.

Relativistic corrections to magnetic moments of nuclear particles. Phys. Rev. 71:400.

Does the electron have an intrinsic magnetic moment? Phys. Rev. 72:1023.

With R.P. Zilsel. Scattering of slow neutrons by bound protons. II. Harmonic binding — neutrons of zero energy. Phys. Rev. 71:232.

With B.T. Darling. Note on the calculation of angular distributions in resonance reactions. Phys. Rev. 71:401.

With I. Bloch. Relativistic correction to the magnetic moment of the deuteron. Phys. Rev. 72:135.

With P.R. Zilsel and B.T. Darling. The scattering of slow neutrons by bound protons. III. Intermediate energies. Phys. Rev. 72:576.

With R.E. Meyerott. Effect of nuclear motion on the hyperfine structure of the ground term of hydrogen. Phys. Rev. 72:1023.

With F.L. Friedman. Calculations on countercurrent electro-migration. J. Res. NBS Research Paper 1836, 39:397.

1948

Erratum: Does the electron have an intrinsic magnetic moment? Phys. Rev. 73:1410.

Some effects of the intrinsic magnetic moment of the electron. Phys. Rev. 74:656.

With I. Bloch. Resonances in Li^7 (p,n) Be^7. Phys. Rev. 74:397.

With A.A. Broyles and M.H. Hull, Jr. Sensitivity of proton-proton scattering to potentials at different distances. Phys. Rev. 73:869.

With G.E. Brown. Effect of nuclear motion on the fine structure of hydrogen. Phys. Rev. 74:1278.

With W.G. Bouricius. A boundary value condition for proton-proton scattering. Phys. Rev. 74:1546.

1949

With W.G. Bouricius. Boundary conditions and range of force for S state of two protons. Phys. Rev. 74:1029.

With R.D. Hatcher and G.B. Arfken. S-wave proton-proton scattering from 0.2 to 40 Mev for the Yukawa and Gauss error potentials. Phys. Rev. 75:1389.

With R.E. Meyerott. The hyperfine structure of the ground term of hydrogen. Phys. Rev. 75:1447.

With G.E. Brown and G.B. Arfken. The effect of nuclear motion on the hyperfine structure of hydrogen. Phys. Rev. 76:1299.

With G.E. Brown. Perturbation Method for Dirac's radial equation. Phys. Rev. 76:1307.

With R.E. Meyerott. A small differential analyzer with ball carriage integrators and selsyn coupling. Revs. Scientific Instr. 20:874.

1950

The Schmidt model and odd-even staggering of the isotopic shift. Phys. Rev. 78:470.

The Fierz and Kopfermann-Brix explanation of the spectroscopic odd-even isotope shift staggering. Phys. Rev. 79:891.

Evidence concerning equality of n-n and p-p forces. Phys. Rev. 80:1110.

With G.B. Arfken and W.W. Clendenin. Spectroscopic isotope shift and nuclear polarization. Phys. Rev. 77:569.

With R.D. Hatcher. Analysis of proton-proton scattering by means of the f function. Phys. Rev. 78:110.

With G.B. Arfken and W.W. Clendenin. Spectroscopic isotope shift and nuclear polarization. Phys. Rev. 78:390.

With M.H. Hull, Jr. Asymptotic expansion of irregular Coulomb function for angular momentum zero. Phys. Rev. 80:392.

With I. Bloch, M.H. Hull, Jr., A.A. Broyles, W.G. Bouricius, and B.E. Freeman. Methods of calculation of radial wave functions and new tables of Coulomb functions. Phys. Rev. 80:553.

With M.H. Hull, Jr. Asymptotic expansion of irregular Coulomb function. Phys. Rev. 80:561.

Sur l'interpretation de la difusion nucleon-nucleon au moyen de courbes de potential. Particules Fondamentales et Noyaux. Paris 24-29 avril 1950 — pp.127-130.

1951

Regarding covariant subtraction of infinites. Science 113:103.

Topics in scattering theory. Revs. Modern Phys. 23:238.

High energy nucleon-nucleon scattering. Phys. Rev. 84:1053.

Electron-neutron interaction. Proc. Nat. Acad. Sci. 37:837.

With M.C. Yovits. Internal excitation and apparent range of nuclear forces in scattering experiments. Phys. Rev. 81:416.

With J.S. McIntosh. Evidence regarding instability of H^4. Phys. Rev. 83:1245.

With I. Bloch, M.H. Hull, Jr., A.A. Broyles, W.G. Bouricius, and B.E. Freeman. Coulomb functions for reactions of protons and alpha particles with the lighter nuclei. Rev. Modern Phys. 23:147.

With M.C. Yovits. Representation of nucleon-nucleon single S scattering. Proc. Nat. Acad. Sci. 37:771.

With H.M. Jones. Limitations on mass changes of scattered nucleons. Phys. Rev. 84:1054.

With M.C. Yovits. Representation of nucleon-nucleon singlet scattering Bull. Am. Phys. Soc. 26:32.

With R.L. Gluckstern. A classical dynamics view of gauge invariance. Science 114:478.

With R.L. Gluckstern and M.H. Hull, Jr. Polarization of Bremsstrahlung radiation. Science 114:480.

1952

Spectroscopic isotope shift and nuclear shell structure. Phys. Rev. 86:254.

Concerning a tentative hypothesis of proton isomers. Bull. Am. Phys. Soc. 27:32.

With M. Yovits, R.L. Smith, M.H. Hull, Jr., J. Bengston. Meson tail effect and treatment of proton-proton scattering data. Phys. Rev. 85:540.

With M.H. Hull, Jr. and R.L. Gluckstern. Possibilities of heavy ion bombardment in nuclear studies. Phys. Rev. 87:74.

With W.W. Clendenin. Effect of intrinsic moment of electron on spectroscopic isotope shift. Phys. Rev. 85:689.

With R.M. Thaler. Relativistic corrections to magnetic moments of nuclear particles. Phys. Rev. 88:1214.

1953

With R.M. Thaler. Relativistic corrections to the magnetic moments of nuclear particles. Phys. Rev. 89:1177.

With R.L. Gluckstern. Advances in nucleon-nucleon scattering experiments and their theoretical consequences. Ann. Revs. Nuclear Sci. 2:365-398.

With M.H. Hull, Jr. Advances in knowledge of nuclear forces. Am. J. Phys. 21:184.

With R.L. Gluckstern and M.H. Hull, Jr. Polarization of Bremsstrahlung Radiation. Phys. Rev. 90:1026.

With M.H. Hull, Jr. and R.L. Gluckstern. Polarization dependence of the integrated Bremsstrahlung cross section. Phys. Rev. 90:1030.

With R.M. Thaler and J. Bengston. Effect of $3P_2$-$^3F_2$ coupling on p-p scattering. Bull. Am. Phys. Soc. 28:30.

1954

Scattering experiments, nuclear. McGraw-Hill Encyclopedia of Science and Technology 12:57.

Longitudinal and transverse stability in linear acceleration systems. Bull. Am. Phys. Soc. 29:31.

Lande-Goudsmit-Fermi-Serge formulae. Rydberg Centennial Conference, June 28-July 1, 1954; Transactions of the Royal Physiographic Society of Lund, Sweden, n.s. 65, no. 21.

With S. Sack and L.C. Biedenharn. The elastic scattering of protons by alpha particles. Phys. Rev. 93:321.

With R.M. Thaler and J. Bengston. Effect of $3P_2$-$^3F_2$ coupling on nucleon-nucleon scattering. Phys. Rev. 93:644.

With H.A. Bethe. Ingoing waves in final state of scattering problems. Phys. Rev. 93:888.

With R.M. Thaler and J. Bengston. Intermediate state coupling and the interpretation of high energy nucleon-nucleon scattering. Phys. Rev. 94:683.

With J.B. Ehrman. Formula for polarization in nucleon-nucleon scattering. Phys. Rev. 96:805.

With J.B. Ehrman, A.M. Saperstein, and M.H. Hull, Jr. Phase shifts for nucleon-nucleon scattering at 280 MeV. Phys. Rev. 96:807.

With P.B. Daitch. Correspondence between semi-classical quantum treatments of Coulomb excitation. Phys. Rev. 96:1447.

With P.B. Daitch, J.P. Lazarus, M.H. Hull, Jr., and F.D. Benedict. Validity of semi-classical treatment of Coulomb excitation. Phys. Rev. 96:1449.

1955

Relativistic corrections for high energy p-p scattering. Phys. Rev. 99:1581.

Relativistic formulation of Coulomb scattering for proton-proton collisions. Science 121:620.

Theory of nucleon transfer. Bull. Am. Phys. Soc. 30:29.

With L.C. Biedenharn, R.L. Gluckstern, and M.H. Hull, Jr. Coulomb functions for large charges and small velocities. Phys. Rev. 97:542.

With M.H. Hull, Jr. Amplitudes in nucleon-nucleon scattering. Phys. Rev. 97:1047.

With J.B. Ehrman and M.H. Hull, Jr. Polarization and amplitudes in nucleon-nucleon scattering. Phys. Rev. 97:1051.

With M.E. Ebel and F.D. Benedict. Angular distribution of gamma-rays in Coulomb excitation. Phys. Rev. 100:429.

With J.P. Lazarus. Effect of finite amplitudes in Coulomb excitation. Phys. Rev. 100:942.

With P.B. Diatch. Redistribution of classical and quantum densities. Proc. Nat. Acad. Sci. 41:653.

1956

Contributions of optical methods to nuclear physics. Invited paper read at Fourth Congress of the International Commission of Optics, March 28, 1956.

Possibilities of heavy-ion bombardment for nuclear theory. Bull. Am. Phys. Soc. 1:162.

With F.D. Benedict and P.B. Daitch. Semiclassical approximation to Coulomb excitation integrals. Phys. Rev. 101:171.

With R.L. Gluckstern and J.P. Lazarus. Comparison of semiclassical and quantum mechanical Coulomb excitation integrals for large 1. Phys. Rev. 101:175.

With M.E. Ebel and J.E. Russell. Gamma-ray angular distribution in Coulomb excitation. Phys. Rev. 101:1504.

Transfer of nuclear particles. Phys. Rev. 102:549.

With M.E. Ebel, Nucleon tunnelling in N^{14} + N^{14} reactions. Phys. Rev. 103:679.

With R.L. Gluckstern and J.E. Russell. Reorientation effect in Coulomb excitation. Phys. Rev. 103:727.

With M.E. Ebel. Nucleon transfer and virtual Coulomb excitation. Phys. Rev. 104:1030.

With M.E. Ebel. Virtual states in transfer reactions. Bull. Am. Phys. Soc.1:303.

1957

Relativistic corrections to p-p scattering. Phys. Rev. 106:314.

Contributions of optical methods to nuclear physics. J. Optical Society of America 47:446.

Energy dependence of reactions at thresholds Phys. Rev. 107:1612.

With R.L. Gluckstern and J.E. Russell. Sources of error in experiments on reorientation effect in Coulomb excitation. Phys. Rev. 105:1121.

With J.S. McIntosh. External Coulomb fields in the Wigner-Eisenbud nuclear reaction theory. Phys. Rev. 106:1246.

With V.W. Hughes. Information obtainable on polarization of μ^+ and asymmetry of e^+ in muonium experiments. Phys. Rev. 106:1293.

1958

Theory of the isotope shift. Revs. Modern Phys. 30:507.

Velocity dependent features of a static nucleon-nucleon potential. Phys. Rev. 111:652.

Effects of atomic electrons on p-p and n-p scattering. Phys. Rev. Letters 1:200.

The theory of transfer reactions. Comptes Rendus du Congres International de Physique Nucleaire. Interactions Nucleaires aux Basses Energies et Structure des Noyaux, Paris, France, 7-12 July 1958. pg. 349.

Second order effects in Coulomb excitation. Comptes Rendus du Congres International de Physique Nucleaire. Interactions Nucleaires aux Basses Energies et Structure des Noyaux, Paris, France, 7-12 July 1958. pg.347.

Theory of heavy ion nucleon transfer reactions. Bull. Am. Phys. Soc. 3:21. Features of nucleon-nucleon interactions suggested by a static model. Bull. Am. Phys. Soc. 3:183.

The theory of transfer reactions. Proceedings of the Conference on Reactions Between Complex Nuclei, Gatlinburg, Tenn., May 5-8, 1958.

1959

Photodisintegration of the deuteron. Gordon Research Conference, New Hampshire, August 24-28, 1959.

Theory of resonance reactions and allied topics. Hanbuch der Physik 41/1, 1-407.

Effects of atomic electrons on p-p and n-p scattering II. Phys. Rev. Letters 2:401.

Calculations on the photodisintegration of the deuteron. International Conference on Nuclear Forces and the Few Nucleon Problem, University College, England, July, 1959. v. 1:23.

Gamma rays in resonance capture theory. Bull. Am. Phys. Soc. 4:473 With J. Shapiro. Metastability of 2s states of hydrogenic atoms. Phys. Rev. 113:179.

With M.H. Hull, Jr. Natural line width for transient excitations. Astrophys. J. 129:456.

With M.L. Rustgi. Electric polarizability of the neutron. Phys. Rev. 114:830.

With W. Zernik and M.L. Rustgi. Note on the photodisintegration of the deuteron. Phys. Rev. 114:1358.

With M.H. Hull, Jr., K.D. Pyatt, Jr., and C.R. Fischer. Nucleon-nucleon spin-orbit interaction potential. Phys. Rev. Letters 2:264.

With M.H. Hull, Jr. Coulomb wave functions. Handbuch der Physik 41/1, 408-466.

With J.S. McIntosh. Polarization of nucleons scattered by nuclei. Hand-

buch der Physik 41/1, 467-495.
With R.L. Gluckstern. Coulomb excitation. Handbuch der Physik 41/1, 496-560.
With J. Shapiro. Metastability of 2s states of hydrogenic atoms. Phys. Rev. 115:1779.

1960

Scattering experiments, nuclear. McGraw-Hill Encyclopedia of Science and Technology 12:57.
The nucleon-nucleon spin-orbit potential. Proc. Nat. Acad. Sci. 46:746.
The nucleon-nucleon spin-orbit interaction and the repulsive core. Phys. Rev. 120:287.
Vector meson coupling in nucleon-nucleon interactions. Science 131:1316.
Nucleon transfer theory. Proceedings of the Second Conference on Reactions Between Complex Nuclei, Gatlinburg, Tenn., May 2-4, 1960. pg. 1 Bull. Am. Phys. Soc. 5:474.
Nucleon-nucleon phase parameters. Proceedings of the Tenth Annual International Conference on High Energy Physics, University of Rochester, August, 1960. pg 132.
The nucleon-nucleon repulsive core and the spin-orbit interaction. Proceedings of the Tenth Annual International Conference on High Energy Physics, University of Rochester, August, 1960. pg. 198.
With M.H. Hull, Jr. Note on one pion exchange potential. Nuclear Phys. 15:216.
With M.H. Hull, Jr., K. Lassila, and K.D. Pyatt, Jr. Tests of charge independence of nucleon-nucleon interactions. Phys. Rev. Letters 4:79.
With R.L. Gluckstern. Effect of finite life of upper level on probability of Coulomb excitation. Nuclear Phys. 20:188.
With M.L. Rustgi, W. Zernik, and D.J. Andrews. Cross section and polarization of photodisintegration of the deuteron. Phys. Rev. 120:1881.
With M.H. Hull, Jr., K.E. Lassila, and K.D. Pyatt, Jr. Phase-parameter representation of proton-proton scattering from 9.7 to 345 MeV Phys. Rev. 120:2227.
With M.H. Hull, Jr., K.E. Lassila, and H.M. Ruppel. Form of the one-pion exchange potential. Phys. Rev. Letters 5:274.
With M.H. Hull, Jr., K.E. Lassila, and H.M. Ruppel. Tests of form on the one-pion exchange potential. Proc. Nat. Acad. Sci. 46:1649.
With M.H. Hull, Jr., K.E. Lassila, and K.D. Pyatt, Jr., and H.M. Ruppel. Nucleon-nucleon phase parameters at energies from 9-340 MeV. Bull. Am. Phys. Soc. 5:268.
With G. Herling and Y. Nishida. Universal curve for neutron transfer

(N^{14}, N^{13}) reactions. Bull. Am. Phys. Soc. 5:293.

With R.L. Gluckstern. Coulomb disintegration of Li^6 by heavy nuclei. Proceedings of the Second Conference on Reactions Between Complex Nuclei, Gatlinburg, Tenn., May 2-4, 1960. pg. 77; Bull. Am. Phys. Soc. 5:474.

1961

On the interpretation of the field theoretical potential in terms of phase shifts. Ann. Phys. 16:346.

Nuclear forces and polarization phenomena. Proceedings of the International Symposium on Polarization Phenomena of Nucleons, Basel, Switzerland, July 4-8, 1960. Helvetica Physica Acta Suppl. VI, pg. 343-358.

D1/1 contribution. Proceedings of the Rutherford Jubilee International Conference, University of Manchester, England, September, 1961. p. 77.

Virtual Coulomb excitation in the nucleon transfer process. Proceedings of the Conference on Electromagnetic Lifetimes and Properties of Nuclear States, Gatlinburg, Tenn., October, 1961. pg. 21.

With K.E. Lassila, H.M. Ruppel, and M.H. Hull, Jr. Two-pion exchange interaction. Phys. Rev. Letters 6:138.

With M.L. Rustgi, W.Zernik, and D.J. Andrews. Erratum: Cross section and polarization in the photodisintegration of the deuteron. Phys. Rev.121:1864.

With M.H. Hull, Jr., K.E. Lassila, H.M. Ruppel, and F.A. McDonald. Phase-parameter representation of neutron-proton scattering from 13.7 to 350 MeV. Phys. Rev. 122:1606.

With W. Zickendraht, D.J. Andrews, M.L. Rustgi, W. Zernik, and A.J. Torruella. Revised calculations on the photodisintegration of the deuteron. Phys. Rev. 124:1538.

With M.H. Hull, Jr., K.E. Lassila, and H.M. Ruppel. Nucleon-nucleon interactions and nucleon structure. Proceedings of the International Conference on Nucleon Structure, June, 1963. pg. 188-203.

With Y. Nishida and G. Herling. Theory of transfer reactions. Proceedings of the Rutherford Jubilee International Conference, University of Manchester, England, September, 1961. pg. 507.

1962

Dispersion theory of nuclear processes. Neutron-pronton scattering. Nuclear forces. Proton-proton scattering. Resonance in nuclear processes. Encyclopadic Dictionary of Physics. N.Y. Pergamon Press. 7 vols.

Aspects on nucleon-nucleon scattering theory. Rev. Modern Phys. 34:766.

Nucleon transfer reactions. Proceedings of the Conference on Direct Interactions and Nuclear Reaction Mechanism. University of Padua, Italy, September, 1962. pg. 480.

Vector mesons and nucleon-nucleon spin-orbit interaction. Eastern Theoretical Physics Conference, University of Virginia, October, 1962. pg. 283.

Nucleon-nucleon interactions. Conference on Advances in Meson and Nuclear Research Below One MeV, Gatlinburg, Tenn., November, 1962.

With M.H. Hull, Jr., K.E. Lassila, H.M. Ruppel, and F.A. McDonald. Phase parameter representation of neutron-proton scattering from 13.7 to 350 MeV. II. Phys. Rev. 128:830.

With M.H. Hull, Jr., F.A. McDonald, H.M. Ruppel. Significance of neutron-proton polarization measurements in the 20 MeV energy range. Phys. Rev. Letters 8:68.

With K.E. Lassila, M.H. Hull, Jr., H.M. Ruppel, and F.A. McDonald. Note on a nucleon-nucleon potential Phys. Rev. 126:881.

With H.M. Ruppel. Wave distortion for magnetic moment effects in nucleon-nucleon scattering. Phys. Rev. 127:2123.

With M.H. Hull, Jr., K.E. Lassila, K.D. Pyatt, Jr., and H.M. Ruppel. Phase parameter representation of proton-proton scattering from 9.7 to 345 MeV. II. Phys. Rev. 128:826.

With H.M. Ruppel. Wave distortion for magnetic moment effects in nucleon-nucleon scattering. Science 136:324.

With M.H. Hull, Jr., F.A. McDonald, and H.M. Ruppel. Form of one-pion exchange potential and charge independence of nucleon-nucleon interaction. Proceedings of the International Conference on High Energy Physics at CERN, July, 1962. pg. 134.

1963

The semiclassical treatment of some quantum problems. Zeit. f. Physik 171:149. Erratum: Aspects of nucleon-nucleon scattering theory. Revs. Modern Phys. 35:734.

Approximations of nucleon transfer theory. Proceedings of the Third Conference on Reactions Between Complex Nuclei, Asilomar, April, 1963, pg. 97.

Nucleon-nucleon interactions and nucleon structure. Proceedings of the International Conference on Nucleon structure, Stanford University, June, 1963. pg. 188.

Definitions of compound states. Conference on Compound Nuclear States, Gatlinburg, Tenn., October, 1963.

Nucleon-nucleon interactions and nucleon structure. Proceedings of the International Conference on Nucleon structure, June, 1963. pg. 188-203.

With H.M. Ruppel. Erratum: Wave distortion for magnetic moment effects in nucleon-nucleon scattering. Phys. Rev. 131:2839.

1964

Limited applicability of the theory of nucleon tunneling. Phys. Rev. 135:1323-1337.

The pursuit of nuclear physics. J.Franklin Institute 278-365.

With A.N. Christakis, M.H. Hull, Jr., H.M. Ruppel, and R.E. Seamon. Phenomenological p-p and n-p phase parameters. Bull. Am. Phys. Soc. 9:378.

With K.W. Chun and H.G. Wahweiler. Neutron transfer in N^{14} (N^{14}, N^{13}) N^{15} at low energies. Phys. Rev. 133:403.

1965

A soluble semiclassical particle transfer problem. Annals of Phys. 34:377.

With M.L. Rustgi. Interaction of nuclei with electromagnetic multipoles. Bull. Am. Phys. Soc. 10:447.

With W. Zickendraht, M.L. Rustgi, and R.G. Brandt. Cross section and polarization in the photodisintegration of the deuteron with arbitrary number of electromagnetic multipoles. Bull. Am. Phys. Soc. 10:447.

With R.G. Brandt, W. Zickendraht, A.J. Torruella, and R. Schrils. Cross section and polarization in the photodisintegration of the deuteron. Bull. Am. Phys. Soc. 10:448.

1966

Nucleon-nucleon scattering. Proceedings of the Williamsburg Conference on Intermediate Energy Physics, February, 1966. pg. 471.

Limitations on the interpretation of electromagnetic form factors of nucleons. proceedings of the XII International Conference on High Energy Physics. pg. 985-987.

With A.N. Christakis, M.H. Hull, Jr., H.M Ruppel, and R.E. Seamon. Phenomenological p-p and n-p phase parameters. Proceedings of the XII International Conference on High Energy Physics. pg. 985-987.

With R.E. Seamon and K.A. Friedman. Electrostatic effects in nucleon-nucleon scattering at intermediate energies. Phys. Rev. 145:779-786.

1967

Phenomenological phase-parameter fits to N-N data at intermediate energies. Rev. Mod. Phys. 39:560.

Virtual Coulomb excitation in nucleon transfer. Proc. Nat. Acad. Sci. 57:849.

Phenomenological phase-parameter fits to N-N data at intermediate energies. Bull. Am. Phys. Soc. 12:990.

With R.D. Haracz. Nucleon-nucleon scattering. High Energy Physics,

Vol. 1, edited by E.H.S. Burhop. Academic Press, Inc., New York, 1967. pg. 21.

With J.A. Polak and D.A. Torchia, ^{14}N (^{14}N, ^{13}N) ^{15}N reaction at low energies and the elastic scattering of ^{14}N by ^{14}N. Phys. Rev. 161:993.

R.G. Newton - scattering theory of waves and particles. J. Frankl. Inst.-Eng. Appl. Math. 284:265.

1968

With R.E. Seamon, K.A. Friedman, R.D. Haracz, J.M. Holt and A. Prakash. Phenomenological phase-parameter fits to N-N data up to 350 MeV. Phys. Rev. 165:1579.

With M.L. Rustgi. The theory of the photodisintegration of the deuteron and of other nuclei. Phys. Rev. 165:1075.

With R.E. Seamon. Polarization in p-p scattering close to 20 MeV. Phys. Rev. Lett. 21:760.

With J.F. Masso, J.A. Polak, D.A. Torchia. Comparison of semiclassical and quantum schematic nucleon- transfer model. J. Phys. Soc. Japan 24:244.

With K.A. Friedman, J.M. Holt, R.E. Seamon. Short- and long-range charge independence. Phys. Rev. 170:1424.

1969

The potential concept in nucleon-nucleon interactions and limitations of existing models. Proceedings of the International Conference on Properties of Nuclear States, Montreal, 1969. pg. 293.

Uncertainties in tests of long-range charge independence. Nuc. Phys. B14:507.

With. J. Lucas and M. Tischler. Concerning N-P scattering cross section at 24-MeV. Phys. Rev. 184:1668.

Uncertainties in tests of long-range charge independence. Proc. Natl. Acad. Sci. U. S. A. 63:223.

1971

With M.L. Rustgi. Proposed polarized/target/beam test for ${}_3S_1$-${}_3S_1$ radiative transitions in thermal $n - p$ capture. Nuc. Phys. A161: 337.

With M. Tischler, S. Mukherje and J. Lucas. Improvements in tests of long-range charge independence. Proc. Nat. Acad. Sci. USA 68: 897.

1973

With M. Tischler, S. Mukherje and G. Pappas. Magnetic-moment effects on tests of charge independence in nucleon-nucleon scattering. Proc. Nat. Acad. Sci. USA 70: 2178.

www.ingramcontent.com/pod-product-compliance
Ingram Content Group UK Ltd.
Pitfield, Milton Keynes, MK11 3LW, UK
UKHW020359150926
8624IPUK00004B/236

9 789810 245535